Greening of Industry Networks Studies

Volume 13

This series aims to improve our understanding of how shifts in industrial regimes, trade, and technology are not only creating significant environmental and social impacts and inequities around the world but also opportunities for sustainable economic growth. It focuses on issues of industrial development, environment and society, and sustainable economy, dealing with the UN Sustainable Development Goals.

The series publishes high-quality peer-reviewed monographs, edited volumes and graduate textbooks related to the theme of sustainable economic development. The topics covered include industry networks, sustainable economic growth, supply chains, supply chain management, sustainable economy, green logistics, green growth, sustainable firms, collaboration for sustainability, corporate sustainability, corporate social responsibility, green banking, operations management, engineering management, sustainable value chains, sustainable business, circular economy, sustainable production, sustainable consumption, waste, resource management, productivity strategies, sustainable management, resource conservation, sustainability strategies, innovation processes, sustainable innovation, renewable resources, among others. Contributors to the series are researchers, business strategists, civil society thought leaders, and government policy makers.

The series is associated with ***The Greening of Industry Network***, an international network of professionals from research, education, business, government and civil society organizations located in more than 50 countries, with the goal of aligning industrial development strategies with sustainable development principles. The network stimulates, coordinates and connects high quality research to policies, strategies and actions in ways that contribute to a more sustainable society. Through a variety of arenas, it provides an open forum for creative debate to engage researchers, business, workers, government, consumers and other actors in developing a shared understanding of the changes required for creating a more sustainable future.

Now indexed by Scopus!

Ani Melkonyan-Gottschalk • Denis Daus
Editors

Transformation Towards Circular Food Systems

Sustainable, Smart and Resilient Citrus Supply Chains in Mediterranean Areas

Springer

Editors
Ani Melkonyan-Gottschalk
Clausthal University of Technology
Clausthal, Germany

Denis Daus
Center for Logistics and Traffic (ZLV)
University of Duisburg-Essen
Duisburg, Germany

ISSN 2543-0246 ISSN 2543-0254 (electronic)
Greening of Industry Networks Studies
ISBN 978-3-031-63792-6 ISBN 978-3-031-63793-3 (eBook)
https://doi.org/10.1007/978-3-031-63793-3

This Springer imprint is published by the registered company Springer Nature Switzerland AG
The registered company address is: Gewerbestrasse 11, 6330 Cham, Switzerland

Foreword

Food supply chains are necessary given modern dietary needs. It is one of the oldest supply chains since the dawn of civilization and agricultural societies. Food sources and supply chains have come a long way. Globalization has become a common characteristic for almost all food and agricultural supply chains. With globalization, growth patterns, and inefficient systems, the amount of food waste has also never been greater. It is in this situation that this edited book by Ani Melkonyan-Gottschalk and Denis Daus offers insight into a potential solution—at least for a regional area of the world, the Mediterranean region, and for citrus products—circularity or closed-loop supply chains.

This edited book in the Greening of Industry Networks series adds to the discourse with a foundation built by over the first dozen or so books in this series that consider the greening of various industries and situations. This book represents the first one dedicated to the agriculture and food industry. Although many other books have chapters on these topics, this compendium brings it together to the fingertips of scholars, students, and professionals.

In this ever-evolving landscape of our global food systems, the need for sustainable and resilient practices has become more important. The journey from farm to table has undergone a profound transformation and this transition continues. The exploration of circular food supply chains, with a specific focus on citrus supply chains, provides a beacon of innovation and responsibility. This book, dedicated to unraveling the complexities and opportunities within circular citrus supply chains, stands as a testament to the collaborative efforts and forward-thinking initiatives that define the future of our food ecosystems and its challenges.

As we navigate the intricate web of challenges facing the agricultural sector, the concept of circularity becomes a guiding principle. For example, when considering the Ellen MacArthur foundation's famous "butterfly" circular economy, they exhibit both technical and biological closed loop systems.

Circular food supply chains, with their emphasis on minimizing waste, optimizing resources, and fostering sustainability, includes both cycles, not many industries have to consider both as viable and important dimensions.

The citrus industry, renowned for its global significance and economic impact, becomes a compelling lens through which to examine the intricacies of the dimensions of circularity and supply chains. The supply chains may seem simple—but it is anything but simple. The book chapters point out local capacity, technology, regulations, and various cultures that exist. The cultural variations in the Mediterranean region and its complex environmental, regulatory, and culture issues add further to this complexity. It is also a cradle to one of the world's civilizations, but there is still much to learn.

The reader will encounter a variety of insights, case studies, and strategies, illuminating a path toward scalable, circular business practices within the citrus industry—but with tools and insights for a variety of agricultural products, other industrial sectors, and supply chains. Learning can be achieved by learning within and between industries, in fact significant competitive and improvement innovations, game changing innovations, can be derived from these insights from within and external to the citrus supply chain.

This book offers a comprehensive examination of the logistics challenges and innovations specific to circular citrus supply chains, shedding light on how effective management can enhance both economic and environmental sustainability.

We are also in an age of digital transformation. Technology adoption becomes a linchpin for progress. Circular citrus supply chains stand at the intersection of traditional farming practices and cutting-edge technologies—and potential tools—as evidenced by a number of examples across chapters. From precision processing technology in agriculture to big data analytics technology, this book explores the various current practice and advancements that are reshaping the citrus industry and catalyzing the transition toward circularity.

Sustainable farming practices appear at the heart of circular food supply chains and are also driven by technological and practice advancements. The book delves into the regenerative and environmentally conscious methods—such as by-products management—that are revolutionizing citrus cultivation. By showcasing sustainable farming models and their positive impact on ecosystems, water use, biodiversity, and soil health, this exploration aims to inspire a broader commitment to responsible agricultural practices.

Environmental and sustainability resilience are needed for unpredictable challenges and is a recurrent theme throughout this book. Circular citrus supply chains, with their inherent adaptability and resourcefulness, serve as a blueprint for building resilient food systems capable of withstanding shocks and uncertainties. Data-driven solutions, forecasting, and management of flows and crops and yields are necessary to carefully control for various environmental, agrometeorological, soil, and topography conditions. The science needs to meet the business practices. The implications in this environment are that there are not only scientific agricultural concerns, but there are technological developments, business concerns, and socio-political concerns. There are significant theories across these areas, but practitioners and stakeholders need to be involved—pointing to a transdisciplinary focus and co-development of solutions. Examples exist throughout, but more can and will be done.

In the pages that follow, readers will embark on a journey through the intricate tapestry of circular citrus supply chains. This book is more than a collection of insights; it is a call to action, a manifesto for change, and a roadmap toward a more sustainable and resilient future for our food systems. As we glean wisdom from the contributors within, let us envision a world where circularity is not just a concept but a way of life. Our food and our lives are dependent on building on this knowledge and then adopting it. I believe this book is instrumental in generating interest, contributing to knowledge, and resulting in action.

Co-Editor Springer-Nature Greening of Industry Networks Book Series
Worcester, MA, USA

Joseph Sarkis

Joseph Sarkis is Professor of Management within Worcester Polytechnic Institute's School of Business. He has published widely on topics of Sustainability, Operations, Supply Chain Management, and Technology. He has over 600 publications in journals, book chapters, conference proceedings, and books. He is currently the co-editor of the Springer-Nature Book Series on Greening of Industry Networks. He is also a founding member and co-chair of the Future Earth Working Group on Circular Economy and an International Program Director of the Greening of Industry Network.

Acknowledgement

The editors and authors express their deep gratitude to the PRIMA Foundation (Partnership for Research and Innovation in the Mediterranean Area) and the German Federal Ministry of Education and Research for funding the Project ImPUlSe (Innovation in the By-ProdUct Supply chain of citrus in Mediterranean area. Period 1 June 2021–31 May 2024).

The contributions in this book are based on this research project, yet including two further papers to strengthen knowledge sharing with other regions and open up further perspectives. For the first time the book establishes an integrated view of the circular food transformation in the Mediterranean areas to make a first step towards increasing resilience and sustainability in food supply chains, and fight food insecurity. The editors expressly thank all the authors for their willingness and commitment to share a variety of highly relevant aspects of their research and to include them in this publication.

Contents

About the Editors

Ani Melkonyan-Gottschalk is Professor for Sustainability and Socio-Technical Transformation at Clausthal University of Technology with focus areas in Environmental Economics, Circular Economy, Sustainable Supply Chains and Logistics/Mobility, as well as in Transformation of Urban Regions. Before she took on this professorship, she was Scientific Director of the Center for Logistics and Traffic at the University of Duisburg-Essen, and international consulting expert in the NASA-funded Project on Land-use and Land Cover Changes in South Caucasus, in frames of which she acted as Visiting Professor at Oklahoma University. Within the frames of the projects ILoNa "Innovative Logistics for Sustainable Lifestyles", CONUS—"Competence Network of Industrial and Rural Interlinkages", PRIMA-funded ImPULSe "Innovative Citrus By-Products Supply Chain in Mediterranean Area", she and her team carried out sustainability assessments throughout the entire value chain of the modern services for stationary and online commerce, while at the same time considering new business models, like sharing and circular economy, drafting transformation pathways along the entire supply chains. Through the projects on "Economic Assessment of Environmental Impact", funded by NASA, "Climate Mitigation and adaptation strategies" for the cities of Duisburg and Essen, "Essen—Green Capital City of Europe, 2017" as well as Mercator Foundation funded project "NEMO—New Emscher Mobility", Prof. Melkonyan-Gottschalk is working with local governmental authorities as well as industrial partners to develop collaborative decision support systems for sustainability transition strategies.

Prof. Melkonyan-Gottschalk has completed her habilitation (Venia Legendi) on Climate Change Influences on Economy. She received her PhD at the University of Duisburg-Essen, on the topic "Statistical Analysis of Long-Term Air Pollution in North Rhine-Westphalia, Germany". Her master's thesis was related to mathematical modelling in Macroeconomy, completed at the Yerevan State University, Armenia. Prof. Dr. Melkonyan-Gottschalk has published more than 30 scientific publications and 5 monographs.

Denis Daus is Researcher and Project Coordinator, currently coordinating an international and interdisciplinary research project at the University of Duisburg-Essen, Germany, focusing on innovations through circular supply chains and sustainable development in the Mediterranean region. His professional path has included roles as a Researcher in supply chain development and strategy at the Fraunhofer Institute for Material Flow and Logistics, Germany, where he contributed to international and German research and industrial projects on digitalization in sustainable supply chain management. Additionally, Daus has been involved in teaching and research at TU Dortmund University, Germany, specializing in circular economy and sustainability in supply chains, while supervising bachelor and master theses. Through exchange programs at Chiang Mai University (Thailand), Milwaukee School of Engineering (USA), and a work placement in Singapore, Daus concluded his studies with two bachelor (B.Sc., B.Sc.) and two master (M.Eng., M.Sc.) degrees, focusing on logistics engineering, international business, and supply chain management.

Chapter 1
Introduction to Transformation of the Agricultural Economy Towards Sustainable and Circular Systems

Ani Melkonyan-Gottschalk and Denis Daus

1 Background and Motivation: Aim of the Book

Incremental climate change adaptation policies in rural areas nowadays often focus on agriculture, agro-technology, and the management of natural resources, while larger trajectories of rural transformation seem to be ignored. Transformation is a way to move from simple understanding of environmental challenges towards addressing them quickly, effectively, and in a more resilient and holistic way (O'Brien 2012). Transformation towards more sustainable economic systems especially in rural areas is of crucial importance given the fact that rural communities are highly vulnerable towards climate change. This is because of their high dependence on natural resources, and the exposure to other processes, like rapid urbanization, trade liberalization, and environmental degradation with increased water scarcity or soil erosion (IPCC 2014; O'Brien et al. 2004). In contrary to the mitigation and adaptation strategies, transformation deals with fundamental changes (Feola 2015), including radical social, economic, technical, environmental, political, and spiritual change, addressing production and consumption systems, entire value and supply chains. It is widely recognized that efficient and effective supply chain management can herald many benefits for all stakeholders along the value chain from raw material suppliers to the end consumers. These benefits are, e.g. supply chain transparency, process/operational efficiency, cost reduction, rapid development of innovative business models, as well as environmental outcomes (Gunasekaran et al. 2017; Marshall et al. 2015). Particularly in the context of environmental pressure,

A. Melkonyan-Gottschalk (✉)
Clausthal University of Technology, Clausthal-Zellerfeld, Germany
e-mail: ani.melkonyan-gottschalk@uni-due.de

D. Daus
University of Duisburg-Essen, Duisburg, Germany

A. Melkonyan-Gottschalk, D. Daus (eds.), *Transformation Towards Circular Food Systems*, Greening of Industry Networks Studies 13,
https://doi.org/10.1007/978-3-031-63793-3_1

increasing market volatility and decreasing public sector investments in agriculture, the capacity of food system actors to apply innovation is crucial for achieving global food security (Bragdon and Smith 2015). Transparency and trust can increase the adoption and spread of technical innovation over larger areas, which will help to achieve long-term impact of digital business transformation within circular economies. The impact will be significant, e.g. in increased agricultural output, higher environmental protection, and risk mitigation along food supply chains (FSCs). Innovation can bring many benefits also to the consumers, such as increased food safety and traceability. Yet, horizontal and vertical fragmentation of the value chain and lack of integration among the actors (efficient industrial symbiosis) limit adoption of innovation. Thus, the core question arising is how innovation adoption can be fostered to transform circular food systems to increase societal benefits, such as food security, regional economic growth and competitiveness, functioning of markets and food value chains, employment, quality of life, and environmental resilience. The paucity of applied research on innovation adoption by the FSC actors, and on the increase in operational efficiency through new business models provides a significant gap which this book aims to address.

Therefore, acknowledging the fact that the complexity and dynamics of rural transformation request multi-layered governance, holistic approaches, and new forms of technical as well as financial assistance, the book investigates socio-economic impact of transformation in agricultural and rural areas using a multi-stakeholder approach. Thus, the main objective of the book is to open horizons for designing innovative supply chains of citrus by-products, taking into consideration current socio-economic situation of the actors, citrus production and distribution infrastructure, consumer preferences, and digitization progress at farmer and production levels. Moreover, the book aims to facilitate an interdisciplinary and systemic understanding of the challenges and opportunities within citrus supply chains (CSCs), promoting collaborative efforts towards a more sustainable future. The book aims to optimize resilience-oriented natural resource use by developing quantitative models to evaluate future climate change impacts on primary/agricultural production across the entire FSC. Additionally, it focuses on creating strategies for circular transformation. Thereby, the book addresses both biological and technological cycles of the circular economy, while focusing on citrus by-products supply chains. Technological circularity and machine/equipment (overhauling, remanufacturing, and repairing) is considered for both production and distribution of investments goods, which provides a solid ground for developing a circular logistics concept. The latter considers system design based on circularity, resource efficiency, and industrial symbiosis. Moreover, food loss and waste generation at each stage of the supply chain is analysed to identify the potential use cases of citrus by-products in the sense of biological circularity. Technological, biological, and logistics circularity across the FSCs in pilot countries presented in this book are reflected in Fig. 1.1. It also shows systematic horizontal and vertical integration of all the stakeholders across the Mediterranean region within an innovation hub to share the relevant knowledge and data, aiming to create innovative business models for citrus by-products supply chains in the region.

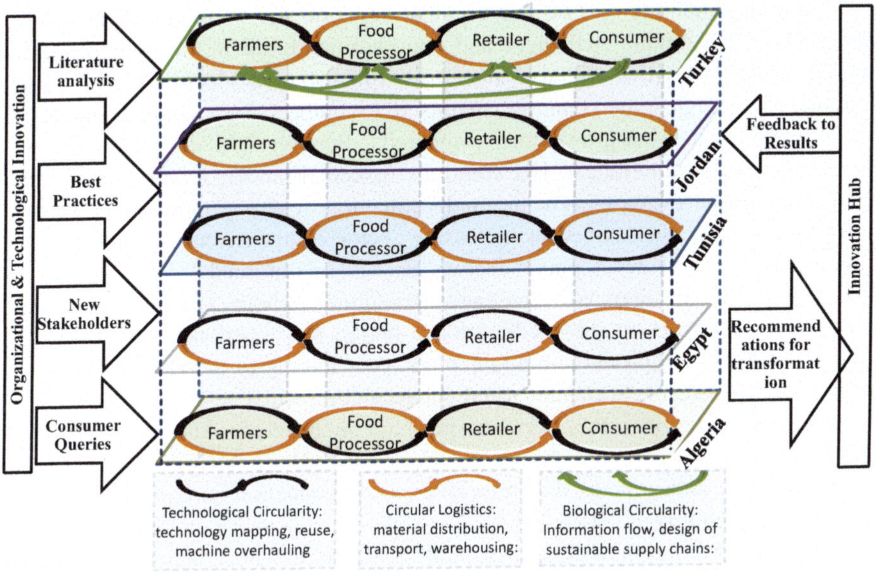

Fig. 1.1 Concept of exploring innovation potential along five citrus (by-products) supply chains in pilot countries (Algeria, Egypt, Tunisia, Jordan, and Turkey)

The arrows in brown and black show the circular characteristics of the value chain through integrating logistics and waste management. Arrows in green show the biological circularity (note, it is the same for all the five pilot countries). This picture also reflects all the structural elements and subobjectives of the book represented below:

1. Efficient adoption of new technologies, ensuring a gender balance in the process of digitization, including ageing population in the digitalization evolution.
2. Enhanced competitiveness and innovation capabilities of smallholders through developing circular economy concepts of citrus by-products (e.g. improved packaging and distribution strategies, use of discarded parts of the fruit, and circular technologies in food processing).
3. Fundamental technical basis for the design of new shortened supply chains with optimized logistics solutions, increasing security and safety by mapping the value chain in terms of structure and involved actors.
4. Increased sustainability of the entire CSC through effective resource management, sustainable primary and processed production, integrated ecological approaches from farm to consumption, waste processing, enhancement of social capital, and resilient distribution infrastructures.
5. Increased trust, understanding, and integration among the main food system actors, meanwhile providing information on EU food safety standards. This measure will undoubtedly accelerate regional development.

While sustainability in FSCs is a global concern, the book specifically concentrates on the unique challenges and opportunities faced in the Mediterranean region. There is a growth trend of the citrus fruit sector in Mediterranean coastal countries, although some countries, mainly EU members, face strong competitive limits, resulting in lower growth rates compared to the overall Mediterranean region. In contrast to Mediterranean EU countries (except France), North African and Near-East countries show dynamic and competitive citrus production. Effective governmental policies, aimed at increasing sector competitiveness, highlight the importance of local development factors as well as the need for reorganization and alignment with demand trends and agricultural policies (Duru et al. 2002). Developing countries in the Mediterranean region are experiencing a significant dynamic increase in production and exports, contributing to the region's competitive power. However, to maintain and enhance competitiveness in citrus exports, breeding activities in citrus varieties need to be fostered. Also strategies that contribute to the economic and social development of North African countries need to be planned, and policies that prioritize citrus production activities need to be established (Schimmenti et al. 2013).

Fostering innovation within sustainable CSCs is essential across this region, given the important relative weight of the sector. Total production of fresh citrus fruit amounted to 60 mln tons in the year 2016/2017, with 75% share being produced in Italy, Spain, Egypt, Morocco, and Tunisia. With 20 mln tons of citrus the Mediterranean region is the largest exporter of the fruit, with 3.6, 1.7, and 1.7 mln tons of export from Spain, Turkey, and South Africa, respectively (http://www.onagri.nat.tn/uploads/veille/Note-de-veille-Agrumes.pdf). With an average price of EUR 61 per 100 kg citrus fruit, the total output of the citrus production was estimated to be EUR 14 billion at basic prices. Besides being directly consumed and traded as raw commodities, fruits are processed into a multitude of food products grouped into frozen and preserved fruits: jams and marmalades (71.7%), juices (19.5%), prepared meals (3.6%), and drained fruits (1.5%). Citrus production activities produce adverse impacts on the environment through GHG emissions, water depletion, and food waste generation. Beyond environmental aspects, there are issues related to citrus quality, poor yield productivity, significant presence of smallholders with low investment in technological and human capital, continuous processes of aging, and growing price and income volatility. Some of these indicators with their current status can be found in Table 1.1.

Table 1.1 shows that Mediterranean countries face many food-related socio-economic challenges and require efficient agricultural policies in the agri-food sector since existing policies have not been successful in addressing socio-economic challenges. To address these challenges and improve the efficiency and effectiveness of material flows, a global value chain analysis approach of the fresh orange export sector, which considers comprehensive value creation rather than solely focusing on cost reduction, should be considered. Actions such as assessing varieties and rootstock combinations, breeding and testing new species, and choosing varieties that meet the preferences of European consumers are suggested (Sausman et al. 2015).

Table 1.1 Measurable indicators of the current situation of citrus supply chains in Mediterranean region

Sustainability indicators	Current situation
Economic indicators	
Output value of citrus production (Schimmenti et al. 2013)	5.3 bln USD in 2009–2010
Efficiency (http://ageconsearch.umn.edu/bitstream/9393/1/sp07la01.pdf)	69% (Spain)
Market share growth	Citrus production reach 10.8 MMT (https://gain.fas.usda.gov/Recent%20GAIN%20Publications/Citrus%20Annual_Madrid_EU-28_12-13-2017.pdf)
Market access (consumption)	Increase of 15.4% in 2016
Operational costs (for conventional and organic Navelina processing in Spain)	2711 €/ha conventional processing 5.355 €/ha for organic processing
Social indicators	
Increased percentage of qualified personnel	13% in the years 2001–2011
Increase percentage of female personnel (Marshall et al. 2015)	8%
Increase in training hours per worker	NA
Horizontal/vertical integration of stakeholders (industrial symbiosis)	NA
Environmental impacts	
Water use (Fader et al. 2016)	Net water requirement of citrus production 10,000 m^3ha^{-1}
CO_2 emissions (Daccache et al. 2014)	12 kg CO_2eq/kg citrus
Waste (Sharma et al. 2017)	12.1 mln tones

bln, billion

2 Theoretical Background: Literature Analysis

To derive circular and sustainable transformation in the agricultural systems, the book points out the necessity to develop dynamic models for sustainable movement of the agricultural goods from the point of production towards consumption and waste. This includes also the steps in between: growing and harvesting, packaging, transporting, warehousing, food processing, distribution and sale of the primary product of citrus and its by-products, such as peels, pulp, and juice. Thereby, the book determines sustainability gaps through the entire FSC, combining the theory of (food) supply chains with the concepts of ecological economics, digitization, circular transformation, and sustainable operations (as transport, warehousing). The overarching aim is to develop sustainable, smart, resilient, and circular farming systems. Integration of environmental risks, planning under uncertainty in terms of resilient and sustainable supply chains with a focus on the essential (agricultural) production have only been considered in the last few years. However, a useful contextualization to economic modelling remained unexplored, although debated on

Table 1.2 State-of-the-art in food supply chains

Entire food supply chain
1. Low efficiency, resilience, and sustainability (http://ageconsearch.umn.edu/bitstream/9393/1/sp07la01.pdf) 2. Low investment rate in new technologies and low adoption rate of innovation (Zude and Kläring 2009) 3. Lack of competitiveness due to high labour costs (Vergamini et al. 2015) and unemployment rate 4. Lack of transparency and coordination: supply and consumption mismatch (Ali and Soosay 2015) 5. Uncertainty in supply chain operations due to regulatory and legal risks (Bachev 2013) 6. Lack of governmental support (Bragdon and Smith 2015)
Farmers
Issues with low crop productivity (Knox et al. 2012), profitability (Bickford and Valverde 2016), high irrigation requirements (Sgarbossa and Russo 2017), innovation (https://www.accenture.com/_acnmedia/Accenture/Conversion-Assets/DotCom/Documents/Global/PDF/Digital_3/Accenture-Digital-Agriculture-Point-of-View.pdf), conventional farming practices, real-time information due to aged population (Ribal et al. 2016)
Food processors
Key issues are related to lack of innovative waste use solutions (biofuel), control systems (Verdouw et al. 2016) and real-time data, traceability, and food quality issues (Jedermann et al. 2014)
Distributors
The distributors of citrus products in the Mediterranean region usually lack access to valuable information and advanced logistics network design and analysis, operations planning, demand forecasting, and data analytics tools due to high cost and educational barriers (Sgarbossa and Russo 2017)
Retailers
Small retailers do not always have quick access to food alerts producing waste (https://ec.europa.eu/eip/agriculture/sites/agri-eip/files/eip-agri_fg_innovative_food_supply_chain_management_final_report_2015_en.pdf). They do not share warehouses: lack/excess of space represent high costs. Furthermore, there is a great disconnection between supplier and retailer in the agricultural sector
Consumers
Consumers generate high quantities of food waste due to the lack of control on the expiration date and less transparency (https://ec.europa.eu/eip/agriculture/sites/agri-eip/files/eip-agri_fg_innovative_food_supply_chain_management_final_report_2015_en.pdf). Consumers are often exposed to innovative products, but only a limited percentage of innovations remain in the market and are successful over time.
Circular engineers
Innovative methods and policies for waste recovery and bioconversion for high-value products are introduced (Laufenberg et al. 2003). There is a lack of systems to strengthen the value-based industrial symbiosis

international level. Surprisingly, the direct supply chain relation to climate change risks stayed weak. Moreover, carrying out a systematic literature analysis across FSCs revealed severe gaps at each stakeholder level across the FSC (Table 1.2).

3 Structure

In the first introductory chapter, the authors and editors Prof. Dr. Ani Melkonyan-Gottschalk and Denis Daus describe the background and the main motivation of this book, stressing out the urgent need for transformation of FSCs towards more circular and sustainable systems to fight poverty, especially in the developing world. The importance of the sustainable supply chain for transformation towards sustainable and circular economy is presented. This is underlined by discussing various theoretical concepts of sustainable supply chain management, contemporary sustainability science, sustainable consumption, delivery methods, food logistics, and resilient infrastructure. Moreover, some transformation pathways are sketched.

Part I analyses sustainable, resilient, and circular FSCs from consumption towards waste. Citrus processing waste is a source of valuable by-products that are crucial for human health and can be a source of biofuel and biogas. Valorizing citrus processing waste promotes efficient use of resources and waste minimization, which advances both circular economy and sustainability objectives. Such a closed-loop system can be referred to as circular CSC. In this part, the authors aim to explore drivers ensuring circular CSC competitiveness. Although wastes are distributed all along the supply chain, the most significant portions are those generated at consumer and processing levels. The focus is put only on wastes generated during the processing of citrus for juice extraction, due to its homogeneous nature, which provides opportunities for better valorization options. More specifically, the authors investigate the existence of drivers related to technology, information and knowledge, economy, market, organizational structure, and regulation. Additionally, an advanced Life Cycle assessment model evaluates the impacts of products flowing through the supply chain and reveals the potential of waste valorization in a circular CSC. Overall, the chapters involved in this part serve as a roadmap for fostering a more sustainable global citrus and citrus by-product supply chain, while identifying opportunities for enhancing resource management, and promoting sustainability within the citrus industry.

Part II analyses best cases in sustainable farming practices starting with yield and quality of citrus fruits. These practices are influenced by several preharvest factors with complex interdependencies, such as genetic, climatic, and cultural factors. Preharvest factors and their interactions may lead to desirable or undesirable results on yield and quality of citrus fruits. By understanding the agro-environmental challenges posed by global warming and drought, strategic decision-making for producers, exporters, and policymakers can be enhanced. This comprehensive examination of citrus growth conditions in the South Mediterranean region holds the key to ensuring the ecological sustainability, cultivation, and management practices of citrus species. The urgency to address these challenges cannot be overstated, and by doing so, the citrus industry can be better equipped to withstand the impacts of climate change and secure a prosperous future. The impacts of climate change are specifically obvious in the Mediterranean region, when it comes to water requirements of the citrus industry driven by frequent drought conditions. FAO software CROPWAT 8.0 has been applied to assess the crop water requirements, while relying on agrometeorological, cultural, and soil texture.

Besides the water management impact on crop growth, pesticides, and their use are also essential to ensure the sustainable growth of the plants. Thus, part II also covers the question of how the different types of pesticide are used in the region, while estimating the risks to human health, the environment, and by-products of the citrus processing chain. Summing up, the influence of preharvest practices in citrus production is discussed, focusing on maximizing yield and fruit quality, fulfilling consumer demands, and mitigating risks.

Moving towards digitization in FSCs, part III is devoted to technology adoption and advanced data analytics. By considering the reuse potential of industrial equipment and developing innovative business models, this discussion explores technological circularity and the overhauling, remanufacturing, and repairing of machinery. These strategies support both the production and distribution of investment goods, laying a solid foundation for the concept of circular logistics. The latter considers system design based on circularity, resource efficiency, and industrial symbiosis. Thus, part III explores the possibilities of collaborative technology adoption as a trigger for circular economy transitions for all the stakeholders involved in the supply chains. Collaborative technologies might significantly reduce technology gaps thanks to the ease of implementation, reduced risk, and better investment conditions. This ensures a better management of the total costs of ownership, especially in the maintenance, reuse, and life extension phases. The efficiency of new technology implementation depends considerably on the skills of the local personnel capacity for maintenance, upgrading, and repair of the technologies as main indicators for eco-efficient consumption. Hence, an ideal profile of a territory is explored in this chapter to allow for a correct and effective adoption of technologies, where the least negative externalities are generated. The key performance indicators of a beneficial circular adoption for both the adopter and the transmitter of the technology are identified, resulting in the development of a technological circularity index as a decision support tool. It integrates elements such as the diagnosis for local capacity competencies for MRO (maintenance, repair, and operations) of the adopted technologies, or the measurement of technological acceptance at difference segments of the society. Moreover, the review of regulatory approaches that facilitate reuse and promote eco-design, as well as the observation of technological achievements, are defined and analysed as study indicators, enabling the design of ideal profiles of study. As a result, concrete operational frameworks are designed to jointly build strategies with a territorial approach, where the appropriate governance can be established for a circular adoption of technologies. Meanwhile different interests of various stakeholders along the FSCs, related with the operation of a certain technology and its transfer, both at local and global level are discussed.

Further, a newly developed technology to optimize orange juice production is described, given the necessity for the juice to undergo pre-treatments that can lead to a change in organoleptic properties. The authors develop and test expanded bed chromatography in a vortex flow reactor, in which the orange juice retains the whole aroma, while avoiding heating the orange juice. The added value of the technology constitutes reduced numbers of technology steps, saving energy costs, and reducing the volume of treated cleaning solution released into the environment.

Moreover, taking the perspective of small-scale farmers, where the technology adoption level is at the lowest stage, the effectiveness of agricultural practices is significantly influenced by the presence or absence of accurate yield forecasting techniques. Therefore, various yield estimation approaches, spanning from traditional statistical methods to advanced machine learning algorithms, are examined with a focus on scalable methods applicable across different croplands. The authors aim to highlight the strengths, limitations, and applicability of these approaches in diverse agricultural contexts, while emphasizing their impact on supply chains and commodity markets.

Given the importance of yield forecast in supply chain planning, different forecasting methods for the future yield projection are analysed and applied. With demand or yield predicted ahead of time, decision-makers can make important decisions about its growth strategies and pricing. The objective of part III is further to compare different basic and advanced time series forecasting models and machine learning algorithms to select the best fitting model to forecast the future citrus crop yields accurately. Using statistical analysis and accuracy metrics, the efficiency of the methods for each dataset is estimated and the results of time series forecasting models and machine learning algorithms are compared. The parameters used in the models of simple moving average, simple exponential smoothing, Holt's and Winter's models are optimized. Further advance models, as automated Auto-Regressive Integrated Moving Average (ARIMA) and Seasonal Auto-Regressive Integrated Moving Average (SARIMA) models are used. The best fitting model is selected based on the results of several statistical and error estimation tests.

Sustainability in food logistics and supply chains is discussed in part IV. The FSC involves various logistics activities and processes that ensure the flow of food products from farms to consumers, such as packaging and transportation. Operational practices in the agricultural supply chain are essential for providing consumers with fresh and high-quality agricultural products while optimizing resource utilization, minimizing food loss, and supporting sustainability development goals. Available literature mainly focuses on the agricultural practices on the farm and operational practices in the processing stages, but rarely on the downstream of the supply chain. Therefore, the authors discuss the key operational practices and provide a catalogue that defines the best operational practices for each stage and the associated logistics activities. Moreover, indicating the fact that closed-loop CSCs eliminate waste, improve profitability of stakeholders within farms and cut the CO_2 emissions released from transportation, a multi-objective System Dynamics model is designed to estimate these positive effects of transportation in closed-loop supply chains. Overall, part IV supports logistics service providers and supply chain managers in selecting the best strategies for efficient and sustainable FSCs based on quantitative analyses.

Considering socio-economic aspects of FSCs, part V discusses innovation potential of circular supply chains. Circular supply chains are seen as a good way to achieve sustainability goals in food systems and developing resilient business models and schemes. However, the deployment of those schemes is a complex process since a multiplicity of factors enter into account. Thus, the identification of main factors that impact the continuity of such schemes is crucial to ensure the sustainability of FSCs. The aim of this chapter is to present the main drivers and limitations of the deployment and exploitation of sustainable and circular supply chain schemes involving by-products in fruits and vegetables. Citrus fruits are among the most consumed fruits worldwide. For this reason, effective management of the CSC is essential to ensure timely product availability while ensuring product quality and minimizing post-harvest losses. The CSC includes several stages, from agricultural production to industrial transformation and final distribution to consumers. In a circular perspective, it must also consider the management of residual materials. Those citrus side streams, the so-called by-products, are generated during the citrus production and processing process. Their valorization is key to circular economy. It describes a process of creation, development, and implementation of innovation practices that generate value with a view on sustainability where economic, environmental, and social dimension are considered. Although side streams or residual materials are common in the food industry, for citrus waste it is mainly technological innovations that have emerged in recent years. Particularly in view of finite raw materials, it will become increasingly important for companies to look at alternative processing methods that save resources and thus create a sustainable circular economy. As eco-systems and their regenerative capacities set the boundaries for human activities, the existing production, transportation, and consumption processes and patterns require profound transformational changes. To support such as transformation, part V explores scenarios for CSCs in the Mediterranean. By addressing potential future developments in Mediterranean citrus production and distribution systems within a circular economy, the authors aim to minimize uncertainties, foster resilience in supply chains, and inform strategic decision-making, while also considering the consumer perspective. The authors explored consumer acceptance of an orange probiotic drink compared to a nectar orange drink and pasteurized Not from Concentrate (NFC) orange juice. The investigation considers the impact of price, packaging, and various sustainability labelling attributes on consumer choice and purchase decisions. To achieve this, a choice-based conjoint survey is conducted using Conjointly in three countries located in Southern Mediterranean (Egypt, Algeria, and Tunisia). The findings aid in setting pricing strategies for orange juice and providing demand insights for the innovative probiotic orange juice, meanwhile acting as a basis to explore drivers for circular CSC competitiveness.

The main conclusions are derived in Chap. 14, describing impacts of the book, providing an outlook, and deriving policy recommendations.

References

Ali I, Soosay C (2015) Understanding risks in the Australian citrus supply chain. In: Proceedings of the 13th operations, supply chain, services management symposium, pp 1–21

Bachev H (2013) Risk management in the agri-food sector. Contemp Econ 7(1):45–62

Bickford R, Valverde C (2016) Spain's citrus report. GAIN Report Number SP 1637

Bragdon S, Smith C (2015) Small-scale farmer innovation. Quaker United Nations Office, Geneva, pp 1–17

Daccache A, Ciurana JS, Rodriguez Diaz JA, Knox JW (2014) Water and energy footprint of irrigated agriculture in the Mediterranean region. Environ Res Lett 9:124014

Duru S, Hayran S, Gül A (2002) The analysis of competitiveness of Mediterranean countries in the world citrus trade. Mediterr Agric Sci 35:21–26. https://doi.org/10.29136/mediterranean.1012466

Fader M, Shi S, von Bloh W, Bondeau A, Cramer W (2016) Mediterranean irrigation under climate change: more efficient irrigation needed to compensate for increases in irrigation water climate change: more efficient irrigation needed to compensate for increases in irrigation water requirements. Hydrol Earth Syst Sci 20:953–973

Feola G (2015) Societal transformation in response to global environmental change: a review of emerging concepts. Ambio 44(5):376–390

Gunasekaran A, Papadopoulos T, Dubey R, Wamba SF, Childe SJ, Hazen B, Akteret S (2017) Big data and predictive analytics for supply chain and organizational performance. J Bus Res 70:308–317

http://ageconsearch.umn.edu/bitstream/9393/1/sp07la01.pdf

http://www.onagri.nat.tn/uploads/veille/Note-de-veille-Agrumes.pdf

https://ec.europa.eu/eip/agriculture/sites/agri-eip/files/eip-agri_fg_innovative_food_supply_chain_management_final_report_2015_en.pdf

https://gain.fas.usda.gov/Recent%20GAIN%20Publications/Citrus%20Annual_Madrid_EU-28_12-13-2017.pdf

https://www.accenture.com/_acnmedia/Accenture/Conversion-Assets/DotCom/Documents/Global/PDF/Digital_3/Accenture-Digital-Agriculture-Point-of-View.pdf

IPCC (2014) Climate change 2014: impacts, adaptation, and vulnerability. Part A: global and sectoral aspects. In: Field CB, Barros VR, Dokken DJ, Mach KJ, Mastrandrea MD, Bilir TE, Chatterjee M, Ebi KL, Estrada YO, Genova RC, Girma B, Kissel ES, Levy AN, MacCracken S, Mastrandrea PR, White LL (eds) Contribution of Working Group II to the fifth assessment report of the intergovernmental panel on climate change. Cambridge University Press, Cambridge, UK and New York, p 1132

Jedermann R, Potsch T, Lloyd C (2014) Communication techniques and challenges for wireless food quality monitoring. Philos Trans R Soc 19:20130304

Knox J, Hess T, Daccache A, Wheeler T (2012) Climate change impacts on crop productivity in Africa and South Asia. Environ Res Lett 7:034032

Laufenberg G, Kunz K, Nystroem M (2003) Transformation of vegetable waste into value added products. Bioresour Technol 87:167–198

Marshall D, McCarthy L, McGrath P, Claudy M (2015) Going above and beyond: how sustainability culture and entrepreneurial orientation drive social sustainability supply chain practice adoption. Supply Chain Manag 20(4):434–454

O'Brien K (2012) Global environmental change II: from adaptation to deliberate transformation. Prog Hum Geogr 36(5):667–676. https://doi.org/10.1177/0309132511425767

O'Brien K, Leichenko R, Kelkar U, Venema H, Aandahl G, Tompkins H et al (2004) Mapping vulnerability to multiple stressors: climate change and globalization in India. Glob Environ Chang 14:303–313

Ribal J, Ramírez-Sanz C, Estruch V, Clemente G, Sanjuán N (2016) Organic versus conventional citrus. Impact assessment and variability analysis in the Comunitat Valenciana (Spain). Int J Life Cycle Assess:1–16

Sausman C, Garcia M, Fearne A, Felgate M, Mekki AAE, Cagatay S, Soliman I, Thabet B, Thabet C, Saïd MB, Laajimi A, Ashkar HA, Hadad-Gauthier FE, Mili S, Martínez C (2015) From value chain analysis to global value chain analysis: fresh orange export sector in Mediterranean partner countries. In: Petit M, Montaigne E, El Hadad-Gauthier F, García Álvarez-Coque JM, Mattas K, Mili S (eds) Sustainable agricultural development: challenges and approaches in southern and eastern Mediterranean countries. Springer International Publishing, Cham, pp 197–225

Schimmenti E, Borsellino V, Galati A (2013) Growth of citrus production among the Euro-Mediterranean countries: political implications and empirical findings. Span J Agric Res 11(3):561–577

Sgarbossa F, Russo I (2017) A proactive model in sustainable food supply chain: insight from a case study. Int J Prod Econ 183:596–606

Sharma K, Mahato N, Cho M, Lee YR (2017) Converting citrus wastes into value-added products: economic and environmentally friendly approaches. Nutrition 34:29–46

Verdouw CN, Wolfert J, Beulens AJM, Rialland A (2016) Virtualization of food supply chains with the internet of things. J Food Eng 176:128e136

Vergamini D, Cuming D, Viaggi D (2015) The integrated management of food processing waste: the use of the full cost method for planning and pricing Mediterranean citrus by-products. Int Food Agribus Manage Rev 18(2):153–172

Zude M, Kläring HP (2009) Two case studies of precision horticulture approaches in citrus production. Tree For Sci Biotechnol 3:139–151

Part I
Sustainable, Resilient and Circular Food Supply Chains

Chapter 2
Drivers for a Competitive Circular Citrus Supply Chain: A Case Study of Citrus Peel Waste Dehydration in Egypt

Heba Yusuf, Hadi Fors, Noha M. Galal, Ahmad E. Elhabashy, and Nermine Harraz

1 Introduction

According to the United States Department of Agriculture (USDA) Foreign Agricultural Service, Egypt is one of the top countries in citrus production globally. Oranges are the major citrus crop in Egypt, representing about 80% of the total cultivated citrus area (Wally 2022). According to the Statistics Portal for Market Data, Egypt was the fifth country among the leading orange-producing countries worldwide in 2023/2024 (Statista 2024).

The export value of Egyptian oranges is substantial, with an estimated value of about 493.28 million dollars during the period 2014–2017, making oranges one of the most important exported fruits in Egypt (Hassanain and Gabr 2020). The excess in production of Egyptian oranges and its high reputation support the transition towards more industrial processing for juice production as a strategic move for Egypt's orange industry (Duru et al. 2022). Industrial processing of orange fruit for juice production adds value to the surplus of oranges, creates employment opportunities, and enhances further the country's position in the global market. Orange processing for juice extraction generates a significant amount of waste as the juice constitutes approximately 55% of the fruit weight, while the remaining is pulp, oil, and Citrus Peel Waste (CPW), with the ratio of 3%, 0.3%, and 41.3%, respectively (Zema et al. 2018a).

H. Yusuf · H. Fors (✉) · A. E. Elhabashy · N. Harraz
Production Engineering Department, Industrial Engineering Division, Faculty of Engineering, Alexandria University, Alexandria, Egypt
e-mail: hadifors@alexu.edu.eg

N. M. Galal
Department of Industrial and Management Engineering, College of Engineering and Technology, Arab Academy for Science, Technology, and Maritime Transport, Alexandria, Egypt

A. Melkonyan-Gottschalk, D. Daus (eds.), *Transformation Towards Circular Food Systems*, Greening of Industry Networks Studies 13,
https://doi.org/10.1007/978-3-031-63793-3_2

By taking into consideration the global situations of food insecurity, economic instability, and recessions, the reuse of food processing waste generated in agri-food production chains is a crucial need (Bhat 2021). In recent years, the term "Circular Economy", which is based on the concept of reducing waste going to landfill, has gained increasing attention. The implementation of Circular Economy (CE) principles in the agri-food industry has the potential to minimize waste, create economic growth, and reduce negative environmental effects by transforming agri-food waste into new materials or products, thus promoting resource efficiency and sustainability (Nattassha et al. 2020).

There are several options for valorizing and transforming agri-food waste into new materials or products. As an example, the CPW generated during orange fruit processing is typically rich in bioactive compounds, including polyphenols, flavonoids, essential oils, and dietary fibers, which have high antioxidant capacity and various health-promoting benefits (Kandemir et al. 2022). Hence, the high nutritional value and the consistency of CPW composition grant it the potential to be utilized as a source of bioactive compounds with high economic value (Putnik et al. 2017; Teigiserova et al. 2021). Many studies highlighted the potential of CPW utilization as a source of bioactive compounds for producing biogas, through cascading industrial processes. This option is called biorefinery, where resources and energy are recovered simultaneously, aiming to achieve net zero waste (Kim et al. 2022).

Another example is valorizing CPW by dehydration. Since CPW is highly perishable in nature, due to its high moist content, dehydrating CPW is a crucial step. Such a step enhances the implementation of the Circular Citrus Supply Chain (CCSC), as it allows the preservation of the bioactive compounds by reducing water content. Adding value to CPW by dehydration turns it from an economic and environmental burden to a driver for economic, environmental, and social competitiveness. Not only does it save land and the environment, but it also transforms the CPW into a valuable resource that can be used in a variety of industries, along with creating new jobs as well as generating revenues. In fact, the selling price for one ton of dehydrated CPW of Egyptian citrus is about $3500 (ITC 2023), which is above the average price for this product by $2000, explained by a rich content of bioactive compounds (Elmaidomy et al. 2023).

Therefore, adding value to orange fruit by juice extraction is not the only benefit, but also the utilization of orange fruit processing waste carries further opportunities supporting the implementation of the CCSC. Although challenges exist for implementing CCSC, recognizing the difficulties encourages stakeholders within citrus supply chains to adopt circular practices. Accordingly, this research aims to explore drivers from the literature for implementing CCSC, in addition to investigating the prospective drivers of CCSC within a specific case study in Egypt, while developing a competitive strategy for implementing CCSC. To achieve this aim, two research questions (RQs) are formulated: RQ1) What are the drivers of a competitive CCSC from the literature? and RQ2) What are the potential drivers and strategies to develop and implement a competitive CCSC in Egypt?

The remaining sections are organized as follows: Section 2 presents the methodology to answer the research questions. Section 3 illustrates the list of drivers extracted from the literature review and the conducted internal and external factors evaluation of the case company. Section 4 discusses the alignment of the company's drivers for implementing CCSC with the reviewed drivers in the literature, along with the developed strategies for supporting the competitiveness of the CCSC. Finally, the research work is concluded in Sect. 5.

2 Methodology

To achieve the research objective, the following methodology is adopted as depicted in Fig. 2.1. After defining the problem by formulating the research questions, a literature review is conducted to identify and categorize the drivers of the CCSC (RQ1). To explore the potential drivers of CCSC in Egypt and judge how they are aligned with the drivers addressed in the literature and suggest strategies for enhancing a CCSC (RQ2), a case-based approach is adopted. The case-based research method is suitable for the exploration stage targeted in this study (Voss et al. 2002). To this end, a case company active in the CCSC in Egypt is investigated. The generated list of drivers and the SWOT matrix are used to support the strategic planning of a CCSC actor in emerging economies. The details of the adopted methodology are described in the following sub-sections.

2.1 Literature Search

The literature search aims to answer the first research question: What are the drivers from the literature of a competitive CCSC? To answer the research question, a set of four keyword groups has been selected.

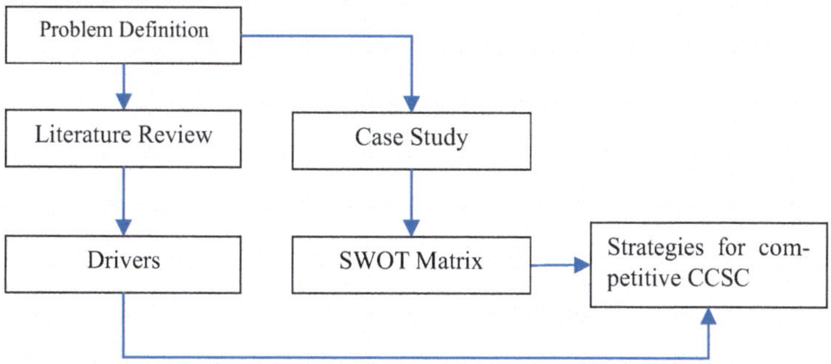

Fig. 2.1 Followed methodology

1. Product type: "citrus" or "orange."
2. Waste: "byproducts" OR "waste" OR "juice" OR "processing" OR "extraction."
3. Circularity: "valorization" OR "valorisation" OR "exploitation" OR "circular economy."
4. Drivers: "driver" OR "opportunity" OR "solution" OR "strength" OR "chance" OR "privilege."

Articles and review papers published in conferences or journals between 2017 and 2023 in the English language were included. Hence, the following query was entered in the advanced search option of the SCOPUS database: TITLE-ABS-KEY ((citrus OR orange) AND (byproducts OR waste OR juice OR processing OR extraction) AND (valorisation OR valorisation OR exploitation OR "circular economy") AND (driver OR opportunity OR solution OR strength OR chance OR privilege)) AND (LIMIT-TO (DOCTYPE,"ar") OR LIMIT-TO (DOCTYPE,"re") OR LIMIT-TO (DOCTYPE,"cp")) AND (LIMIT-TO (PUBYEAR,2023) OR LIMIT-TO (PUBYEAR,2022) OR LIMIT-TO (PUBYEAR,2021) OR LIMIT-TO (PUBYEAR,2020) OR LIMIT-TO (PUBYEAR,2019) OR LIMIT-TO (PUBYEAR,2018) OR LIMIT-TO (PUBYEAR,2017)) AND (LIMIT-TO (LANGUAGE, "English")).

This search was conducted in late September of 2023 and resulted in 93 papers, which have been checked for accessibility and relevance. The full text of two papers was not available to the authors. A relevance check was conducted by examining the paper title and abstract to determine whether the scope of the paper is relevant to the current research. This process resulted in 63 papers included in the current review and 28 publications being omitted due to irrelevance. Each paper has been read in full to identify the drivers for valorizing citrus waste. Drivers are then categorized into economical, environmental, social, technological, and legal drivers. The categorization scheme is inspired by the PESTEL analysis, which is a tool commonly used for strategic management. While PESTEL analysis is used to identify external factors affecting the market, in this work the categories include all types of internal and external drivers, relevant for the supply chain. Due to their similarity and limited discrimination found in the reviewed literature, the political and legal categories have been merged into one category. The different categories are delineated as follows:

- Political/Legal: Related to governmental policies and initiatives, specific laws, and regulations.
- Economical: Related to cost, profit, income, and taxes.
- Social: Related to societal health and safety.
- Technological: Related to the process development and implementation.
- Environmental: Related to the effect on the ecosystem including earth, air, animals, and water resources.

The number of papers included in the review has been equally divided between three of the authors. Initially, the authors have individually identified the drivers from the reviewed papers and categorized them. For some drivers it was feasible to assign them to multiple categories, in such cases a main category was identified based on the relevance to the context of the paper, and the other category was considered as an additional category. Furthermore, several review meetings and discussions have been conducted including all three authors to check the consistency of the category assignment and the appropriateness of the driver description.

2.2 *Case Study Design*

Case-based research is selected in this work to investigate the possible drivers of CCSC in Egypt, assess how well they match the drivers discussed in the literature, and offer recommendations for improving the competitiveness of the case company, which is a primary actor in the CCSC. One in-depth case study on a leading company that dehydrates fruits and vegetables in Egypt has been conducted. The company offers high-quality organic dry orange peels, and they are considered one of the main products offered by the company for international food and pharmaceutical manufacturers. The company acts as the first step for CPW utilization, as a source of valuable bioactive materials. The company specializes in dehydrating fruits and vegetables, with dehydrated CPW accounting for 10% of the total sales of the company. The company's market share for dehydrated CPW is approximately 90% of the local producer's market. The CPW are purchased from juice processors, while the dehydrated CPW are mainly exported to pharmaceuticals and food manufacturers in Germany, along with being sold domestically for local exporters and farms as animal feed. Figure 2.2 shows the case study company's supply chain for the dehydrated CPW.

Data are gathered through semi-structured interviews with the factory manager, financial manager, operation manager, and marketing manager. Aiming to conduct External Factor Evaluation (EFE) and Internal Factor Evaluation (IFE), to assess internal and external factors affecting the industry (David 2011), the interviews took place in two sessions, each of which was 3 hours long, these evaluations along with a SWOT matrix help in formulating strategies to support the competitiveness of the company.

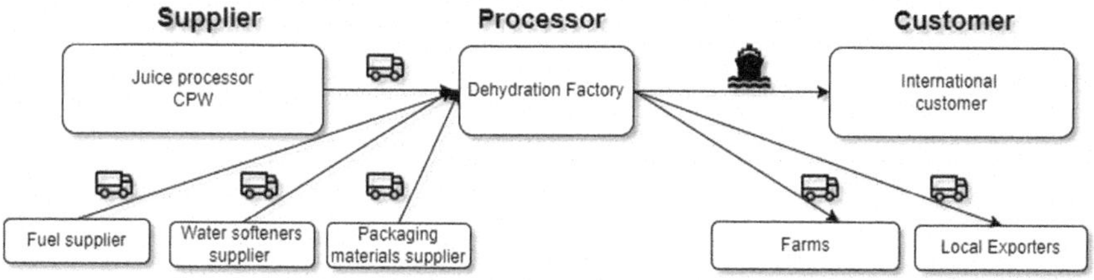

Fig. 2.2 The case study factory's supply chain for the dehydrated CPW

The developed IFE matrix summarizes the company's internal factors evaluation to measure business competitiveness, through identifying key strengths and weaknesses. Each internal factor is weighted and scored to obtain a weighted score; with the weights ranging from 0.0 (low importance) to 1.0 (high importance). Scores from 1 to 4 are used to rate each factor, the strengths receive a 3 or 4 rating, and the weaknesses receive a 1 or 2 rating. More specifically, a minor strength receives a rating of 3, while a major strength receives a rating of 4; whereas a major weakness receives a rating of 1, while a minor weakness receives a rating of 2. The sum of the weighted scores is the total weighted score for the organization.

The developed EFE matrix is to measure the industry's attractiveness, by weighing and scoring the threats and the opportunities that affect the industry to obtain a total weighted score for all factors. Similarly, each factor is assigned a weight that ranges from 0.0 (not important) to 1.0 (very important), indicating the relative importance of that factor to industry. Then, a score is also assigned to each factor between 1 and 4 to each key external factor (opportunities and threats), which indicates how effectively the company is responding to that factor. More specifically, a score of 4 indicates the response is superior, 3 indicates the response is above average, 2 indicates the response is average, and 1 indicates the response is poor. Each factor's weight is multiplied by its rating to determine a total weighted score for the organization's response to the external factors.

After conducting the IFE and EFE, the total weighted scores obtained from both evaluations are plotted on the Internal-External (IE) matrix. The IE Matrix is a strategic management tool used in analyzing the current position of a business and providing appropriate future strategies based on both the IFE and EFE. It was developed by General Electric to position an organization's business units in a nine-cell matrix (David 2011). The nine cells of the IE matrix are divided into three regions corresponding to three types of strategy: Growth and Build, Hold and Maintain, and Harvest or Divest.

To formulate possible strategies based on the company's strengths, weaknesses, opportunities, and threats, a SWOT Matrix is used to formulate four categories of strategies:

- SO (strengths-opportunities) strategies address the internal strengths with the external opportunities.
- WO (weaknesses-opportunities) strategies tackle the internal weaknesses with the external opportunities.
- ST (strengths-threats) strategies handle the internal strengths with the external threats.
- WT (weaknesses-threats) strategies deal with the internal weaknesses with the external threats.

3 Results

3.1 Drivers for a Competitive Circular Citrus Supply Chain

The economical drivers of the CCSC listed in Table 2.1 present initiatives to several actors in the supply chain. Farmers may benefit from circularity through the reduction in operating expenses as well as the generation of new revenue streams. Additionally, on the agricultural level, the valorized waste reduces the need for fertilizer usage and provides a source for ruminant feed. For the manufacturer, economic competitiveness may be improved through the reduction of the disposal cost (Raimondo et al. 2018) and the generation of new revenue streams resulting from the valorized waste (Zema et al. 2018b).

Table 2.1 Economical drivers category

Driver	Additional category	References
Enhancing economic competitiveness	–	Raimondo et al. (2018), Zema et al. (2018b)
Providing lower feeding costs for ruminants	–	Raimondo et al. (2018), Tayengwa and Mapiye (2018)
Providing income to farmers	Social	Zema et al. (2018b), Tayengwa and Mapiye (2018)
A cost-effective valorization technique	–	Ramya and Balachandran (2021), Alwazeer and Elnasanelkasim (2023), Tayengwa and Mapiye (2018), Lucia et al. (2022)
Saving the cost of Citrus Waste Water (CWW) treatment and displacement	–	Lucia et al. (2022)
Reducing the farmers' operating costs	–	Lucia et al. (2022)
Reducing the use of fertilizers	–	Lucia et al. (2022)
Extracting valuable compounds and chemicals	Social	Lucia et al. (2022), Xu et al. (2023), Stone et al. (2019)
Developing the bioeconomy	Technological	Ortiz-Sanchez et al. (2023)
Increasing efficiency of the food supply chain	Social	Gonzalez-Canche et al. (2021)
Promoting investments to mitigate negative socioeconomic and environmental impacts	Social/Environmental	Gonzalez-Canche et al. (2021)
Material to be valorized is inexpensive	–	Castaño et al. (2022)
Developing low-cost adsorbent/biosorbents	Environmental	Fernández-López et al. (2021), Campagnolo et al. (2019)
Cheap source of carbon	–	Vaishnav et al. (2022)

Table 2.2 Social drivers category

Driver	Additional category	References
Extracting dietary fiber (pectin)	–	Raimondo et al. (2018), Perez-Pirotto et al. (2022)
Contributing to food and nutrition safety	–	Tayengwa and Mapiye (2018)
Low-calorie food	–	Xu et al. (2023)
Reducing food loss during growth and storage phases	–	Xu et al. (2023)
Enhancing the nutritional value of food products	–	Xu et al. (2023), Iqbal et al. (2021), Eliopoulos et al. (2022), Rosales and Fabi (2023), Molina-Montero et al. (2023)
Enriched content of bioactive materials	–	Cassano et al. (2021), Ningrum et al. (2018), Iqbal et al. (2021), Kandemir et al. (2022), Tropea et al. (2022), Zuin and Ramin (2018), Khan et al. (2023), Ndayishimiye and Chun (2017), Myrtsi et al. (2022)
Using byproducts in the cosmetics industry	–	Khan et al. (2023), Eliopoulos et al. (2022)
Providing additional employment opportunities	Technological	Boccia et al. (2021)

Actors involved in the valorization process are also driven by the cost-effectiveness of the valorization techniques (Ramya and Balachandran 2021; Alwazeer and Elnasanelkasim 2023; Tayengwa and Mapiye 2018; Lucia et al. 2022). The recovery of valuable material from citrus waste is another important driver for a competitive CCSC (Lucia et al. 2022; Xu et al. 2023; Stone et al. 2019).

Supply chain-wide drivers include the promotion of bioeconomy (Ortiz-Sanchez et al. 2023) and increasing the efficiency of the food supply chain (Gonzalez-Canche et al. 2021) through the generation of value-added products from waste.

The social drivers for a competitive CCSC are summarized in Table 2.2. These address the health, safety, and welfare of humans. The main driver affecting the health and safety of humans is the enriched nutritional content (Raimondo et al. 2018; Cassano et al. 2021; Ningrum et al. 2018; Iqbal et al. 2021; Kandemir et al. 2022; Tropea et al. 2022; Zuin and Ramin 2018; Khan et al. 2023; Ndayishimiye and Chun 2017; Myrtsi et al. 2022; Perez-Pirotto et al. 2022) of the CPW that can be deployed in food (Iqbal et al. 2021; Eliopoulos et al. 2022; Rosales and Fabi 2023; Molina-Montero et al. 2023) and cosmetic industries (Khan et al. 2023; Eliopoulos et al. 2022). Additionally, the valorization of waste creates new job opportunities (Boccia et al. 2021), thus promoting societal welfare.

The technological drivers, summarized in Table 2.3, are the ones related to process development and implementation. Perhaps the most significant driver in this category is valorization techniques that could be easily implemented in general (Fazzino et al. 2021) and, more preferably, on a large scale (Ramya and Balachan-

Table 2.3 Technological drivers category

Driver	Additional category	References
Extracting essential oil (D-limonene)	–	Raimondo et al. (2018)
Producing biogas and ethanol	Environmental	Raimondo et al. (2018), Zema et al. (2018b)
A scalable valorization technique	–	Ramya and Balachandran (2021), Fazzino et al. (2021)
Development of sustainable life, pollution control, and energy storage material	Environmental	Xu et al. (2023), Gonzalez-Canche et al. (2021)
Green and sustainable separation methods for agro-industrial waste	Environmental	Zuin and Ramin (2018), Kaur et al. (2022)
Medicine development	Social	Boccia et al. (2021), Jamsheera and Pradeep (2023)
Ease of valorization technique implementation	–	Fazzino et al. (2021)
Producing activated carbon	Environmental	Súarez-Velázquez et al. (2022), Grima-Olmedo et al. (2023)
Helping with fertilizer development	Environmental	Pérez et al. (2022), Piscitelli et al. (2022), Tolisano and Del Buono (2023)
Developing healthier meat products	Environmental	Nieto et al. (2021)
Producing food packaging films	Environmental	Papadaki et al. (2022)
Helping with food preservation	Environmental	Khalil et al. (2023), Jorge et al. (2023)
Increasing CH4 production; boosting anaerobic digestion	–	Chen et al. (2023)
Promising alternative to lignocellulosic biomass	–	Jeong et al. (2021)
Preparing biodegradable mulch films	Environmental	Merino and Athanassiou (2022)
Enhancing the recyclability of other products	Environmental	Carrodeguas et al. (2020)

dran 2021; Fazzino et al. 2021). Being able to use the citrus waste to produce other valuable products is also a common technological driver for competitive CCSC. A wide range of products could be manufactured using the waste, including biogas and ethanol (Raimondo et al. 2018; Zema et al. 2018b), essential oil (Raimondo et al. 2018), activated carbon (Súarez-Velázquez et al. 2022; Grima-Olmedo et al. 2023), medicines (Boccia et al. 2021; Jamsheera and Pradeep 2023), fertilizers (Pérez et al. 2022; Piscitelli et al. 2022; Tolisano and Del Buono 2023), meat products (Nieto et al. 2021), and packaging films (Papadaki et al. 2022).

Furthermore, some valorization techniques have an additional favorable impact on the environment, beyond just recycling citrus waste. One example is techniques

Table 2.4 Legal drivers category

Driver	Additional category	References
Complying with regulations and strategies	–	Ortiz-Sanchez et al. (2023), Eriksson and Spångberg (2017)
Achieving "zero waste" policies	–	Gullón et al. (2020)

that can separate the waste in a sustainable way (Zuin and Ramin 2018; Kaur et al. 2022). Another example is producing environmentally friendly products, such as energy storage material (Xu et al. 2023; Gonzalez-Canche et al. 2021) and food packaging materials that help with preserving the food (Khalil et al. 2023; Jorge et al. 2023).

Table 2.4 presents the legal drivers for CCSC. There are only less than a handful of regulations and policies that promote circularity and sustainability in the related literature. For example, valorizing citrus waste while considering various transformation pathways promotes adherence to the Farm-to-Fork (F2F) strategy (Ortiz-Sanchez et al. 2023), the United Nation's Sustainable Development Goals (SDG) (Ortiz-Sanchez et al. 2023), and the European Waste Framework Directive (WFD) (Eriksson and Spångberg 2017). In addition, utilizing byproducts in the agri-food industry is considered a motive for the adoption of "zero waste policies" (Gullón et al. 2020).

Table 2.5 lists the environment-related drivers for implementing a CCSC. These drivers revolve around saving the environment, enhancing soil and plant properties, managing feedstock, and providing environmentally friendly energy sources. Valorizing the large quantities of generated citrus byproducts helps in protecting the environment. Utilizing these byproducts leads to the reduction of fruit wastes (Kumar et al. 2022) and the prevention of accumulating wastes in landfills (Grima-Olmedo et al. 2023). As a result of waste valorization, municipal and agro-industrial waste management is reduced (Súarez-Velázquez et al. 2022). In addition, by adding these byproducts to ruminants' diets, problems related to their disposal into the environment can be avoided (Tayengwa and Mapiye 2018). Disposal of CPWs has a negative effect on soil and groundwater because they produce leachate, and have an adverse effect on air, as they emit odors and greenhouse gases (Zema et al. 2018b). Furthermore, the presence of oil in citrus peels reduces the rate of decomposition (Raimondo et al. 2018).

Citrus Waste Water (CWW) is derived from fruit washing and processing. Utilizing this wastewater for irrigation could prevent the natural water sources from being significantly depleted, enabling full use of the water for commercial and industrial uses (Lucia et al. 2022). In addition, CWW could provide organic matter and nutrients to the soil, thus improving its fertility (Lucia et al. 2022) and enhancing carbon fixation (Xu et al. 2023). Another waste obtained from fruit processing is the Orange Peel Waste (OPW), this waste can be used in removing phosphorus or nitrate in wastewater, thus obtaining high-value material rich in phosphorus or nitrogen, which could be used for agricultural purposes (Pérez et al. 2022; Moreno et al.

Table 2.5 Environmental drivers category

Driver	Additional category	References
A valorization technique that reduces the negative impacts on the environment	–	Raimondo et al. (2018), Zema et al. (2018b), Ramya and Balachandran (2021), Alwazeer and Elnasanelkasim (2023), Tayengwa and Mapiye (2018), Lucia et al. (2022), Xu et al. (2023), Castaño et al. (2022), Ndayishimiye and Chun (2017), Eliopoulos et al. (2022), Grima-Olmedo et al. (2023), Pérez et al. (2022), Piscitelli et al. (2022), Khalil et al. (2023), Carrodeguas et al. (2020), Kumar et al. (2022), Moreno et al. (2022), Abdullahi et al. (2022), Belorkar and Kausar (2023), Satari and Karimi (2018), Zema et al. (2019), Bendaoud-Boulahlib and Ghizellaoui (2017)
The negative environmental impact of landfilling waste compared to waste valorization	–	Raimondo et al. (2018), Zema et al. (2018b), Tayengwa and Mapiye (2018), Xu et al. (2023), Ortiz-Sanchez et al. (2023), Ningrum et al. (2018), Iqbal et al. (2021), Kandemir et al. (2022), Tropea et al. (2022), Khan et al. (2023), Grima-Olmedo et al. (2023), Jamsheera and Pradeep (2023), Kaur et al. (2022), Panuccio et al. (2022), Mariana et al. (2021), Antonio et al. (2021), Valenti et al. (2017)
Using byproducts as animal feed	–	Raimondo et al. (2018), Tayengwa and Mapiye (2018), Stone et al. (2019), Eliopoulos et al. (2022), Súarez-Velázquez et al. (2022)
Material to be valorized is abundant	–	Zema et al. (2018b), Tayengwa and Mapiye (2018), Lucia et al. (2022), Xu et al. (2023), Ortiz-Sanchez et al. (2023), Fernández-López et al. (2021), Kandemir et al. (2022), Ndayishimiye and Chun (2017), Molina-Montero et al. (2023), Nieto et al. (2021), Kaur et al. (2022), Gullón et al. (2020), Abdullahi et al. (2022), Antonio et al. (2021), Teigiserova et al. (2019), Bousios and Worrell (2017), Tamelová et al. (2018), Calabrò et al. (2019)
Saving water use by providing water for crop irrigation	Economical	Lucia et al. (2022)
Providing organic matter and nutrients to the soil	–	Lucia et al. (2022), Xu et al. (2023)
Providing an alternate source of energy	–	Xu et al. (2023), Zuin and Ramin (2018)
Abundancy in waste: high residue-to-product rate	–	Kandemir et al. (2022), Antonio et al. (2021)

(continued)

Table 2.5 (continued)

Driver	Additional category	References
Reducing municipal and agro-industrial waste management	–	Súarez-Velázquez et al. (2022)
Helping with wastewater management	–	Pérez et al. (2022), Moreno et al. (2022), Zema et al. (2019)
Enhancing agronomic attributes of plants	–	Tolisano and Del Buono (2023)
Reducing food and fruit waste	Social	Gullón et al. (2020), Kumar et al. (2022), Bendaoud-Boulahlib and Ghizellaoui (2017), Sabater et al. (2022), Singh and Singh (2022)
Producing green diesel	Technological	Abdullahi et al. (2022)

2022). Furthermore, Tolisano and Del Buono (2023) mentioned that nanoparticles obtained from orange peels can enhance the agronomic attributes of tomato plants (Tolisano and Del Buono 2023).

Agro-industrial wastes could be utilized in several applications. Citrus pulp could be used as animal feed (Eliopoulos et al. 2022). Citrus peel could also be valorized in producing environmentally friendly diesel (Abdullahi et al. 2022). As dietary supplements, citrus byproducts can improve ruminants' nutrition, health, productivity, and meat quality (Tayengwa and Mapiye 2018). Zuin and Ramin (2018) mentioned that agro-industrial waste can be considered as an alternate source of energy leading to minimizing the dependence on non-renewable materials, such as fossil fuels, and reducing greenhouse gas emissions (Zuin and Ramin 2018).

3.2 Internal and External Factors Evaluation

3.2.1 Internal Factor Evaluation (IFE)

Table 2.6 shows the resulting IFE matrix for the case study company, where the evaluation shows a relatively strong internal position. A description is provided for each point of strength and weakness in the table.

3.2.2 External Factor Evaluation (EFE)

Table 2.7 shows the resulting EFE matrix for the case study company, the evaluation shows that the factory is doing slightly below average for the dry CPW. Such a performance means that the factory is not taking full advantage of external opportunities and some threats are significant and is not properly handled, such as the

Table 2.6 The resulting IFE matrix for the dehydration company

Factors		Weight	Rate	Weighted score	Description
Strengths	S1: High product quality	0.1	4	0.4	The company follows a set of quality assurance measures to verify that the dehydrated peels are of very high quality complying to the European Union (EU) specifications
	S2: High market share	0.1	4	0.4	The company's market share is almost 90% of the local exporter's market
	S3: Using food dehydrator	0.03	3	0.09	Using food dehydrators helps in controlling the specifications of the product
	S4: Reputation	0.2	4	0.8	The company has an excellent reputation in both local and international markets, especially in the German market
	S5: High profitability of production	0.1	4	0.4	The return on investment is very high
Weaknesses	W1: Using outdated obsolete technologies	0.2	2	0.4	The facility is too old, causing too many stops during production, increasing waste and costs
	W2: Diesel is the main source of energy	0.03	2	0.06	Diesel fuel implies high cost and an increase in CO_2 emissions
	W3: Location of the facility	0.02	1	0.02	The facility is located far from the main suppliers for CPW, increasing the transportation cost and the negative environmental impact of the product
	W4: Shortage of conforming biomass	0.2	2	0.4	Current farming practices lead to few suppliers having oranges complying with food safety standards
	W5: Water circulation system	0.02	2	0.04	The facility does not apply any wastewater management system, causing high conferral costs
	Total	**1**		**3.01**	

Table 2.7 The resulting EFE matrix for the dehydration company

Factors		Weight	Rate	Weighted score	Description
Opportunities	O1: Proximity from international markets	0.15	3	0.45	Egypt has a unique geographical location near three major markets: African, European, and Asian markets linked through the Suez Canal, making it a favorable supplier for perishable food products
	O2: High quality of Egyptian orange peels	0.06	3	0.18	Egyptian oranges are branded worldwide with high nutritional content
	O3: High demand	0.1	2	0.2	There is a high demand from developed countries for citrus peels of high quality to be utilized as a source of valuable food additives like pectin, as well as other pharmaceutical compounds
	O4: Trade agreements with nearshoring markets	0.2	2	0.4	Egypt is involved in many trade agreements with the neighboring countries that support the ease of trade
	O5: Production seasons are in market gaps	0.01	2	0.02	The production seasons of Egyptian orange products and byproducts are off seasons for the key market players, such as Brazil and the USA
Threats	T1: Excess use of pesticides	0.2	4	0.8	The pesticide residues are one of the key determinants for any food product to be commercialized locally or internationally
	T2: Food safety regulations	0.05	2	0.1	Each year the national and international food safety regulations become tighter, requiring the need to be updated regularly
	T3: The potential of suppliers for forward integration	0.1	1	0.1	Increasing the practices of adding value to orange fruit by juice extraction, increases the supplier's potential for forward integration
	T4: Increase in raw material price	0.03	2	0.06	Increasing demand due to the raised awareness of the value of CPW may surge the price of biomass
	T5: Environmental regulations	0.1	1	0.1	Carbon footprint certifications will be a need for commercializing products nationally and internationally. Recently, it has been a requirement for entrance of some EU markets, such as Germany
	Total	**1**		**2.41**	

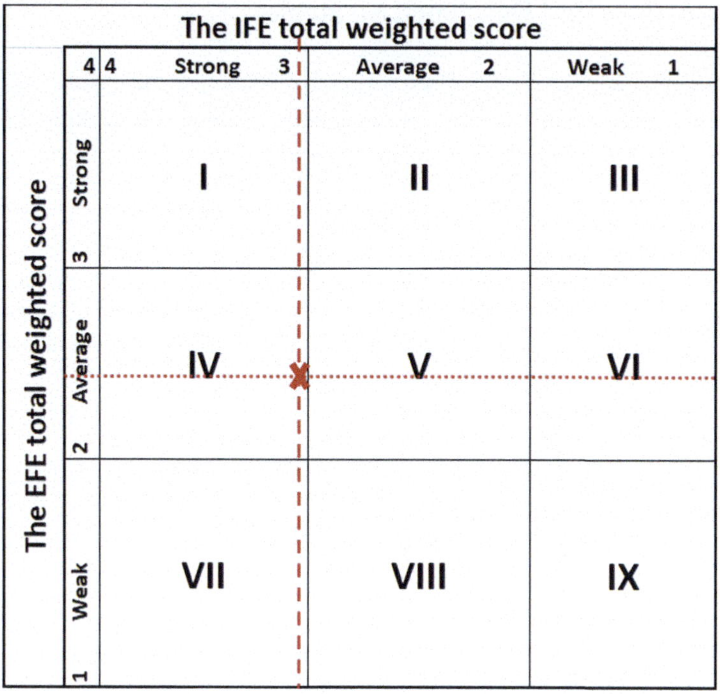

Fig. 2.3 IE matrix for the dehydration company

suppliers' potential for forward integration and compliance with the environmental regulations.

3.2.3 Internal-External (IE) Matrix

To obtain a strategic business strategy at the corporate level, the IFE and EFE total weighted scores are used in the IE Matrix, which was developed from the General Electric (GE) Business Screen Matrix. As shown in Fig. 2.3, the weighted scores lie in cell IV, which corresponds to the strategy "Growth and Build." This strategy typically includes intensive (market penetration, market development, and product development) and integration (backward integration, forward integration, and horizontal integration) actions (David 2011). Accordingly, the supply chain integration strategy is recommended, to gain more strength for market development.

4 Discussions and Developed Strategies

Based on the internal and external factor analyses, it is evident that the company has been motivated by multiple drivers for circularity that conform with the literature review results (Sect. 3.1). From an economic and environmental perspective, dehydration is a cost-effective valorization technique, relying on inexpensive material to extract valuable compounds, chemicals, and dietary fibers. The enriched content of bioactive materials drives the use of valorized products (dried citrus peels) to enhance the nutritional value of food products and their use in the cosmetics industry. The company further benefits from the abundant availability of CPW and its high residue-to-product rate. Moreover, the used valorization technique reduces the negative impacts on the environment, as it reduces the disposal of CPW. Regarding the technological aspects, the ease of valorization technique implementation has shaped the strengths of the company.

Nevertheless, more opportunities are still to be explored to ensure the competitiveness of the company within the CCSC. Explicitly, the following drivers are identified to have a potential contribution to increase the company's competitiveness:

- Producing biogas and ethanol.
- Green and sustainable separation methods for agro-industrial waste.
- Achieving "zero waste" policies.
- Complying with regulations and strategies.
- Helping with wastewater management.

To formulate possible strategies in which internal strengths were used to capitalize on external opportunities, the SWOT Matrix was used as a matching tool for key external and internal factors developed in IFE and EFE, while considering the above-listed potential drivers elicited from the literature. Table 2.8 shows possible strategies which are categorized as: SO (strengths-opportunities), WO (weaknesses-opportunities), ST (strengths-threats), and WT (weaknesses-threats) strategies.

The formulated strategies are directed towards gaining more strength for market development through supply chain integration, as recommended through the IE matrix. The penetration of new markets is currently constrained by environmental and food safety regulations. Hence, to overcome this barrier, a stringent quality assurance system for the farmers needs to be pursued, to ensure the compliance of the CPW to environmental regulations. The improvement of the current technology used and the efficient use of clean energy at the dehydrator facility may also reduce the environmental impact. Furthermore, the exploitation of the seasonality market gap of large producers of oranges is also a means for entering new markets. Horizontal and backward integration will contribute to market penetration and will also reduce the bargaining power of the supplier and mitigate the problems related to CPW quality. The company's good reputation and high profitability provide the potential for horizontal integration through strategic alliances.

Table 2.8 The SWOT matrix of the dehydration company

	S1: High product quality *S2: High market share* *S3: Using food dehydrator* *S4: Reputation* *S5: High profitability of production*	*W1: Using outdated obsolete technologies* *W2: Diesel is the main source of energy* *W3: Location of the facility* *W4: Shortage of conforming biomass* *W5: Water circulation system*
O1: Proximity from international markets *O2: High quality of Egyptian orange peels* *O3: High demand* *O4: Trade agreements with nearshoring markets* *O5: Production seasons are in market gaps*	Penetrate new markets by exploiting trade agreements and proximity to demanding markets (S1, S4, O1, O2, and O4). Seize new markets with high demand during the seasonal market gaps by offering competitive prices (S5, O3, and O5)	Adopt new green technologies to improve economic efficiency and reduce the environmental impact of the product, to exploit most of the opportunities through developing new markets (W1, O1, O2, O3, and O4). Relocate the current facility or locate a new facility dedicated to CPW dehydration in proximity to biomass suppliers (W3, W4, O3, and O5). Use a cleaner source of energy, or produce biogas from CPW, to reduce the cost and carbon footprint of the products. (W2, O1, and O4)
T1: Excess use of pesticides *T2: Food safety regulations* *T3: The potential of suppliers for forward integration* *T4: Increase in raw material price* *T5: Environmental regulations*	Apply high-quality assurance measures in biomass supply selection to ensure compliance with health and environmental regulations (S1, T1, and T2). Apply backward or horizontal integration through strategic alliances (S3, S4, T3, and T4). Diversify the revenue streams (S1, S2, S3, S4, S5, T3, and T4)	Introduce a new plant dedicated to citrus products and by-products through backward or horizontal integration (W3, W4, T3, and T4). Upgrade the existing plant to use more efficient and greener state-of-the-art technologies (W1, W2, and T5). Introduce a water management system to reduce cost and improve the environmental impact and product's competitiveness in new markets (W5 and T5)

Despite the profitability of the dehydration process, the company currently valorizes the CPW partially. The potential value gained through the valorization of citrus wastewater which includes valuable nutritional values remains unused. Also, the generation of biogas is unexplored. Hence, the company may diversify its revenue streams by addressing more than one valorization path.

5 Conclusion

Circular supply chains transform waste materials to valuable products to promote sustainability. The use of CPW supports the implementation of the CCSC. Despite the difficulties in putting CCSC into practice, identifying the key factors that motivate its adoption helps in encouraging citrus supply chain stakeholders to embrace circular practices. This study has examined the drivers for CCSC implementation found in the literature and investigated a practical case study of a CPW dehydrating plant in Egypt to identify competitive strategies to support a CCSC via a set of strategic management tools.

From the reviewed literature, a total of 53 drivers were found and classified into five categories: economical, social, technological, legal, and environmental drivers. Key identified drivers for a competitive CCSC included valorization techniques that could be easily implemented in general, are cost-effective, and have the potential to reduce the negative impacts on the environment. Many of the drivers found in the literature were also identified within the practical case study, such as having a cost-effective valorization technique.

Internal and external factors influencing a leading Egyptian company that dehydrates orange peels were assessed. Such an evaluation recommended implementing an integration strategy to gain more strength for market development. Furthermore, a SWOT matrix was created to assist in formulating strategies, such as: (1) implementing strict quality assurance systems, to comply with environmental regulations; (2) using greener technologies, to reduce the negative environmental impacts; (3) exploiting the seasonality market gaps of large orange producers, to potentially enter new markets; (4) building strategic alliances for horizontal integration, based on the company's excellent reputation; and (5) exploring additional valorization paths, to increase revenue.

While the conducted review elicited drivers for all actors in the CCSC, the case study was limited to the consideration of one actor, the dehydrating plant. It is believed that the consideration of the whole supply chain may lead to better exploitation of several drivers, leading to the design and operation of more competitive CCSCs. Furthermore, the review did not categorize the drivers according to geographic location, while the case study has revealed some challenges and opportunities related to territory-based regulations.

Hence, the following suggestions may constitute future research areas. First, the adoption of a multi-case study to investigate similar companies or a complete CCSC to allow for generalization of conclusions to be drawn to the CCSC. Second, analyzing the challenges and opportunities faced by global CCSC encompassing emerging and developed economies.

Acknowledgments This work was supported by the Academy of Scientific Research & Technology (ASRT)—Egypt, PRIMA call 2020 section 2 project no.26, entitled "Innovation in the By-Product Supply chain of citrus in Mediterranean area" (ImPUISe). Also, the authors would like to thank the managers of Agromisr factory for dehydrated food ingredients and essential oils in Egypt for their valuable time and information with regard to the case study.

References

Abdullahi TM, Said MFM, Othman N, Malik SA (2022) Potentials of citrus peel waste valorisation for green diesel production: a mini review. Chem Eng Trans 97:475–480 SE-Research Articles. https://doi.org/10.3303/CET2297080

Alwazeer D, Elnasanelkasim MA (2023) Hydrogen-rich water as a green solvent for the extraction of phytochemicals from agri-food wastes. Sustain Chem Pharm 33:101035. https://doi.org/10.1016/j.scp.2023.101035

Antonio AS, Wiedemann LSM, Galante EBF et al (2021) Efficacy and sustainability of natural products in COVID-19 treatment development: opportunities and challenges in using agro-industrial waste from Citrus and apple. Heliyon 7:e07816. https://doi.org/10.1016/j.heliyon.2021.e07816

Belorkar SA, Kausar H (2023) Valorization parameters to determine the fermentative applicability of selected fruit peels for Xylanase production. Waste Biomass Valoriz 14:185–196. https://doi.org/10.1007/s12649-022-01858-y

Bendaoud-Boulahlib Y, Ghizellaoui S (2017) Study of inhibiting effect of orange peel on hard growth water. J Mater Environ Sci 8:1076–1081

Bhat R (2021) Chapter 1 - Sustainability challenges in the valorization of agri-food wastes and by-products. In: Bhat R (ed) Valorization of agri-food wastes and by-products. Academic Press, pp 1–27

Boccia F, Di Pietro B, Covino D (2021) Food waste and environmental-sustainable innovation: a scenario for the Italian citrus market. Qual Access Suc 22:145–153

Bousios S, Worrell E (2017) Towards a Multiple Input-Multiple Output paper mill: opportunities for alternative raw materials and sidestream valorisation in the paper and board industry. Resour Conserv Recycl 125:218–232. https://doi.org/10.1016/j.resconrec.2017.06.020

Calabrò PS, Fazzino F, Folino A et al (2019) Semi-continuous anaerobic digestion of orange peel waste: effect of activated carbon addition and alkaline pretreatment on the process. Sustain For 11:3386. https://doi.org/10.3390/su11123386

Campagnolo L, Morselli D, Magrì D et al (2019) Silk Fibroin/Orange Peel foam: an efficient biocomposite for water remediation. Adv Sustain Syst 3:1800097. https://doi.org/10.1002/adsu.201800097

Carrodeguas LP, Chen TTD, Gregory GL et al (2020) High elasticity, chemically recyclable, thermoplastics from bio-based monomers: carbon dioxide, limonene oxide and ε-decalactone. Green Chem 22:8298–8307. https://doi.org/10.1039/D0GC02295K

Cassano A, Conidia C, Drioli E (2021) Integrated membrane systems as an innovative approach for the recovery of high value-added compounds from agro-food by-products. Chem Eng Trans 87:361–366. https://doi.org/10.3303/CET2187061

Castaño FC, Igal K, Arreche R, Vázquez P (2022) Synthesis of silica-based solids by sol-gel technique using lemon bio-waste: juice, peels and ethanolic extract. Curr Res Green Sustain Chem 5:100322. https://doi.org/10.1016/j.crgsc.2022.100322

Chen L, Fang W, Liang J et al (2023) Biochar application in anaerobic digestion: performances, mechanisms, environmental assessment and circular economy. Resour Conserv Recycl 188:106720. https://doi.org/10.1016/j.resconrec.2022.106720

David FR (2011) Strategic management concepts and cases, 13th edn. Pearson

Duru S, Hayran S, Gül A (2022) The analysis of competitiveness of Mediterranean countries in the world citrus trade. Mediterr Agric Sci 35:21–26. https://doi.org/10.29136/mediterranean.1012466

Eliopoulos C, Markou G, Langousi I, Arapoglou D (2022) Reintegration of food industry by-products: potential applications. Foods 11:3743. https://doi.org/10.3390/foods11223743

Elmaidomy AH, Abdel-Maqsoud NMR, Tammam Omar Y et al (2023) Egyptian mandarin peel oil's anti-scabies potential via downregulation-of-inflammatory/immune-cross-talk: GC–MS and PPI network studies. Sci Rep 13:14192. https://doi.org/10.1038/s41598-023-38390-5

Eriksson M, Spångberg J (2017) Carbon footprint and energy use of food waste management options for fresh fruit and vegetables from supermarkets. Waste Manag 60:786–799. https://doi.org/10.1016/j.wasman.2017.01.008

Fazzino F, Mauriello F, Paone E et al (2021) Integral valorization of orange peel waste through optimized ensiling: lactic acid and bioethanol production. Chemosphere 271:129602. https://doi.org/10.1016/j.chemosphere.2021.129602

Fernández-López JA, Doval Miñarro M, Angosto JM et al (2021) Adsorptive and surface characterization of Mediterranean Agrifood processing wastes: prospection for pesticide removal. Agronomy 11:561. https://doi.org/10.3390/agronomy11030561

Gonzalez-Canche NG, Carrillo JG, Escobar-Morales B et al (2021) Physicochemical and optical characterization of Citrus aurantium derived Biochar for solar absorber applications. Materials 14:4756. https://doi.org/10.3390/ma14164756

Grima-Olmedo C, Valle-Falcones LM, Galindo DG, Esparver RR (2023) Production of AC from bamboo, orange, and paulownia waste–influence of activation gas and biomass maturation. Materials 16:3498. https://doi.org/10.3390/ma16093498

Gullón P, Gullón B, Romaní A et al (2020) Smart advanced solvents for bioactive compounds recovery from agri-food by-products: a review. Trends Food Sci Technol 101:182–197. https://doi.org/10.1016/j.tifs.2020.05.007

Hassanain HT, Gabr RH (2020) An economic study on Egyptian Orange exports and its competitiveness in the international markets. Zagazig J Agric Res 47:623–639. http://doi.org/10.21608/zjar.2020.94500

Iqbal A, Schulz P, Rizvi SSH (2021) Valorization of bioactive compounds in fruit pomace from agro-fruit industries: present insights and future challenges. Food Biosci 44:101384. https://doi.org/10.1016/j.fbio.2021.101384

ITC (2023) List of importing markets for the product exported by Egypt in 2023. Available online: https://www.trademap.org/Country_SelProductCountry.aspx?nvpm=1%7c818%7c%7c%7c%7c0814%7c%7c%7c4%7c1%7c1%7c2%7c1%7c1%7c2%7c1%7c1%7c1

Jamsheera CP, Pradeep BV (2023) Antibacterial activity of silver functionalized bacterial cellulose isolated from Acetobacter Tropicalis Jcp OK356208.1 from rotten orange. Res J Biotechnol 18:105–120. https://doi.org/10.25303/1807rjbt1050120

Jeong D, Park H, Jang BK et al (2021) Recent advances in the biological valorization of citrus peel waste into fuels and chemicals. Bioresour Technol 323:124603. https://doi.org/10.1016/j.biortech.2020.124603

Jorge AMS, Gaspar MC, Henriques MHF, Braga MEM (2023) Edible films produced from agrifood by-products and wastes. Innovative Food Sci Emerg Technol 88:103442. https://doi.org/10.1016/j.ifset.2023.103442

Kandemir K, Piskin E, Xiao J et al (2022) Fruit juice industry wastes as a source of bioactives. J Agric Food Chem 70:6805–6832. https://doi.org/10.1021/acs.jafc.2c00756

Kaur B, Rana P, Sridhar K (2022) Mass transfer kinetics and process optimization of osmotic dehydration of Kinnow mandarin (Citrus reticulata) peel. J Food Process Preserv 46:e16318. https://doi.org/10.1111/jfpp.16318

Khalil RKS, Sharaby MR, Abdelrahim DS (2023) Novel active edible food packaging films based entirely on citrus peel wastes. Food Hydrocoll 134:107961. https://doi.org/10.1016/j.foodhyd.2022.107961

Khan ZU, Khan T, Khan H et al (2023) Eco-friendly sustainable nanocarriers to treat oxidative stresses and skin aging-related ailments, valorization of a by-product. Bioengineering 10. https://doi.org/10.3390/bioengineering10070798

Kim IJ, Jeong D, Kim SR (2022) Upstream processes of citrus fruit waste biorefinery for complete valorization. Bioresour Technol 362:127776. https://doi.org/10.1016/j.biortech.2022.127776

Kumar A, Bhattacharya T, Shaikh WA et al (2022) Valorization of fruit waste-based biochar for arsenic removal in soils. Environ Res 213. https://doi.org/10.1016/j.envres.2022.113710

Lucia C, Laudicina VA, Badalucco L et al (2022) Challenges and opportunities for citrus wastewater management and valorisation: a review. J Environ Manag 321:115924. https://doi.org/10.1016/j.jenvman.2022.115924

Mariana O-S, Alzate C, Ariel C (2021) Comparative environmental life cycle assessment of orange peel waste in present productive chains. J Clean Prod 322:128814. https://doi.org/10.1016/j.jclepro.2021.128814

Merino D, Athanassiou A (2022) Biodegradable and active mulch films: hydrolyzed lemon peel waste and low methoxyl pectin blends with incorporated biochar and neem essential oil. ACS Sustain Chem Eng 10:10789–10802. https://doi.org/10.1021/acssuschemeng.2c01539

Molina-Montero C, Vicente-Jurado D, Igual M et al (2023) Fiber enrichment of 3D printed Apricot gel snacks with orange by-products. Gels 9:569. https://doi.org/10.3390/gels9070569

Moreno N, Bocachica D, Marimon-Bolivar W et al (2022) Application of agro-industrial waste biomass of plantain (Musa paradisiaca) and Tangerine (Citrus reticulata) for selective removal of nitrates from agricultural wastewater. Environ Eng Manag J 21:1513–1524. https://doi.org/10.30638/eemj.2022.134

Myrtsi ED, Koulocheri SD, Evergetis E, Haroutounian SA (2022) Agro-industrial co-products upcycling: recovery of carotenoids and fine chemicals from Citrus sp. juice industry co-products. Ind Crop Prod 186:115190. https://doi.org/10.1016/j.indcrop.2022.115190

Nattassha R, Handayati Y, Simatupang TM, Siallagan M (2020) Understanding circular economy implementation in the agri-food supply chain: the case of an Indonesian organic fertiliser producer. Agric Food Secur 9:10. https://doi.org/10.1186/s40066-020-00264-8

Ndayishimiye J, Chun BS (2017) Optimization of carotenoids and antioxidant activity of oils obtained from a co-extraction of citrus (Yuzu ichandrin) by-products using supercritical carbon dioxide. Biomass Bioenergy 106:1–7. https://doi.org/10.1016/j.biombioe.2017.08.014

Nieto G, Fernández-López J, Pérez-Álvarez JA et al (2021) Valorization of citrus co-products: recovery of bioactive compounds and application in meat and meat products. Plants 10:1069. https://doi.org/10.3390/plants10061069

Ningrum A, Supriyadi AS et al (2018) Valorization of food by product from selected tropical fruits pomace. IOP Conf Ser Earth Environ Sci 205:12034. https://doi.org/10.1088/1755-1315/205/1/012034

Ortiz-Sanchez M, Omarini AB, González-Aguirre J-A et al (2023) Valorization routes of citrus waste in the orange value chain through the biorefinery concept: the Argentina case study. Chem Eng Proc Proc Intensif 189:109407. https://doi.org/10.1016/j.cep.2023.109407

Panuccio MR, Marra F, Maffia A et al (2022) Recycling of agricultural (orange and olive) bio-wastes into ecofriendly fertilizers for improving soil and garlic quality. Resour Conserv Recycl Adv 15:200083. https://doi.org/10.1016/j.rcradv.2022.200083

Papadaki A, Manikas AC, Papazoglou E et al (2022) Whey protein films reinforced with bacterial cellulose nanowhiskers: improving edible film properties via a circular economy approach. Food Chem 385. https://doi.org/10.1016/j.foodchem.2022.132604

Pérez S, Giraldo S, Forgionny A et al (2022) Eco-friendly reuse of agricultural wastes to produce biocomposites with high potential in water treatment and fertilizers. Biomass Convers Biorefinery. https://doi.org/10.1007/s13399-022-02948-6

Perez-Pirotto C, Moraga G, Quiles A et al (2022) Techno functional characterization of green-extracted soluble fibre from orange by-product. LWT 166. https://doi.org/10.1016/j.lwt.2022.113765

Piscitelli L, Bennani Z, El Chami D, Mondelli D (2022) A circular economy model to improve phosphate rock fertiliser using agro-food by-products. Sustainability (Switzerland) 14. https://doi.org/10.3390/su142316228

Putnik P, Bursać Kovačević D, Režek Jambrak A et al (2017) Innovative "green" and novel strategies for the extraction of bioactive added value compounds from citruswastes - a review. Molecules 22:680. https://doi.org/10.3390/molecules22050680

Raimondo M, Caracciolo F, Cembalo L et al (2018) Making virtue out of necessity: managing the citrus waste supply chain for bioeconomy applications. Sustainability (Switzerland) 10:4821. https://doi.org/10.3390/su10124821

Ramya AV, Balachandran M (2021) Valorization of agro-industrial fruit peel waste to fluorescent nanocarbon sensor: ultrasensitive detection of potentially hazardous tropane alkaloid. Front Environ Sci Eng 16:27. https://doi.org/10.1007/s11783-021-1461-z

Rosales TKO, Fabi JP (2023) Valorization of polyphenolic compounds from food industry by-products for application in polysaccharide-based nanoparticles. Front Nutr 10. https://doi.org/10.3389/fnut.2023.1144677

Sabater C, Villamiel M, Montilla A (2022) Integral use of pectin-rich by-products in a biorefinery context: a holistic approach. Food Hydrocoll 128. https://doi.org/10.1016/j.foodhyd.2022.107564

Satari B, Karimi K (2018) Citrus processing wastes: environmental impacts, recent advances, and future perspectives in total valorization. Resour Conserv Recycl 129:153–167. https://doi.org/10.1016/j.resconrec.2017.10.032

Singh A, Singh A (2022) Microbial degradation and value addition to food and agriculture waste. Curr Microbiol 79:119. https://doi.org/10.1007/s00284-022-02809-5

Statista (2024) Leading orange producing countries worldwide in 2023/2024. Available online: https://www.statista.com/statistics/1044840/major-orange-producers-worldwide/

Stone J, Garcia-Garcia G, Rahimifard S (2019) Development of a pragmatic framework to help food and drink manufacturers select the most sustainable food waste valorisation strategy. J Environ Manag 247:425–438. https://doi.org/10.1016/j.jenvman.2019.06.037

Súarez-Velázquez GG, Pech-Rodríguez WJ, León JAR-D et al (2022) Orange peel as substrate to synthesize conductive carbon nanostructures by a green thermal process. Revista Internacional de Contaminacion Ambiental 38:34–47. https://doi.org/10.20937/RICA.54242

Tamelová B, Malaťák J, Velebil J (2018) Energy valorisation of citrus peel waste by torrefaction treatment. Agron Res 16(1):276–285. http://dx.doi.org/10.15159/ar.18.029

Tayengwa T, Mapiye C (2018) Citrus and winery wastes: promising dietary supplements for sustainable ruminant animal nutrition, health, production, and meat quality. Sustainability (Switzerland) 10:3718. https://doi.org/10.3390/su10103718

Teigiserova DA, Hamelin L, Thomsen M (2019) Review of high-value food waste and food residues biorefineries with focus on unavoidable wastes from processing. Resour Conserv Recycl 149:413–426. https://doi.org/10.1016/j.resconrec.2019.05.003

Teigiserova DA, Tiruta-Barna L, Ahmadi A et al (2021) A step closer to circular bioeconomy for citrus peel waste: a review of yields and technologies for sustainable management of essential oils. J Environ Manag 280:111832. https://doi.org/10.1016/j.jenvman.2020.111832

Tolisano C, Del Buono D (2023) Biobased: biostimulants and biogenic nanoparticles enter the scene. Sci Total Environ 885:163912. https://doi.org/10.1016/j.scitotenv.2023.163912

Tropea A, Ferracane A, Albergamo A et al (2022) Single cell protein production through multi food-waste substrate fermentation. Fermentation 8:91. https://doi.org/10.3390/fermentation8030091

Vaishnav A, Upadhyay K, Koradiya M et al (2022) Valorisation of fruit waste for enhanced exopolysaccharide production by Xanthomonas campestries using statistical optimisation of medium and process. Food Biosci 46:101608. https://doi.org/10.1016/j.fbio.2022.101608

Valenti F, Porto SMC, Chinnici G et al (2017) Assessment of citrus pulp availability for biogas production by using a GIS-based model: the case study of an area in southern Italy. Chem Eng Trans 58:529–534. https://doi.org/10.3303/CET1758089

Voss C, Tsikriktsis N, Frohlich M (2002) Case research in operations management. Int J Oper Prod Manag 22:195–219. https://doi.org/10.1108/01443570210414329

Wally A (2022) Citrus Annual (Report No. EG2022-0033). United States Department of Agriculture (Foreign Agricultural Service). Available online: https://fas.usda.gov/data/egypt-citrus-annual-6

Xu G, Zhao J, Shi K et al (2023) Trends in valorization of citrus by-products from the net-zero perspective: green processing innovation combined with applications in emission reduction. Trends Food Sci Technol 137:124–141. https://doi.org/10.1016/j.tifs.2023.05.012

Zema DA, Calabrò PS, Folino A et al (2018a) Valorisation of citrus processing waste: a review. Waste Manag 80:252–273. https://doi.org/10.1016/j.wasman.2018.09.024

Zema DA, Fòlino A, Zappia G et al (2018b) Anaerobic digestion of orange peel in a semi-continuous pilot plant: an environmentally sound way of citrus waste management in agro-ecosystems. Sci Total Environ 630:401–408. https://doi.org/10.1016/j.scitotenv.2018.02.168

Zema DA, Calabro PS, Folino A et al (2019) Wastewater management in citrus processing industries: an overview of advantages and limits. Water (Switzerland) 11. https://doi.org/10.3390/w11122481

Zuin VG, Ramin LZ (2018) Green and sustainable separation of natural products from agro-industrial waste: challenges, potentialities, and perspectives on emerging approaches. Top Curr Chem 376:3. https://doi.org/10.1007/s41061-017-0182-z

Chapter 3
Evaluating the Environmental Impact of Orange Juice Production in Egypt Through Life Cycle Assessment

Sk. Riad Bin Ashraf, Ani Melkonyan-Gottschalk, Denis Daus, Emad Alzubi, and Ahmed Kassem

1 Introduction

The 2015 Paris Agreement (The Paris Agreement|UNFCCC 2023) and other global initiatives emphasize the importance of reducing the adverse effects of climate change on all aspects of the community, including the food supply chain (FSC) (Beske et al. 2014). Greenhouse gas (GHG) emissions and resource use are highly impacted by the FSC, especially in the citrus supply chain (Cambeses-Franco et al. 2021; Esposito et al. 2020; Gerber et al. 2013; Filho et al. 2022). The LCA approach has been the focus of several researchers (Beccali et al. 2009a; Vono et al. 2023; Pergola et al. 2013; Cabot et al. 2022; Ribal et al. 2019), particularly important in the Mediterranean regions (Knoema 2020; Statista 2023; Zacarias et al. 2020). It has become clear that moving toward a circular economy (CE) is crucial since it encourages an economic model in which resources are reused to reduce waste and emissions (Haupt and Hellweg 2019; Secondi 2021). Research on the olive oil industry, among others in the agri-food sector, has shown that implementing CE approaches within the FSC is feasible and environmentally beneficial (Gallo et al. 2023; Nikkhah et al. 2021).

In the Mediterranean region, circular approaches to orange juice production are gaining attention, mainly through the lens of the LCA. The concept of a circular bioeconomy is vividly illustrated in research focusing on the LCA of orange peel waste for sustainable production practices (Teigiserova et al. 2022). Similarly, the

S. R. B. Ashraf (✉) · A. Melkonyan-Gottschalk · D. Daus
Zentrum für Logistik und Verkehr (ZLV), Universität Duisburg-Essen, Duisburg, Germany
e-mail: riad.ashraf@uni-due.de

E. Alzubi · A. Kassem
Transportssysteme und -logistik, Fakultät für Ingenieurwissenschaften, Universität Duisburg–Essen, Duisburg, Germany

A. Melkonyan-Gottschalk, D. Daus (eds.), *Transformation Towards Circular Food Systems*, Greening of Industry Networks Studies 13,
https://doi.org/10.1007/978-3-031-63793-3_3

potential of biofuel production from orange juice industrial waste is explored, presenting a novel approach within the framework of CE (Christofi et al. 2022). Although not directly related to orange juice, another research provides valuable insights into sustainable agricultural practices that can be applied to the orange juice sector to enhance circularity (Guinée et al. 2006). Further, one more paper delves into the broader implications of sustainable agri-food processes, emphasizing the importance of circular pathways in the agricultural sector, including orange juice production (Stillitano et al. 2021). Lastly, research conducted in the Italian citrus supply chain offers a comprehensive assessment of the environmental impacts associated with citrus production, underlining the significance of LCA in understanding and improving the sustainability of orange juice production (Beccali et al. 2009b). These studies highlight the importance of integrating CE concepts and LCA methodologies to enhance the sustainability of orange juice production in the Mediterranean region.

However, the absence of similar research in the Egyptian region, particularly regarding orange juice production focusing on the CE principles, is noticeable. This gap signifies an unexplored area where transitioning from a linear to a circular supply chain model could offer significant environmental benefits. This research aims to illuminate the ecological advantages of this transition in the Egyptian orange juice industry. Specifically, it explores the integration of a composting unit to reduce fertilizer use, thereby contributing to a more sustainable and circular production model. Such an approach is expected to enhance the industry's environmental sustainability and align with global trends towards more circular and eco-friendly production.

2 Objectives and Research Question

This study aims to conduct a comprehensive LCA to evaluate the environmental impacts of NFC orange juice production's current linear supply chain model and contrast it with a proposed circular approach. The research leverages the advanced analytical capabilities of the LCA simulation software Eco chain to offer detailed insights into the environmental effects of both the linear and circular models. The existing research predominantly focuses on LCA methodologies. Still, there is a distinct lack of studies applying a circular strategy tailored explicitly to orange juice production in Egypt. This research aspires to bridge this gap, concentrating on the selected pilot case by providing a novel analysis of the current linear and proposed circular supply chain approaches. This exploration is anticipated to deliver a perspective that could serve as a benchmark for future research in sustainable orange production in Egypt and potentially extend to other agricultural sectors.

Therefore, the goal of this research can be accomplished by addressing the following research question:

How can transitioning from a linear to a circular supply chain model impact the environmental sustainability of NFC orange juice production?

3 Literature Review

The Paris Agreement influences the FSC by encouraging sustainable practices, which is crucial for the citrus supply chain to adapt to climate change and maintain productivity (Hussain et al. 2023). It emphasizes the need for environmentally friendly approaches in agriculture, impacting citrus production through adopting sustainable farming practices (Meadu et al. 2015). The Agreement's focus on reducing GHG emissions is directly relevant to the citrus industry, which must innovate to reduce its carbon footprint (Abdullah et al. 2017; Smith 2017).

LCA research by Rasines et al. (2023) provides significant insights into food waste and environmental impact, specifically focusing on the storage of broccoli. Their findings indicate that maintaining bioplastic bags at a temperature of 5 °C can notably reduce food waste and its carbon footprint (CF). This storage method for 16 days has the potential to save approximately 4.63 kg of broccoli and 3.16 kg of CO_2 eq. per functional unit, offering a sustainable solution compared to less effective scenarios. Building upon these sustainable practices, Tsouti et al. (2023) delve into innovative packaging solutions for fresh tomatoes. Their research highlights that while the production of novel packaging films has a higher CF impact, these films effectively extend the shelf life of tomatoes, thus reducing the overall CF in the FSC.

Transitioning from packaging to agricultural practices, Keikha et al. (2023) examine the environmental impacts of wheat cultivation in arid climates. Their study reveals a 9% higher potential for ecological damage in such conditions, primarily due to increased nitrogen inputs and diesel use. In a similar vein of technological innovation, Da Costa et al. (2023) explore the integration of Internet of Things (IoT) technologies in a frozen meals company. Despite a slight increase in environmental impact by about 0.4%, the implementation of IoT shows promise in significantly reducing food waste, suggesting a beneficial trade-off for sustainability.

Further emphasizing the importance of strategic waste management, Albizzati et al. (2022) critically analyze food waste prevention in the EU's FSC. Their research underscores the necessity of including rebound effects in the LCA to understand the net environmental benefits of such strategies fully. Complementing this perspective, Boenzi et al. (2022) investigate the environmental impacts of using fresh fruit versus semi-finished products in apricot jam production. Their findings indicate that fresh apricots lead to 50–65% lower environmental impacts, making a solid case for fresh produce, provided transport factors are favorable.

Expanding the scope to a broader range of food products, Vidergar et al. (2021) conducted an in-depth analysis of 49 LCA studies on the FSC. Their research points out the need for enhanced methodologies in objective setting, scope definition, inventory accuracy, and impact assessment. In line with this need for comprehensive approaches, Notarnicola et al. (2015) and Notarnicola et al. (2022) present an extensive overview of agri-food LCA. They discuss global initiatives and the evolving understanding of the interplay between consumer behavior and the product and production chain in achieving food sustainability.

The research by Alzubi et al. (2022) concludes that polypropylene crates, reducing citrus loss by 60% and having lower environmental impacts due to their reusability and recyclability, are a more sustainable choice for transporting citrus fruits in Jordan compared to polystyrene crates. Another study by Negro et al. (2017) evaluates the environmental and economic impact of ten orange peel waste management scenarios in Southern Italy, finding that anaerobic co-digestion strategies are the most favorable in terms of global warming and resource depletion, despite a potential eutrophication risk and are economically preferable compared to traditional methods like landfilling, composting, and animal feeding. Another literature by Omolayo et al. (2021) related to food loss and waste across the FSC reveals a research gap in studies focusing on prevention and the differentiation between avoidable and unavoidable waste streams, with most studies being Europe-centric, often lacking in spatial and temporal specificity and data transparency, and primarily assessing global warming potential with less focus on energy, water demand, and human toxicity.

Focusing on fruit production, Cerutti et al. (2015) offer an overview of LCA in fruit systems. Their discussion includes various criteria and the environmental influence of nurseries and orchards, thereby highlighting the complexities in fruit production systems. Echoing the need for specificity in LCA studies, Notarnicola et al. (2022) address the adequacy of commercial Life Cycle Inventory (LCI) databases for the Italian agri-food sector. Their research identifies gaps in inventory datasets, particularly for critical Italian products like olive oil, wine, wheat, and citrus fruits.

Connecting these broader themes to specific case studies, Ribal et al. (2019) demonstrate the environmental benefits of organic farming for Valencian oranges, explicitly noting the reduction in CF. Complementing this, Esturo et al. (2023) and Liao et al. (2022) focus on the CF impact of packaging materials in orange juice production. Their studies advocate for waste reduction and transforming agricultural byproducts into commercially valuable products, thus contributing to the sector's sustainability.

4 Comparison Between the Current Linear and Circular Case

4.1 Current Linear Supply Chain Scenario

The current linear case we selected focuses on the NFC orange juice supply to the European market. NFC orange juice is obtained by directly extracting juice from oranges and then bottled without undergoing any concentration processes.

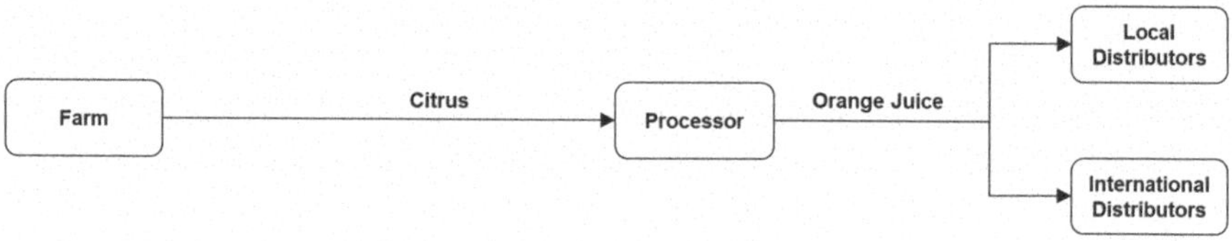

Fig. 3.1 Overall linear supply chain case

This process maintains a flavor profile that consumers frequently perceive as superior in freshness, resembling freshly squeezed oranges. NFC juice is subjected to pasteurization, a necessary thermal step to eliminate any potential microbial contamination (Abedelmaksoud et al. 2019; Li et al. 2021; Rodrigues and Fernandes 2022).

To get a better understanding of the current linear supply chain, different steps associated with the overall process are given below:

- *Agricultural production:* The selected case owns a farm where fruits are harvested, sorted, and packaged before being packed in boxes and loaded onto trucks for transportation. No pesticides are used, but synthetic fertilizer is applied.
- *Processing and packaging:* Harvested oranges are transported to processing units, which undergo washing, peeling, and pasteurization. This juice is directly squeezed from the oranges and bottled. Given its direct nature, it does not experience concentration. However, to ensure safety and extend its shelf life, NFC orange juice typically undergoes pasteurization. This process heats the juice to kill potential pathogens without significantly affecting its taste or nutritional content. The juice is packaged in materials designed to preserve quality and extend shelf life.
- *Distribution and sales:* Packaged juice is distributed in the local and international markets. In global markets, shipping companies ship refrigerated containers.

Figure 3.1 demonstrates the selected case's overall linear supply chain scenario.

4.2 Proposed Circular Model

The Mediterranean's unique climate significantly influences orange production, with countries like Egypt leading in this industry due to optimal growing conditions and resource-intensive processes (Ashraf and Daus 2022; Delrio and Cocco 2012; Dimico et al. 2017; Fraga and Santos 2021; Orhan 2021). Traditional orange juice production follows a linear supply chain, often leading to substantial waste and resource misuse, prompting a shift towards the CE principles for environmental sustainability (Korhonen et al. 2018; Miranda-Ackerman et al. 2017; Ortiz et al. 2020; Ortiz-Sanchez et al. 2023). CE practices in orange production emphasize resource optimization and waste reduction, notably through converting orange peels into compost, reducing environmental impacts like global warming potential and ozone depletion (Hanes et al. 2021; Korhonen et al. 2018; Siles et al. 2016). This work advocates for a circular approach, utilizing waste from orange juice production as compost to replace traditional fertilizers, illustrated through detailed calculations of environmental impact reductions (Hanes et al. 2021; Siles et al. 2016). Implementing composting units represents a pivotal move towards a more circular system in the industry, enhancing soil quality, promoting sustainable cultivation, and offering environmental benefits (Bansode et al. 2016; Eremeeva et al. 2023; Mahapatra et al. 2022).

All assumptions and key calculations related to the transition to the circular model are discussed for a better understanding.

- *Juice production efficiency*: It takes 2 kg of oranges to produce 1 liter of juice.
- *Pomace*: During juice production, approximately 50% of the used oranges turn into pomace, a mixture of peels and pulps. This byproduct is often overlooked but has significant potential.
- *Composting—turning waste into resources*: Remarkably, 25% of this pomace can be transformed into compost through efficient composting practices. This result means that a quarter of what is wasted once becomes a valuable resource.
- *Reducing fertilizer use with compost*: The compost produced from pomace can be used as a natural fertilizer. Using this compost reduces the need for synthetic fertilizers to just 7.65 grams per liter of orange juice. This reduction of fertilizer is a substantial decrease, cutting down fertilizer use by approximately 82.37% compared to traditional methods, which require 43.4 grams of fertilizer.

The overall process of the proposed circular supply chain model is discussed briefly.

- *Agricultural production and waste collection:* The process begins on the farm, where oranges grow without pesticides but with reduced fertilizers obtained from the current linear scenario. Organic waste, including unusable fruit and plant debris, is diligently collected for composting.
- *Processing, packaging, and waste diversion:* Following harvest, oranges are processed into juices. Byproducts, such as peels, pulp, and wastewater, are carefully segregated during this phase. The juice was packaged and prepared for distribution.
- *Waste processing and compost production:* Organic waste from farms and processing facilities is transported to composting sites. It is converted into nutrient-rich compost, effectively closing the supply chain loop.
- *Compost utilization in farming:* The final step involved compost produced at the orange farm. This way, compost replaces synthetic fertilizers, enhancing soil health and reducing chemical dependence, thus completing the circular model.

Figure 3.2 shows the proposed circular model.

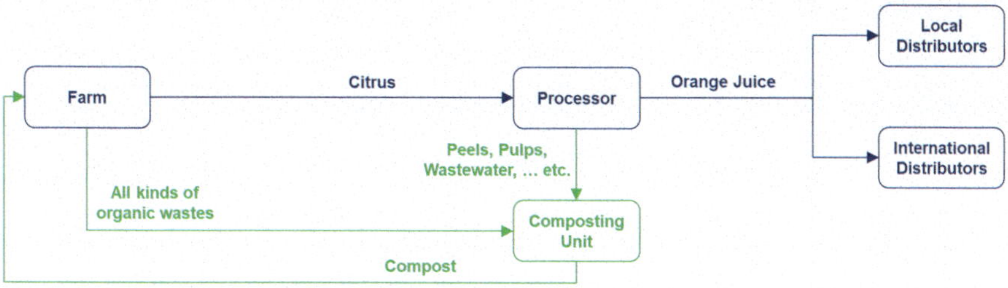

Fig. 3.2 Proposed circular case by utilizing a composting unit

4.3 Justification of the Selection of the Composting Unit as a Valorization Unit

Various studies have supported the importance of composting in orange production and its role in CE production. These studies highlight composting as a method for transforming orange production waste into valuable resources, emphasizing sustainability and resource efficiency. For example, composting can recycle orange-processing waste as a soil conditioner, with acceptable humification parameters, phytotoxicity, and low phenolic compounds, and is suitable for organic fertilization, nursery crops, and horticultural crops (Gelsomino et al. 2010). In addition, when utilized in the composting process, orange peel waste demonstrates effective degradation, transforming it into quality and mature compost, which is beneficial for environmental sustainability. Including orange peels in composting units not only aids in reducing industrial waste but also enhances the nutrient content, especially total nitrogen, making it a valuable resource for soil amendment (Medina-Salas et al. 2020).

On the other hand, from the LCA perspective, the research evaluates the environmental impacts of composting, revealing significant potentials such as reduced global warming, acidification, eutrophication, and ozone depletion (Serafini et al. 2023). Also, composting as a sustainable method for managing agricultural byproducts positively impacts soil quality, as demonstrated by an LCA analysis (Pergola et al. 2020).

Finally, composting units as a circular approach is economically and environmentally viable (Maskova et al. 2023). Incorporating composting into waste management systems helps decrease the use of landfills and the release of GHG while also generating valuable compost for agricultural and landscaping applications. This strategy aligns with the ideas of sustainability and CE, as it transforms waste into a valuable resource and promotes the growth of local economies (Sulewski et al. 2021; Vaverková et al. 2020). Therefore, selecting a composting unit is a rational choice for our research.

5 Basics of the LCA

An LCA study has four phases (Muralikrishna and Manickam 2017):

1. *Goal and scope definition* (Curran 2017; Muralikrishna and Manickam 2017):
   - Establishes the LCA's objectives and boundaries, including choosing a suitable functional unit for comparison standards.
   - Outlines the specific life cycle stages and sets the criteria for system comparisons and time considerations.
2. *Inventory analysis* (Chen and Wang 2017; Muralikrishna and Manickam 2017):
   - Provides a detailed account of resources and energy used, focusing on environmental exchanges, raw material consumption, and emissions.
   - Involves a comprehensive quantification of inputs and outputs across all lifecycle stages, highlighting the environmental aspects linked to the product or activity.
3. *Impact assessment* (Bradley et al. 2023; Muralikrishna and Manickam 2017):
   - Utilizes inventory data to evaluate environmental repercussions, categorizing impacts and assessing their significance.
   - The process includes selecting relevant models and indicators, classifying inventory results into impact categories, and calculating the category-specific impacts.
   - Additional steps involve comparing impacts against standard references (normalization), categorizing them based on importance (grouping), and consolidating results through a scoring system influenced by value choices (weighting).
4. *Interpretation* (Singal et al. 2023):
   - Analyzes findings from the inventory and impact assessments, focusing on understanding environmental outcomes.
   - Concludes the study by discussing implications, acknowledging study limitations, and suggesting forward-looking recommendations.

6 LCA for the Selected Case

This chapter discusses a methodological discussion of the simulation process, software and database selection, and a breakdown of the LCA stages.

6.1 Simulation Software

- Using Eco-chain Mobius software owing to its ease of use, scientifically endorsed outcomes, and valuable insights, aiding businesses in enhancing sustainability.
- This software demonstrated effectiveness in real-world cases, such as reducing Vodafone's packaging carbon emissions and assisting EMMA Safety Footwear in sustainable production practices.

6.2 Database Selection

- Adopting the Ecoinvent v3.8 database for its comprehensive, current data is especially beneficial for the Egyptian context.
- The selection is justified by the database's extensive datasets, support for various environmental impact methodologies, and relevance in providing realistic simulation inputs in Egypt.

6.3 LCA Stages for the Selected Case

All four stages of the analysis of the selected case are discussed in this subsection.

- *Goal and Scope Definition:*

  I. *Aim:* To assess the ecological impacts of existing linear and proposed circular supply chains.
  II. *Scope:* It encompasses the life cycle from cultivation to distribution, following a "cradle to gate" strategy.
  III. *Functional unit:* One liter of orange juice production.
  IV. *System boundary:* All required inputs and outputs involving all three supply chain stages of the proposed circular scenario are as follows:

Stage 1: Crop Production
- Input: water, energy, fertilizers, pesticides, and packages.
- Outputs: Wastewater, fruit waste, emissions, and packaging waste.

Stage 2: Producer (Pack House)
- Input: water, energy, and packaging.
- Outputs: wastewater, fruit waste, emissions, and packaging waste.

Stage 3: Transportation to the Distributor
- Inputs: energy and packages.
- Outputs: fruit waste, processed fruit waste, product waste, emissions, and packaging waste.

Table 3.1 LCI data for the proposed circular case

Supply chain stages	Sub stages	Input	Quantity
Crop production	Water	Primary data	20.9 liters
	Energy	Primary data	0.12 kWh
	Fertilizer	Primary and Secondary data	7.65 grams
	Pesticides	Primary data	None (organic farming)
	Packages	Primary data	1.65 grams of Polypropylene (PP) crate
1-1 Producer (processed fruit)	Water	Primary data	0.75 liters
	Energy	Primary data	0.05 kWh
	Packages	Secondary data	40 grams tetra pack
1-4 Transportation to the distributor	Energy (boiler and refrigeration)	Primary data	0.01 kWh
	Weight times distance by road	Primary data	1.73 ton-km
	Weight times distance by ship	Primary data	14.72 ton-km
	Packages	Secondary data	40 grams tetra Pak

- *Lifecycle Inventory Analysis*

Gather quantitative data regarding the resources used (material and energy flow) at each supply chain phase. The following Table 3.1 shows the inventory input data for all three stages, and the measurement is shown in terms of the functional unit of 1 liter of orange juice production.

- *Life Cycle Impact Assessment*

The environmental impact categories were examined using the ReCiPe 2016 assessment midpoint (E) method. The selected impact categories were global warming (GWP100a), ozone depletion (OD), eutrophication (freshwater), and eutrophication (marine).

Global warming is a critical focus in agricultural LCAs due to its direct connection with GHG emissions, highlighting the pressing need to assess and mitigate climatic impacts in agricultural practice (McClelland et al. 2018). While not as prevalent as global warming, ozone depletion remains an important impact category in agricultural LCAs, underlining the necessity of evaluating its effects to ensure comprehensive environmental assessments (Owens 1996). Lastly, eutrophication, particularly in aquatic environments, is a significant concern in agricultural LCAs, emphasizing the importance of evaluating nutrient runoff and its effects on freshwater and marine ecosystems (Møller et al. 2023).

- *Life Cycle Interpretation*

Critical evaluation and comparison of linear and circular supply chain models based on LCIA findings.

7 Results Analysis

This section compares the linear and circular cases for the four selected impact categories of the GWP100a, ODP, eutrophication (freshwater), and eutrophication (marine) (Table 3.2).

The following subsection discusses the detailed analysis for each impact category.

7.1 *Global Warming (GWP100a)*

In assessing the impact of orange juice production on global warming over a hundred years, this research found a significant difference between the traditional linear supply chain and a more sustainable circular model. The linear approach, which follows a straightforward "produce, use, dispose" method, results in about 0.674 kg of CO_2 eq. emissions. However, when moving to a circular model, where waste is utilized in a composting unit, these emissions drop to 0.63 kg CO_2 eq. This change marks a notable reduction of 6.53% in GHG emissions (Fig. 3.3).

7.2 *Ozone Depletion*

The OD category quantifies the detrimental effects of specific chemicals on the ozone layer, represented in (kg CFC11 eq). The impact within a linear supply chain is 1.49E−06 kg CFC11 eq. Conversely, this impact drops markedly to 6.26E−07 kg

Table 3.2 Comparative impact assessment: linear vs circular

Impact category	Unit	Linear supply chain	Circular supply chain	Overall % reduction in the circular case
GWP100a	kg CO_2 eq	6.74E−01	6.30E−01	6.53%
OD	kg CFC11 eq	1.49E−06	6.26E−07	58.05%
Eutrophication (freshwater)	kg P eq	7.24E−06	5.17E−06	28.70%
Eutrophication (marine)	kg N eq	1.08E−05	5.06E−06	53.15%

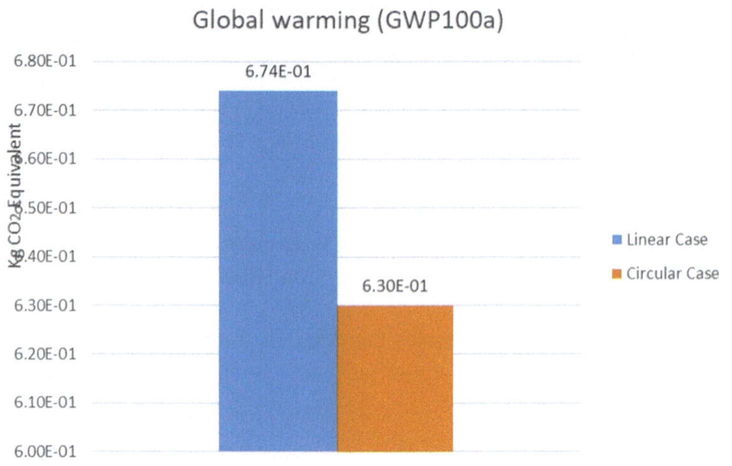

Fig. 3.3 Comparing linear and circular models' impact on the GWP100a

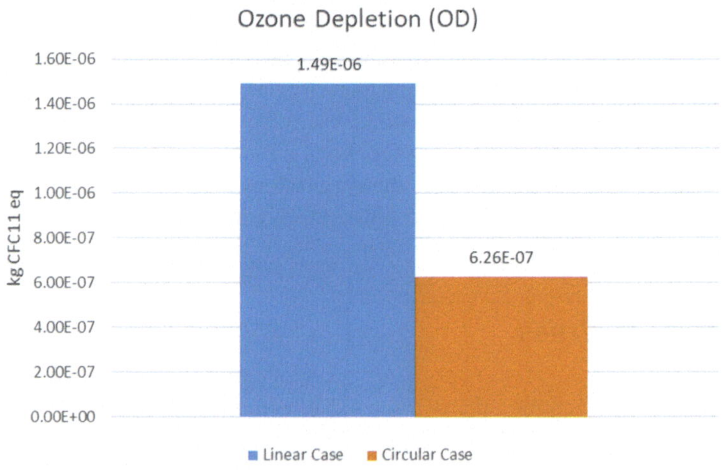

Fig. 3.4 Comparing linear and circular models' impact on the OD

CFC11 eq, indicating a significant decrease of 58.05% in potential ozone layer damage (Fig. 3.4).

7.3 *Eutrophication (Freshwater)*

Eutrophication, a phenomenon in freshwater cases triggered by an overabundance of nutrients that spur excessive plant proliferation and subsequent oxygen scarcity, is assessed in kg P eq. The impact is identified in a linear supply chain context at

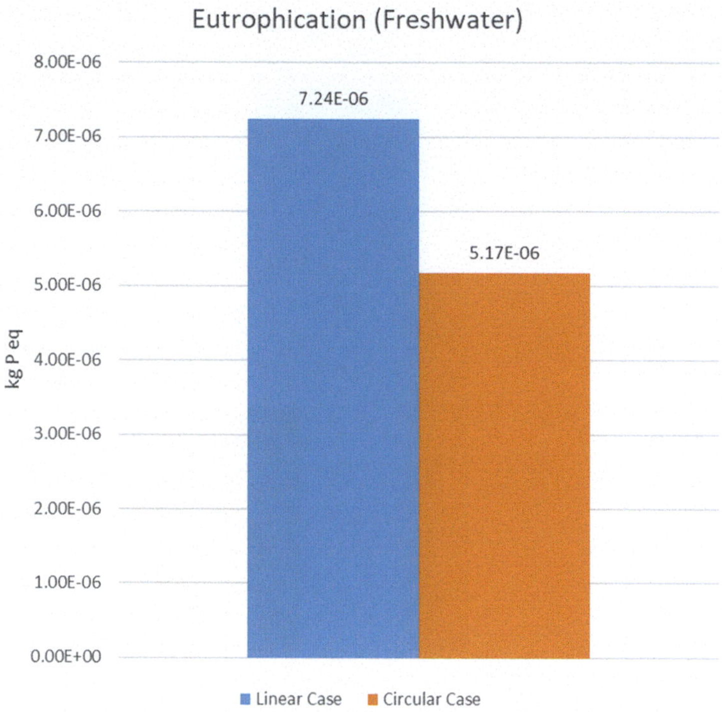

Fig. 3.5 Comparing linear and circular models' impact on eutrophication (freshwater)

7.24E−06 kg P eq. However, transitioning to a circular methodology curtails this impact to 5.17E−06 kg P eq, representing a 28.70% reduction in eutrophication potential (Fig. 3.5).

7.4 Eutrophication (Marine)

Like the freshwater variant, marine eutrophication transpires in oceanic ecosystems and is evaluated in kg N eq. The detrimental effect of this impact in a linear supply chain is 1.08E−05 kg N eq. Conversely, this impact diminished to 5.06E−06 kg N eq, signifying a substantial reduction of 53.15% in this impact category (Fig. 3.6).

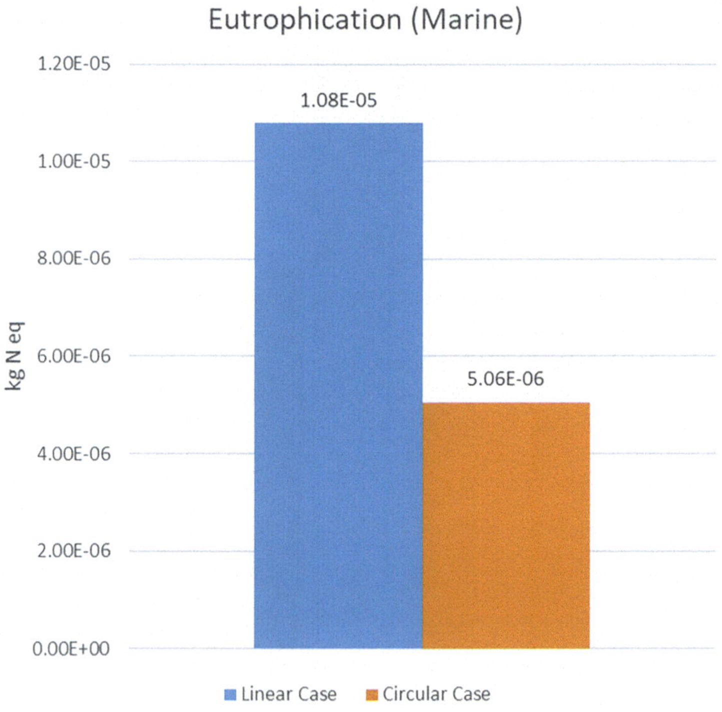

Fig. 3.6 Comparing linear and circular models' effects on eutrophication (marine)

8 Summary and Future Works

This study uses a selected case study to examine the environmental impacts of the orange juice supply chain in Egypt. The traditional linear model is compared with the proposed circular scenario. The research revealed that transitioning to a circular model by integrating a composting unit reduced fertilizer use by 82.37%, significantly improving all four environmental impact categories. The detailed analysis shows a 6.53% reduction in GWP100a, a 58.05% decrease in ODP, and a substantial decrease of 28.70% and 53.15%, respectively, for fresh and marine eutrophication. The findings suggest that adopting a circular system is a strategic pivot towards sustainability, serving as a valuable blueprint for future sustainable agricultural practices in regions beyond Egypt.

For further research on CE in orange juice production, an in-depth life cycle cost analysis from an economic perspective would be valuable for assessing the viability of improving the composting unit.

Exploring other waste management facilities, such as biogas plants, could also provide further opportunities to reduce waste and generate renewable energy. Embracing renewable energy sources, such as solar or wind energy, can significantly enhance the overall circularity and sustainability of the process. Collaborations with technology farms and enterprises can alleviate the financial burden of adopting these innovative solutions. Moreover, extensive awareness campaigns highlighting the benefits of CE would foster greater consumer understanding and encourage the broader adoption of sustainable practices.

References

Abdullah SNA, Ho CL, Wagstaff C (eds) (2017) Crop improvement: sustainability through leading-edge technology. Springer, Cham

Abedelmaksoud TG, Mohsen SM, Duedahl-Olesen L, Elnikeety MM, Feyissa AH (2019) Impact of ohmicsonication treatment on pectinmethylesterase in not-from-concentrate orange juice. J Food Sci Technol 56:3951–3956. https://doi.org/10.1007/s13197-019-03834-2

Albizzati PF, Rocchi P, Cai M, Tonini D, Astrup TF (2022) Rebound effects of food waste prevention: environmental impacts. Waste Manag 153:138–146. https://doi.org/10.1016/j.wasman.2022.08.020

Alzubi E, Kassem A, Noche B (2022) A comparative life cycle assessment: polystyrene or polypropylene packaging crates to reduce citrus loss and waste in transportation? Sustain For 14:12644. https://doi.org/10.3390/su141912644

Ashraf SRB, Daus D (2022) Integration of Blockchain Technology in the citrus food supply chain: a systematic literature review. https://www.researchgate.net/publication/362413028_Integration_of_Blockchain_Technology_in_the_Citrus_Food_Supply_Chain_A_systematic_literature_review/citations. Accessed 8 Jan 2024

Bansode R, Randolph P, Hassan O, Rehrah D, Ahmedna M (2016) Mixed solid municipal waste-based biochar for soil fertility and greenhouse gas mitigation. In: Qatar Foundation annual research conference proceedings volume 2016 issue 1. Hamad bin Khalifa University Press (HBKU Press)

Beccali M, Cellura M, Iudicello M, Mistretta M (2009a) Resource consumption and environmental impacts of the agrofood sector: life cycle assessment of italian citrus-based products. Environ Manag 43:707–724. https://doi.org/10.1007/s00267-008-9251-y

Beccali M, Cellura M, Iudicello M, Mistretta M (2009b) Resource consumption and environmental impacts of the agrofood sector: life cycle assessment of Italian citrus-based products. Environ Manag 43:707–724. https://doi.org/10.1007/s00267-008-9251-y

Beske P, Land A, Seuring S (2014) Sustainable supply chain management practices and dynamic capabilities in the food industry: a critical analysis of the literature. Int J Prod Econ 152:131–143. https://doi.org/10.1016/j.ijpe.2013.12.026

Boenzi F, Digiesi S, Facchini F, Silvestri B (2022) Life cycle assessment in the agri-food supply chain: fresh versus semi-finished based production process. Sustain For 14:13010. https://doi.org/10.3390/su142013010

Bradley T, Ling-Chin J, Maga D, Speranza LG, Roskilly AP (2023) Life Cycle Assessment (LCA) of algae biofuels. In: Yang X-S (ed) Benchmarks and hybrid algorithms in optimization and applications. Springer Verlag, Singapor, [S.l.], pp 387–404

Cabot MI, Lado J, Clemente G, Sanjuán N (2022) Towards harmonised and regionalised life cycle assessment of fruits: a review on citrus fruit. Sustain Prod Consum 33:567–585. https://doi.org/10.1016/j.spc.2022.07.024

Cambeses-Franco C, González-García S, Feijoo G, Moreira MT (2021) Encompassing health and nutrition with the adherence to the environmentally sustainable New Nordic Diet in Southern Europe. J Clean Prod 327:129470. https://doi.org/10.1016/j.jclepro.2021.129470

Cerutti AK, Beccaro GL, Bosco S, Luca AI de, Falcone G, Fiore A, Iofrida N, Giudice A, Lo Strano A (2015) Life cycle assessment in the fruit sector. https://www.springerprofessional.de/life-cycle-assessment-in-the-fruit-sector/2297868. Accessed 13 Nov 2023

Chen HE, Wang L (2017) Technologies for biochemical conversion of biomass, 1st edn. Elsevier, London

Christofi A, Tsipiras D, Malamis D, Moustakas K, Mai S, Barampouti EM (2022) Biofuels production from orange juice industrial waste within a circular economy vision. J Water Proc Eng 49:103028. https://doi.org/10.1016/j.jwpe.2022.103028

Curran MA (2017) Overview of goal and scope definition in life cycle assessment. In: Goal and scope definition in life cycle assessment. Springer, Dordrecht, pp 1–62

Da Costa TP, Gillespie J, Pelc K, Adefisan A, Adefisan M, Ramanathan R, Murphy F (2023) Life cycle assessment tool for food supply chain environmental evaluation. Sustain For 15:718. https://doi.org/10.3390/su15010718

de Medina-Salas L, Giraldi-Díaz MR, Castillo-González E, Morales-Mendoza LE (2020) Valorization of orange peel waste using precomposting and vermicomposting processes. Sustain For 12:7626. https://doi.org/10.3390/su12187626

Delrio G, Cocco A (2012) The peach fruit fly, Bactrocera Zonata: a major threat for Mediterranean fruit crops? Acta Hortic:557–566. https://doi.org/10.17660/actahortic.2012.940.80

Dimico A, Isopi A, Olsson O (2017) Origins of the Sicilian mafia: the market for lemons. J Econ Hist 77:1083–1115. https://doi.org/10.1017/s002205071700078x

Eremeeva NA, Savoskina OA, Poddymkina LM, Abdulmazhidov KA, Gamidov AG (2023) Analysis of anthropogenic impact on the environment, measures to reduce it, and waste management. Front Bioeng Biotechnol 11:1114422. https://doi.org/10.3389/fbioe.2023.1114422

Esposito B, Sessa MR, Sica D, Malandrino O (2020) Towards circular economy in the agri-food sector. A systematic literature review. Sustain For 12:7401. https://doi.org/10.3390/su12187401

Esturo A, Lizundia E, Sáez de Cámara E (2023) Fruit juice industry's transition towards sustainability from the viewpoint of the producers. Sustain For 15:3066. https://doi.org/10.3390/su15043066

Filho WL, Setti AFF, Azeiteiro UM, Lokupitiya E, Donkor FK, Etim NANA, Matandirotya N, Olooto FM, Sharifi A, Nagy GJ, Djekic I (2022) An overview of the interactions between food production and climate change. Sci Total Environ 838:156438. https://doi.org/10.1016/j.scitotenv.2022.156438

Fraga H, Santos JA (2021) Assessment of climate change impacts on chilling and forcing for the main fresh fruit regions in Portugal. Front Plant Sci 12:689121. https://doi.org/10.3389/fpls.2021.689121

Gallo F, Manzardo A, Camana D, Fedele A, Scipioni A (2023) Integration of a circular economy metric with life cycle assessment: methodological proposal of compared agri-food products. Int J Life Cycle Assess. https://doi.org/10.1007/s11367-022-02130-0

Gelsomino A, Abenavoli MR, Princi G, Attinà E, Cacco G, Sorgonà, A (2010) Compost from fresh orange waste: a suitable substrate for nursery and field crops? Compost Sci Util 18(3):201–210.

Gerber PJ, Steinfeld H, Henderson B, Mottet A, Opio C, Dijkman J, Falcucci A (2013) Tackling climate change through livestock: a global assessment of emissions and mitigation opportunities. Food and Agriculture Organization of the United Nations (FAO), Rome

Guinée JB, Oers V, de Koning A, Tamis WLM (2006) Life cycle approaches for conservation agriculture. In: SETAC-Europe 16th annual meeting at the Hague

Hanes R, Ghosh T, Key A, Eberle A (2021) The circular economy lifecycle assessment and visualization framework: a case study of wind blade circularity in Texas. Front Sustain 2. https://doi.org/10.3389/frsus.2021.671979

Haupt M, Hellweg S (2019) Measuring the environmental sustainability of a circular economy. Environ Sustain Indic 1–2:100005. https://doi.org/10.1016/j.indic.2019.100005

Hussain S, Khalid MF, Ali MA, Ahmed N, Hasanuzzaman M, Ahmad S (eds) (2023) Citrus production: technological advancements and adaptation to changing climate. CRC PRESS, Boca Raton

Keikha M, Darzi-Naftchali A, Motevali A, Valipour M (2023) Effect of nitrogen management on the environmental and economic sustainability of wheat production in different climates. Agric Water Manag 276:108060. https://doi.org/10.1016/j.agwat.2022.108060

Knoema (2020) World— Citrus fruit production quantity

Korhonen J, Nuur C, Feldmann A, Birkie SE (2018) Circular economy as an essentially contested concept. J Clean Prod 175:544–552. https://doi.org/10.1016/j.jclepro.2017.12.111

Li Z, Wu C, Huang J, Zhou R, Jin Y (2021) Membrane fouling behavior of forward osmosis for fruit juice concentration. Membranes (Basel) 11. https://doi.org/10.3390/membranes11080611

Liao CY, Guan YJ, Bustamante-Román M (2022) Techno-economic analysis and life cycle assessment of pineapple leaves utilization in Costa Rica. Energies 15:5784. https://doi.org/10.3390/en15165784

Mahapatra S, Ali MH, Samal K (2022) Assessment of compost maturity-stability indices and recent development of composting bin. Energy Nexus 6:100062. https://doi.org/10.1016/j.nexus.2022.100062

Maskova AR, Shayakhmetov RZ, Aminova AF, Klyavlina YM, Mazitova AK (2023) Reduced carbon footprint through targeted recycling. ogbus:93–124. https://doi.org/10.17122/ogbus-2023-3-93-124

McClelland SC, Arndt C, Gordon DR, Thoma G (2018) Type and number of environmental impact categories used in livestock life cycle assessment: a systematic review. Livest Sci 209:39–45. https://doi.org/10.1016/j.livsci.2018.01.008

Meadu V, Coche I, Vermeulen SJ, Friis AE (2015) The Paris Climate Agreement: what it means for food and farming

Miranda-Ackerman MA, Azzaro-Pantel C, Aguilar-Lasserre AA (2017) A green supply chain network design framework for the processed food industry: application to the orange juice agrofood cluster. Comput Ind Eng 109:369–389. https://doi.org/10.1016/j.cie.2017.04.031

Møller H, Lyng K-A, Röös E, Samsonstuen S, Olsen HF (2023) Circularity indicators and added value to traditional LCA impact categories: example of pig production. Int J Life Cycle Assess:1–13. https://doi.org/10.1007/s11367-023-02150-4

Muralikrishna I, Manickam V (2017) Environmental management: science and engineering for industry. Butterworth-Heinemann, Oxford

Negro V, Ruggeri B, Fino D, Tonini D (2017) Life cycle assessment of orange peel waste management. Resour Conserv Recycl 127:148–158. https://doi.org/10.1016/j.resconrec.2017.08.014

Nikkhah A, Firouzi S, Dadaei K, van Haute S (2021) Measuring circularity in food supply chain using life cycle assessment; refining oil from Olive Kernel. Food Secur 10. https://doi.org/10.3390/foods10030590

Notarnicola B, Tassielli G, Renzulli PA, Lo Giudice A (2015) Life Cycle Assessment in the agri-food sector: an overview of its key aspects, international initiatives, certification, labelling schemesand methodological issues. https://www.springerprofessional.de/life-cycle-assessment-in-the-agri-food-sector-an-overview-of-its/2297820. Accessed 13 Nov 2023

Notarnicola B, Tassielli G, Renzulli PA, Di Capua R, Saija G, Salomone R, Primerano P, Petti L, Raggi A, Casolani N, Strano A, Mistretta M (2022) Life cycle inventory data for the Italian Agri-food sector: background, sources and methodological aspects. Int J Life Cycle Assess. https://doi.org/10.1007/s11367-021-02020-x

Omolayo Y, Feingold BJ, Neff RA, Romeiko XX (2021) Life cycle assessment of food loss and waste in the food supply chain. Resour Conserv Recycl 164:105119. https://doi.org/10.1016/j.resconrec.2020.105119

Orhan O (2021) Land suitability determination for citrus cultivation using a GIS-based multi-criteria analysis in Mersin, Turkey. Comput Electron Agric 190:106433. https://doi.org/10.1016/j.compag.2021.106433

Ortiz DL, Batuecas E, Orrego CE, Rodríguez LJ, Camelin E, Fino D (2020) Sustainable management of peel waste in the small-scale orange juice industries: a Colombian case study. J Clean Prod 265:121587. https://doi.org/10.1016/j.jclepro.2020.121587

Ortiz-Sanchez M, Omarini AB, González-Aguirre J-A, Baglioni M, Zygadlo JA, Breccia J, D'Souza R, Lemesoff L, Bodeain M, Cardona-Alzate CA, Pejchinovski I, Fernandez-Lahore MH (2023) Valorization routes of citrus waste in the orange value chain through the biorefinery concept: the Argentina case study. Chem Eng Proc Proc Intensif 189:109407. https://doi.org/10.1016/j.cep.2023.109407

Owens JW (1996) LCA impact assessment categories. Int J Life Cycle Assess 1:151–158. https://doi.org/10.1007/BF02978944
Pergola M, D'Amico M, Celano G, Palese AM, Scuderi A, Di Vita G, Pappalardo G, Inglese P (2013) Sustainability evaluation of Sicily's lemon and orange production: an energy, economic and environmental analysis. J Environ Manag 128:674–682. https://doi.org/10.1016/j.jenvman.2013.06.007
Pergola M, Persiani A, Pastore V, Palese AM, D'Adamo C, de Falco E, Celano G (2020) Sustainability assessment of the green compost production chain from agricultural waste: a case study in Southern Italy. Agronomy 10:230. https://doi.org/10.3390/agronomy10020230
Rasines L, Castillejo N, San Miguel G, Aguayo E (2023) Can household storage conditions reduce food waste and environmental impact? A broccoli case study. Sci Total Environ 892:164779. https://doi.org/10.1016/j.scitotenv.2023.164779
Ribal J, Estruch V, Clemente G, Fenollosa ML, Sanjuán N (2019) Assessing variability in carbon footprint throughout the food supply chain: a case study of Valencian oranges. Int J Life Cycle Assess 24:1515–1532. https://doi.org/10.1007/s11367-018-01580-9
Rodrigues S, Fernandes FAN (2022) Glow discharge plasma processing for the improvement of pasteurized orange juice's aroma and off-flavor. PRO 10:1812. https://doi.org/10.3390/pr10091812
Secondi L (2021) A regression-adjustment approach with control-function for estimating economic benefits of targeted circular economy practices: evidence from European SMEs. SAE 39:10.25115/eea.v39i2.4070
Serafini LF, Feliciano M, Rodrigues MA, Gonçalves A (2023) Systematic review and meta-analysis on the use of LCA to assess the environmental impacts of the composting process. Sustain For 15:1394. https://doi.org/10.3390/su15021394
Siles JA, Vargas F, Gutiérrez MC, Chica AF, Martín MA (2016) Integral valorisation of waste orange peel using combustion, biomethanisation and co-composting technologies. Bioresour Technol 211:173–182. https://doi.org/10.1016/j.biortech.2016.03.056
Singal SK, Goel V, Nautiyal H, Papantonis DE (2023) Chapter 10 - Environmental impacts of small hydropower system. In: Singal SK, Goel V, Nautiyal H, Papantonis D (eds) Small hydropower: design and analysis. Elsevier, pp 259–269
Smith R (2017) Sustainable supply chains: utilisation of whole crops for new product development. In: Abdullah SNA, Ho CL, Wagstaff C (eds) Crop improvement: sustainability through leading-edge technology. Springer, Cham, pp 271–286
Statista (2023) Fruit: citrus production worldwide by region 2021 | Statista. https://www.statista.com/statistics/264002/production-of-citrus-fruits-worldwide-by-region/. Accessed 24 July 2023
Stillitano T, Spada E, Iofrida N, Falcone G, de Luca AI (2021) Sustainable agri-food processes and circular pathways in a life cycle perspectives: where is applicative research going? Sustain For. https://doi.org/10.20944/preprints202102.0034.v1
Sulewski P, Kais K, Gołaś M, Rawa G, Urbańska K, Wąs A (2021) Home bio-waste composting for the circular economy. Energies 14:6164. https://doi.org/10.3390/EN14196164
Teigiserova DA, Hamelin L, Tiruta-Barna L, Ahmadi A, Thomsen M (2022) Circular bioeconomy: life cycle assessment of scaled-up cascading production from orange peel waste under current and future electricity mixes. Sci Total Environ 812:152574. https://doi.org/10.1016/j.scitotenv.2021.152574
The Paris Agreement | UNFCCC (2023) https://unfccc.int/process-and-meetings/the-paris-agreement. Accessed 23 July 2023
Tsouti C, Papadaskalopoulou C, Konsta A, Andrikopoulos P, Panagiotopoulou M, Papadaki S, Boukouvalas C, Krokida M, Valta K (2023) Investigating the environmental benefits of novel films for the packaging of fresh tomatoes enriched with antimicrobial and antioxidant compounds through life cycle assessment. Sustain For 15:7838. https://doi.org/10.3390/su15107838
Vaverková MD, Adamcová D, Winkler J, Koda E, Petrželová L, Maxianová A (2020) Alternative method of composting on a reclaimed municipal waste landfill in accordance with the

circular economy: benefits and risks. Sci Total Environ 723:137971. https://doi.org/10.1016/j.scitotenv.2020.137971
Vidergar P, Perc M, Lukman RK (2021) A survey of the life cycle assessment of food supply chains. J Clean Prod 286:125506. https://doi.org/10.1016/j.jclepro.2020.125506
Vono G, Falcone G, Mistretta M, Fazari A, Strano A (2023) Life cycle inventory of the Italian citrus fruit supply chain: modelling the agricultural phase inventory through statistical data processing. Clean Environ Syst 10:100131. https://doi.org/10.1016/j.cesys.2023.100131
Zacarias L, Cronje PJ, Palou L (2020) Postharvest technology of citrus fruits. In: The genus citrus, p 421–446. https://doi.org/10.1016/B978-0-12-812163-4.00021-8

Part II
Best Cases in Sustainable Farming Practices

Chapter 4
Citrus Production Conditions in Algeria: Drought and Irrigation Issues

Lotfi Mustapha Kazi-Tani

1 Introduction

Citrus has a long history and has accompanied the development of agriculture since the Babylonian era. Indeed, citrus fruits bear witness to historical epochs and great civilizations, captivating admiration wherever they have been cultivated. Their allure quickly elevated them to the forefront of literature, arts, markets, and gastronomy. Among these fruits, the orange, in particular, symbolizes the convergence of European Islamic agricultural wisdom and sensory pleasure. Its influence radiated throughout Andalusia and Sicily during the transition of the first millennium CE and subsequent centuries, leaving an indelible mark on the cultural and culinary landscapes of these regions. On the other hand, citrus fruit emerged as an enduring symbol of the Italian Renaissance (Inglese and Sortino 2019).

In Algeria, citrus cultivation dates back to the Roman era. The country is home to a wide variety of citrus fruits, including oranges, lemons, grapefruits, mandarins, and pomelos. The most famous native variety is Clementine.

In agronomy, citrus refers to a group of evergreen fruit-bearing trees and shrubs from to the Rutaceae family. These plants are cultivated for their edible fruits, which are known for their acidic, juicy, and aromatic characteristics. Citrus fruits are highly valued for their nutritional content, high vitamin C content, and refreshing flavours.

Taxonomically, citrus fruits are classified into six botanical genera of the tribe Citreae, which are sexually compatible with each other: *Clymenia*, *Eremocitrus*, *Microcitrus*, *Poncirus*, *Fortunella*, and *Citrus*. The first three are of Oceanian origin, while the last three are Asian and include the most well-known, cultivated, and

L. M. Kazi-Tani (✉)
University of Tlemcen, SNV-STU Department of Agronomy, Tlemcen, Algeria

A. Melkonyan-Gottschalk, D. Daus (eds.), *Transformation Towards Circular Food Systems*, Greening of Industry Networks Studies 13,
https://doi.org/10.1007/978-3-031-63793-3_4

consumed citrus fruits (Rebour 1950; Ortiz 2002; Khan and Khan 2021; Luro et al. 2022).

A. Genus *Citrus*: is the most important genus and it comprises the following main species (Rebour 1950; Ortiz 2002; Luro et al. 2022):

   1. *Citrus sinensis* (Sweet Orange): This species includes varieties such as Valencia, Navel, and Blood oranges.
   2. *Citrus limon* (Lemon): Varieties include Eureka and Lisbon lemons.
   3. *Citrus reticulata* (Mandarin): Varieties include Clementine, Satsuma, and Tangerine.
   4. *Citrus paradisi* (Grapefruit): Varieties include Ruby Red, White, and Pink grapefruits.
   5. *Citrus aurantium* (Bitter Orange): This species is used for its peel in marmalades and essential oils.

B. Genus *Fortunella*:

   1. *Fortunella japonica*: the round kumquat.
   2. *Fortunella margarita*: the oval kumquat.

C. Genus *Poncirus:*

   1. *Poncirus trifolia.*

The genus *Citrus* includes economically significant species such as sweet orange (*Citrus sinensis*), lemon (*Citrus limon*), lime (*Citrus aurantifolia*), and grapefruit *(Citrus paradisi*). These species are not only valued for their nutritional and culinary uses but also contribute to local and global economies through agricultural production and international trade. Hence, the global citrus industry generates substantial revenue through exports and international trade. Citrus production in the Mediterranean region accounts for about one-fifth of the world's total citrus production. This production predominantly comprises sweet oranges and mandarin-like fruits, which are primarily intended for fresh consumption and, to a lesser extent, for use in agro-industrial processes. As a result, citrus farming plays a significant role in generating income for numerous individuals across Mediterranean countries, providing jobs at all levels from production to distribution. Additionally, traditional citrus cultivation is labour-intensive and plays a key role in driving the economy of the Mediterranean region (Duarte et al. 2016), especially in North Africa.

Some of the major citrus producers in the world include the North African countries such as Egypt, Morocco, Algeria then Tunisia (Fig. 4.1).

Referring to the FAO data (FAO 2021), the total worldwide citrus exports were estimated to be more than 17.4 million tons in 2019. This represents a market value of more than $14 billion. By 2021, the global market value had increased to over $16 billion. Spain was the primary exporter, with $4.12 billion, while Germany was the main importer, with $1.74. Among the North African countries, Egypt leads

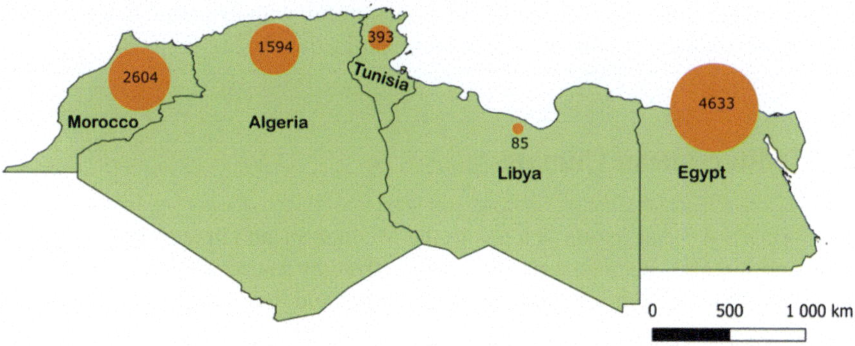

Fig. 4.1 Citrus production of the North African countries in thousand tonnes. (Data source: FAO 2021)

with a total of $982 million, followed by Morocco with $566 million, representing a combined 9.24% of the global citrus exports billion (OEC 2023).

The above mentioned data highlight the significance of studying citrus, as they provide insights into market dynamics, supply chains, trade policies, and consumer preferences. This knowledge can inform strategic decision-making for producers, exporters, and policymakers, fostering efficient trade practices and facilitating market access.

In this chapter, we focus on the agro-environmental conditions for citrus growth in the southern Mediterranean region in the context of climate change. The emerging climate realities of this region, characterized by rising temperatures, frequent heat waves, and prevailing drought conditions, necessitate a comprehensive examination of citrus species.

Despite their adaptation capacities, likely linked to their geographic origin, all citrus species will have to face the consequences of climate change. However, there have been limited efforts to explore the impact of climate change on citrus (Rasera et al. 2023).

Nevertheless, several researches on the relationship between citrus cultivation and climate change (Rosenzweig et al. 1996; Quiroga and Iglesias 2009; Albrigo et al. 2019; Ziogas et al. 2021) suggests that climate change is likely to have a significant impact on citrus production. In the Mediterranean context, excessive heat during the winter heat and drought are the main driving forces that lead to decline of the yield.

Environmental factors significantly impact citrus growth, productivity, and overall ecological sustainability. Hence, this study aims to investigate citrus climatic conditions in Algeria and to determine the irrigation requirements of citrus trees. To achieve this goal, the FAO software CROPWAT 8.0 has been used to determine irrigation requirements according to different meteorological stations. This method necessitates the incorporation of climatic, cultural, and soil data. Soil texture is the primary focus of this study due to its close relationship with irrigation requirements, followed by climatic conditions with a future projection for 2050.

2 Mediterranean Climate

The northern African region belongs to the Mediterranean climate type where the summer drought is the main trait. It is characterized by a succession of two seasons: winter mild to cool and wet, summer warm to hot and dry. Autumn and spring are considered as transitional seasons. Average rainfall is low in these areas (generally, 350 to 750 mm/yr) and generally falls in the winter. According to the Köppen-Geiger climate types classification and updated by Peel (Peel et al. 2007), the Mediterranean climate is one in which winter rainfall is more than three times summer rainfall. It is coined as warm temperate, dry summer (Cs), and the main subtype in Algeria is Csa Mediterranean hot summer climates, where the bulk of the citrus surfaces is concentrated.

2.1 Climate Change "Hotspot"

Climate change is already happening and must be considered; some regions are more affected than others, known as "hotspots". These are areas where the physical and ecological effects of climate change are strongest, especially for agriculture (Cos et al. 2022). All climate models agree that the western Mediterranean region will become much drier in the coming decades. Winter rainfall (December-January-February, or DJF), which is essential for agriculture, could decrease by up to 40% in the west and 20–30% in the east. The importance of winter precipitation to the region's agriculture and economy cannot be overstated, rendering its future a matter of paramount significance for Mediterranean countries. Furthermore, these models indicate a notable elevation in summer temperatures (June-July-August, JJA) by 3–4 °C over the ensuing century (Tuel and Eltahir 2020; Cos et al. 2022).

Mediterranean climate is also known for its seasonal and inter-annual variability linked to the North Atlantic Oscillation (NAO) (Barnston and Livezey 1987). For instance, Zenata station in the region of Tlemcen (North West of Algeria) has received approximately 474 mm in 2013 and 60% less in 2021 (175 mm).

An ongoing decrease in rainfall has been observed in the Mediterranean climate in the past three decades and is still current, with frequent and intense droughts. According to data from the Zenata station, annual rainfall has decreased by 25% over the past 40 years (Fig. 4.2). Similarly, the region of Boufarik in the Mitidja plain (North centre of Algeria) has also seen a decline in annual rainfall, with an estimated 25% loss from 1951 to 2008 (Drouiche et al. 2019).

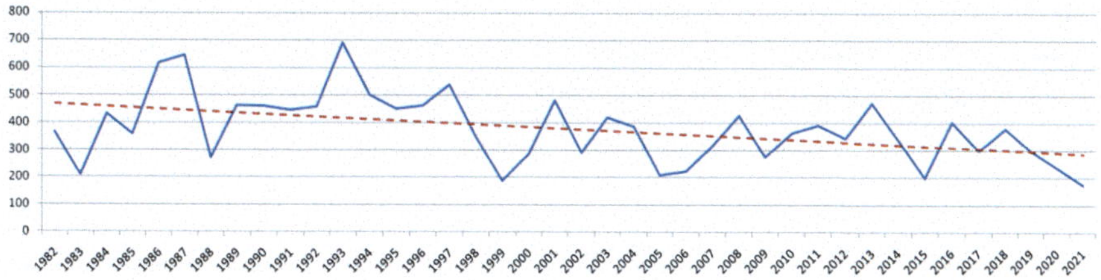

Fig. 4.2 Precipitations variation in the region of Tlemcen (Zenata station NW of Algeria). The red dashed line represents the trend line

2.2 *Effect of Climate Change on Citrus*

In the Mediterranean region, farmers perceive climate as a significant risk, likely due to the recurrence of severe weather events. Among the various climate risks, drought is perceived to be the most damaging, primarily because of its long duration and extensive spatial coverage (Quiroga and Iglesias 2009; Balfagón et al. 2022). Additionally, irregularities in temperature and rainfall affect phenological stages such as blossoming and ripening, which, in turn, impact the harvest cycle and fruit quality (Verner et al. 2018). Since citrus, as a perennial crop, offers limited opportunities for short-term adaptation to climate change, the development of new irrigation and farming techniques is essential to maintain yield and fruit quality within acceptable limits.

3 Mediterranean Soils

Mediterranean soils refer to a diverse group of soils found in the regions surrounding the Mediterranean Sea. Their evolution is a unique combination of natural and anthropogenic soil forming factors (Yaalon 1997; Kazi-Tani and Gaouar 2015; Lagacherie et al. 2017). These soils are influenced by a Mediterranean climate, characterized by hot, dry summers and mild, wet winters. The Mediterranean region is characterized by extensive coverage of carbonaceous rocks, such as limestone, dolomites, and marls. This occurrence is considered somewhat exceptional on a global scale and significantly affects soil formation processes in the region, with specific soil calcium dynamic.

The combination of climate, geology, and vegetation in these areas contributes to the distinctive characteristics of Mediterranean soils and their agronomic potentialities. Pedodiversity is probably higher in the Mediterranean region than in any other climatic zone (Ibáñez et al. 1998, 2013; Kazi-Tani and Gaouar 2015, 2016), which may represent an asset in crop production by ensuring various ecosystem services. In contrast, Mediterranean soils, particularly those in North Africa under arid and semi-arid conditions, exhibit low soil organic carbon content, which hinders several key soil functions, such as nutrient retention, water holding capacity, and soil physical quality (Romanyà and Povira 2011; Di Prima et al. 2018). These issues ultimately lead to soil degradation and negatively impact agriculture.

4 Ecological Requirements of Citrus

4.1 Climatic Requirements

Citrus plants have specific climatic requirements for optimal growth, flowering, and fruit production. Understanding these requirements is crucial for successful cultivation and maximizing yield. While different citrus species and cultivars may exhibit some variation in their tolerance to specific climate conditions, there are certain key factors that influence their growth and productivity.

Citrus species originated in tropical and subtropical regions, so they require a warm climate for high-quality production (Khan and Khan 2021). The optimal climate extends from the equator to above 40° latitude in both the Northern and Southern Hemispheres, where the risk of prolonged freezing temperatures is low. Most of the world's citrus production comes from countries located between 20° latitude and the equator (Davenport 1990; Albigo et al. 2019; Khan and Khan 2021).

It is within Mediterranean and semi-tropical climates that citrus fruits attain their most vibrant hues, smoothest peel texture, and the ideal balance between sugars and acidity, rendering them perfectly suited for the fresh fruit market. Varieties such as Navel and blood oranges, alongside lemons, are primarily confined to Mediterranean climates. The production of citrus, both in terms of quantity and quality, exhibits a close interdependence with climatic elements, particularly low temperatures and drought (Albigo et al. 2019; Ghada et al. 2019; Khan and Khan 2021). Low temperatures limit the geographic extent of citrus cultivation to coastal regions with elevations below 400 meters above sea level. The Mediterranean region grapples with recurring drought episodes, resulting in highly variable rain-fed production and conflicts related to water resources for irrigation. Consequently, with the onset of climate change, the suitability of areas for citrus cultivation is expected to decline (Vincent et al. 2020).

Increases in irrigation in response to these temperature and precipitation patterns are expected to impact soil water availability and quality. In semiarid areas, the groundwater extraction for irrigation led to the lowering of groundwater.

Generally, citrus trees prefer temperatures between 15 and 30 °C. Freezing temperatures can severely damage or kill citrus trees, especially young or tender growth. Frost protection measures, such as using protective covers or heaters, may be necessary in regions prone to frost events.

High temperature is the main trend of North African climates, and frost is not the main threat to the citrus trees. Like other fruit trees with the C3 photosynthetic pathway, the most favourable temperature for photosynthesis ranges from 25 to

Table 4.1 Citrus key temperatures by comparison with corresponding temperatures in citrus areas cultivation in Algeria

Key citrus temperatures (Rebour 1950; Albrigo et al. 2019)	Corresponding temperatures in citrus regions in Algeria
Mean annual temperature must be around 14 °C	Between 18 and 20 °C
Mean summer temperature must be around 22 °C	Around 25 °C
Mean winter temperature must be equal or higher than 10 °C	Around 13 °C
Minimum temperatures of −4 °C	Very few regions reach this temperature mainly in the coastal area.

35 °C. The following Table 4.1 summarizes key citrus temperatures in comparison with corresponding temperatures in citrus cultivation areas in Algeria.

It is important to note that different citrus species and cultivars have varying degrees of adaptability to different climates. Some species, such as oranges and grapefruits, are more tolerant of subtropical and tropical climates, while others, like lemons and limes, can tolerate slightly cooler conditions. Additionally, specific citrus rootstocks are selected for their adaptability to different climates and soil conditions.

Successful citrus cultivation often involves selecting appropriate citrus varieties or cultivars that are well-suited to the local climate conditions, implementing proper irrigation practices, managing pest and disease pressures, and providing necessary protection during extreme weather events.

4.2 Soil Conditions

As per the FAO/UNESCO global soil map, the principal soil types underpinning citrus orchards in North Africa encompass chromic cambisols, calcic cambisols, calcaric fluvisols, and eutric fluvisols. While the FAO soil classification provides valuable insights from an agronomic perspective, it is insufficient in isolation. For a more comprehensive assessment, physical and chemical attributes of the topsoil (≤30 cm) warrant consideration. These attributes encompass the soil's textural class, pH level, cation exchange capacity, and calcium carbonate content.

Soil physical quality is strongly influenced by its texture, which is determined by the relative proportions of sand, silt, and clay. Texture, in turn, exerts a profound impact on soil structure, which holds a pivotal role in the soil's capacity to receive, retain, and convey water, as well as in the distribution of chemical fertilizers, soil aeration, and drainage (FAO–IIASA 2021). This aspect assumes particular significance for citrus species, given their roots' heightened oxygen requirements, prompting them to extend their feeder roots in close proximity to the soil surface

(Calabrese 2002). Consequently, the practice of shallow tillage demands meticulous attention, especially in soils of greater density.

Nevertheless, a recent investigation by Di Prima et al. (2018) has underscored that the most effective approach for enhancing soil physical quality in citrus cultivation is the application of sheep and goat manure, while tilling and herbicide treatments should be avoided. Generally, citrus exhibits higher productivity and fruit quality in sandy soils.

5 Citrus in Algeria

5.1 Geography of Citrus Area in Algeria

Broadly, Algeria is divided into two main regions: the Northern part, known as the Tell, is the most fertile and humid, while the Southern part is more arid and falls within the Sahara (precipitation $\leq$100 mm/year), thereby exhibiting a negative rainfall gradient from the North to the South.

The Tell region has also a negative rainfall gradient from the East (800 mm/year) to the West (350 mm/year), and the terrain is predominantly flat, featuring more plains in the centre and western parts. As a result, citrus cultivation in Algeria is limited to the coastal regions and is distributed across three main areas:

- Eastern Region, representing 10.76% of the total citrus area.
- Central Region, representing 63.17% of the total citrus area.
- Western Region, representing 26.10% of the total citrus area.

Clearly, the central part of Algeria is most important for citrus cultivation, primarily due to the Mitidja Plain (Fig. 4.3). The western region (approximately 26%) is also important because it has several plains, with the Sig Plain being the most significant. Irrigation is essential in this area due to insufficient rainfall. Conversely, the eastern part (approximately 11%), despite being the wettest region in Algeria, is limited in large-scale orchard expansion because of its difficult terrain and few plains.

Close examination of the production statistics (Fig. 4.4) reveals that the Wilaya of Blida (Wilaya is an administrative department) is by far the most productive citrus producer in Algeria. There are three main reasons for this:

- Blida contains the largest portion of the Mitidja Plain, with the remainder shared between Tipaza, Algiers, and Boumerdes.
- Blida is close to the Mediterranean Sea.
- Blida receives an average annual rainfall of over 600 mm.

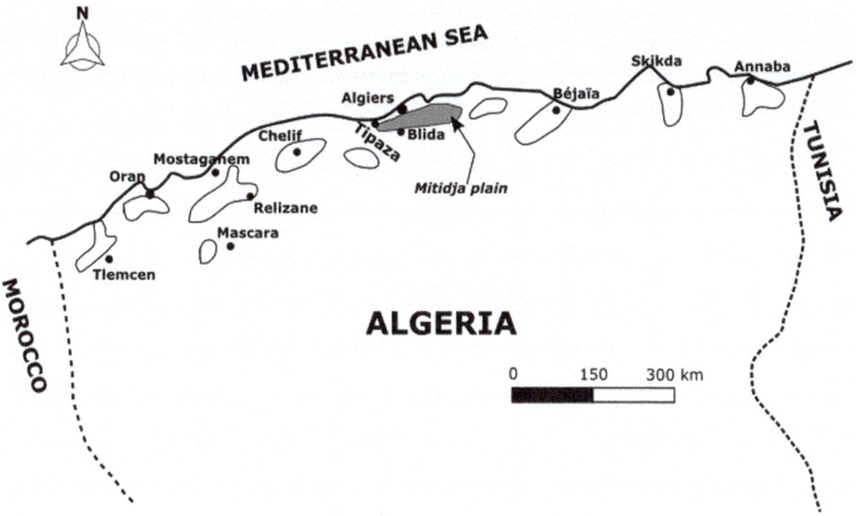

Fig. 4.3 Main citrus-growing regions in Algeria, with a focus on the Mitidja Plain study area

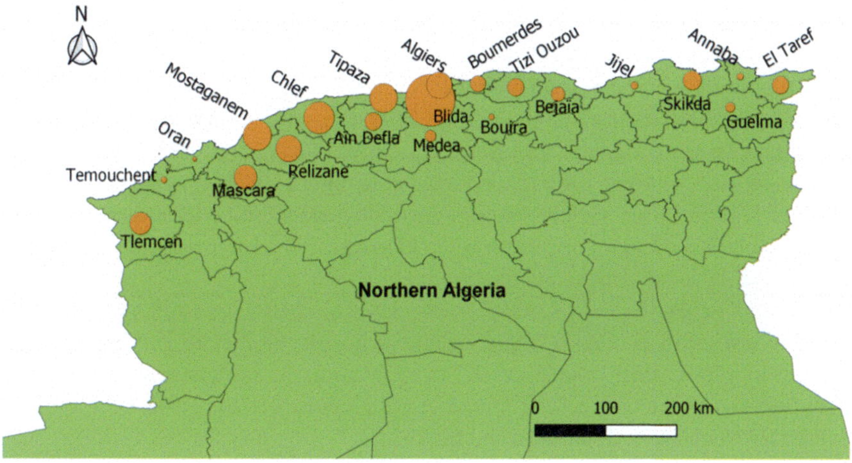

Fig. 4.4 Citrus production in Algeria's main departments

5.2 Citrus Varieties in Algeria

The nomenclature of citrus varieties has long been a challenge, due to a number of factors:

- Environmental factors can induce variation.
- The abundance of citrus trees in gardens and orchards throughout the Mediterranean region has facilitated cross-pollination, resulting in the emergence of new

hybrids. The most famous of these is the Clementine tree, which originated in the late nineteenth century. Father Clément fostered its development at the orphanage in Misserghin, a small village about 20 kilometres southwest of Oran, Algeria (Rebour 1950; Luro et al. 2022).

Citrus rootstock plays a crucial role in the cultivation of citrus trees, influencing their growth, health, and overall productivity. Rootstock selection is a strategic decision for citrus growers, as it can determine the tree's vigour, adaptability to specific soils, and resistance to pests, diseases, as well as drought and salinity. In Algeria, 10 species and varieties of citrus rootstocks are approved (CNCC 2015). The main citrus rootstocks are (Blondel 1974; Ortiz 2002; ITAFV 2012; CNCC 2015):

- *Citrus aurantium* (L.), also known as bitter orange or Seville orange, is the most widely used rootstock or rootstock progenitor due to its good compatibility with the main cultivated varieties, ease of propagation and grafting, adaptability to soil conditions (especially salinity and limestone), and drought tolerance. Resistant to Phytophthora and gummosis. However, sensitive to nematodes, mal secco, and tristeza, except when grafted with lemon trees.
- *Poncirus trifoliata* (L.) Raf., also known as the trifoliate orange, is used as a rootstock or rootstock progenitor because it has desirable resistance and tolerance traits for citrus cultivation, including resistance to tristeza, nematodes, Phytophthora gummosis, and cold. However, it is sensitive to limestone, salinity, and the viroid disease exocortis (CEV Citrus exocortis viroid).
- Citrange Troyer *(Poncirus trifoliata* L. × *Citrus sinencis* L. Osbeck), is a hybrid citrus rootstock known for its exceptional cold hardiness. It imparts strong vigour to the grafted citrus variety and demonstrates adaptability to a variety of soil types, tolerates moderately damp soils. Citrange Troyer is generally resistant to some soil-borne diseases, and to tristeza.
- Citrange Carrizo (*Citrus sinensis* L. Osbeck × *Poncirus trifoliate* L.) is a versatile and robust citrus rootstock, appreciated for its adaptability to varying climates and soil types. With moderate to strong vigour, it imparts resilience to the grafted citrus variety. Known for relative cold hardiness and resistance to specific diseases, such as Phytophthora and tristeza.
- *Citrus limonia* "volkameriana", highly compatible with lemon and pomelo. It exhibits robust resistance to cold conditions; very tolerant to calcareous soils and high pH. It is ideally suited to well-drained and aerated soils, and showcases resilience to soil salinity. Tolerant to tristeza and exocortis. Very resistant to gummosis.
- *Citrus macrophylla* Wester, highly compatible with lemon and pomelo, cold, sensitive, and humid soils. Tolerates soil salinity and calcareous soils. Sensitive to tristeza, and tolerant to gummosis and to exocortis.
- Cleopatra mandarin *(Citrus reshni* HORT. *ex* TAN.*)* presents poor graft recovery. Tolerates soil salinity and calcareous soils. Tolerant to tristeza, xylophorosis, cachexia. Resistant to gummosis.

Table 4.2 The most important varieties of citrus in Algeria

Species	Groups	Varieties
Orange	Orange Blond	Cadenera Hamlin Maltaise Salustiana Shamouti N°85 (Jaffa) Valentia Late
	Orange Navel	Washington Navel Washington Navel Bernard Thomson Navel
	Orange Sanguine	Double fine Portugaise Sanguine Sanguinelli
Mandarins and hybrids	Clementine	Aïn Taoujdat Clementine 2749 (seedless) Messerghine 48 Montreal Nova Ragheb Trabut
	Mandarin	Kara N°165 Mandarine commune Mandarine de Blida Mandarine ortanique Wilking vivier Satsuma
Lemon	Lemon	Eureka N°4 Feminello Lisbon Santa Tereza Villa Franca
Pomelo	Pomelo white flesh	Duncan Marsh seedless
	Pomelo pink flesh	Foster Redblush Star ruby

Algeria boasts 273 varieties of citrus, among them 97 are approved (CNCC 2015), 34 are local varieties, and 9 are the most used (INRAA 2006). Some of the most important varieties of citrus fruits grown in Algeria are presented in Table 4.2, based on CNCC (2015) study.

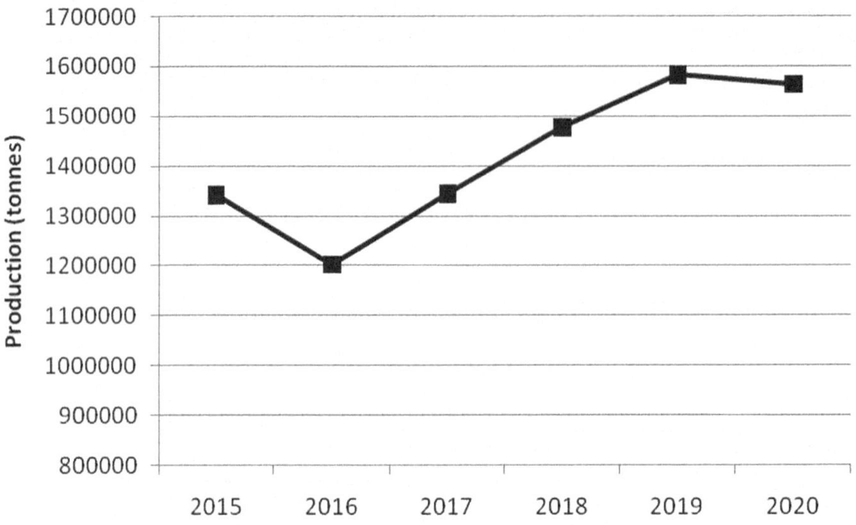

Fig. 4.5 Evolution of citrus production 2015 and 2020. (Data source: AMARD)

5.3 Citrus Production in Algeria

The main citrus species distributed worldwide have been previously introduced. However, only a few of them hold economic importance in Algeria. Figure 4.5 depicts the evolution of citrus production in Algeria from 2015 to 2021. According to the 2021 statistics provided by the Algerian Ministry of Agriculture and Rural Development (AMARD), the national citrus production comprises: 89% of orange, 7% of lemons, 3% of mandarins, <1% of Clementine and <1% of pomelos.

The citrus production in Algeria faces a number of challenges, including water scarcity, pests and diseases, and a lack of investment in post-harvest infrastructure including cold stores. Citrus cultivation is primarily focused on fresh consumption, resulting in substantial waste that can be recycled but poses collection challenges. However, there are also a number of opportunities for the industry, including growing demand for citrus fruits in international markets, the development of new varieties, and the adoption of new technologies.

6 Methodology

To identify citrus water stress during the flowering period, which is the most sensitive phenological stage, the agriculture stress index (ASI) was employed. This study also used the FAO software CROPWAT to calculate irrigation requirements and scheduling for citrus crops based on climate data. The results will be useful for estimating agricultural water demand and comparing it to irrigation citrus data to calculate irrigation efficiency particularly in Tipaza. Subsequently, global climate change data for the Mediterranean region was used to predict irrigation requirements for 2050.

6.1 Study Area

As quoted above, the plain of the Mitidja is the largest and among the most fertile plains of Algeria. It stretches over 100 km from East-North-East to West-North-West and 5–25 km from North to South, in the center of the Tell (Fig. 4.3) and covers a surface area of about 1400 km^2. The Mitidja plain is divided into two physical units: the lower Mitidja in the East and the upper Mitidja in the West. All the plain receives about 600 mm/year of rainfall.

Our study area covers a part of the upper Mitidja in Tipaza department, and citrus cultivation is the dominant activity.

6.2 Computerized Crop Water Use Simulations (CROPWAT)

The FAO CROPWAT program (Smith 1992) was developed to calculate reference crop evapotranspiration and crop water requirements (CWR). It enables the simulation of crop water use under various climate, crop, soil texture, and soil depth conditions.

In the realm of irrigation planning, computer programs have become essential tools for estimating reference crop evapotranspiration (ET_0) from climate data. This advancement has resulted in the development of standardized information and criteria for the purpose of planning and managing both rainfed and irrigated agricultural systems.

One notable contribution in this field is the decision support system known as CROPWAT, developed by the Food and Agriculture Organization of the United Nations (FAO). CROPWAT incorporates a range of agrometeorological parameters, including the geographical coordinates of the specific area, maximum and minimum surface temperatures, maximum and minimum relative humidity, wind speed, and solar radiation/sunshine hours. It can also be used to calculate ET_0 and crop evapotranspiration (ET_c), for different climates, crops, and soil types. Notably, CROPWAT calculates CWR and crop irrigation requirements as well as irrigation schedules across diverse management.

In our specific context, CROPWAT 8.0 for Windows has been deployed for the purpose of calculating CWR and irrigation requirements. This calculation is based on existing climate data sourced from CLIMWAT 2.0 for CROPWAT, complemented by relevant citrus crop data, such as the crop coefficient (Kc). Moreover, the application CLIMWAT, designed to retrieve climate data, has been used in this study. We have harnessed adjusted data derived from climatic models to project CWR trends up to the year 2050.

Climatic Models Data

As previously mentioned, climate models project significant warming in summer (JJA) of 3–4 °C by the end of the century (Tuel and Eltahir 2020; Cos et al. 2022). The same models also project a decrease in winter (DJF) precipitation of 30–40% in North Africa. Due to the large number of scenarios and models available, we have chosen to use the worst-case scenario in this study, with a 4 °C increase in annual temperature and a 40% decrease in rainfall, to estimate future CWR and citrus production. The input climate data will be modified according to these projections. Assuming that both temperature increase and precipitation decrease follow a linear progression throughout the century, we will reach +2 °C and −20% of precipitation by 2050.

Estimation of Crop Water Requirements (CWR)

The term CWR pertains to the volume of water needed to offset the loss through evapotranspiration from a cultivated area. Although the numerical values for crop evapotranspiration under standard conditions (ET_c) and CWR may be the same, the distinction lies in their interpretation (FAO 1977; FAO–IIASA 2021). CWR signifies the amount of water that must be provided, whereas crop evapotranspiration denotes the quantity of water lost due to evapotranspiration (Eq. 4.1).

$$\mathrm{CWR} = \mathrm{ET}_0 \times \mathrm{Kc} \times \%\mathrm{Planted\ area} \tag{4.1}$$

Irrigation Requirements

Represents the quantity of water that must be supplied to a crop to meet the plant's needs, compensate for evaporation and other losses, and ensure optimal crop growth. These requirements depend on various factors such as the type of crop, climate, soil conditions, and growth stage of the plants. Irrigation requirement is the most important parameter considered in this paper (FAO 1977; FAO–IIASA 2021).

Equation 4.1 can be used to estimate irrigation requirement (IR):

$$\mathrm{IR} = \mathrm{CWR} \times \mathrm{Effecive\ rain} \tag{4.2}$$

Where effective rain was calculated using the FAO/AGLW empirical formula; it is based on analysis for different arid and sub-humid climates. This formula is as follows (Brouwer and Heibloem 1986):

- Effective rain = 0.6 × Total rainfall − 10

  For Total Rainfall <70 mm

- Effective rain = 0.8 × Total rainfall − 24

  For Total Rainfall >70 mm

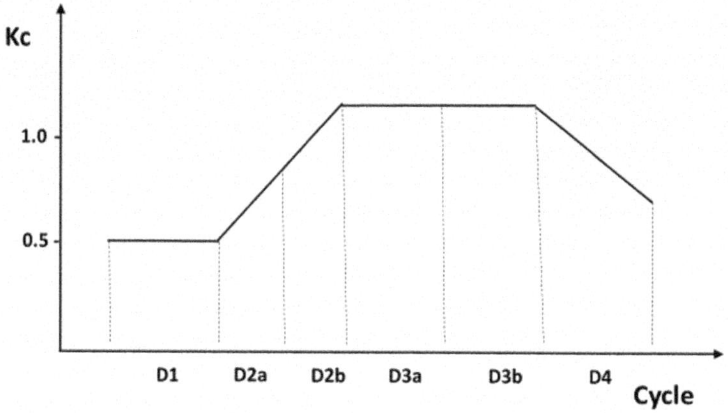

Fig. 4.6 Schematic representation of Kc values for different crop development stages

Crop Evapotranspiration Under Standard Conditions (ETc)
Is the amount of water lost through evapotranspiration from healthy, well-fertilized crops grown in large fields under optimal soil water conditions and achieving maximum production potential under the given climatic conditions. ETc is calculated by multiplying reference potential evapotranspiration (ET_0) by a crop coefficient (Kc) (Eq. 4.3). The values of Kc vary depending on the crop type and growth stage and are given as input parameters in Fig. 4.6 (Allen et al. 1998).

$$ETc = Kc \times ET_0 \tag{4.3}$$

Crop Coefficient (Kc)
The Crop coefficient (Kc) is an integrating factor that takes into account the characteristics distinguishing a particular crop from the reference crop. In accordance with the crop coefficient approach, the calculation of crop evapotranspiration under standard conditions (ETc) involves multiplying the reference evapotranspiration (ET_0) by the appropriate Kc value specific to the given crop.

- *D1:* Initial phase: from planting to 10% ground cover (from planting/germination/emergence to establishment).
- *D2a:* early crop development stage.
- *D2b:* late crop development stage.
- *D3a:* early mid-season stage (flowering).
- *D3b*: late mid-season stage (reproductive stage).
- *D4:* late season stage (start maturation to full maturity).

Kc is dependent on crop type and varies according to the phenological stages.

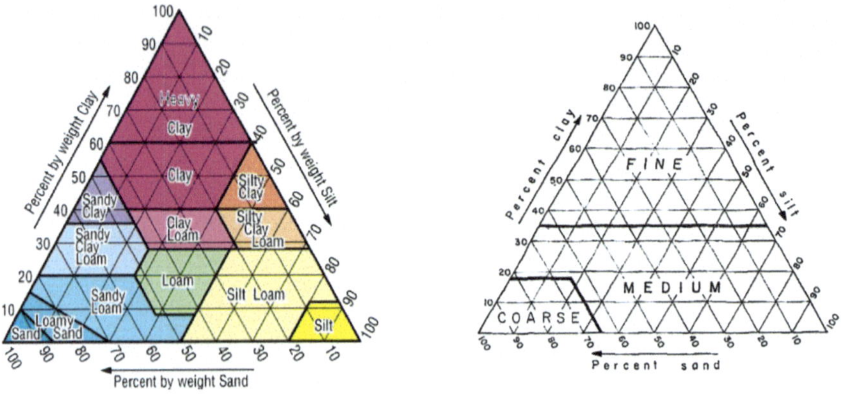

(a) *(b)*

Fig. 4.7 The USDA soil textural triangle (**a**) and soil textural groupings (**b**) (Fisher et al. 2021)

Reference Evapotranspiration (ET_0)
The reference evapotranspiration (ET_0) represents evapotranspiration of well-watered grass of uniform height (12 cm), actively growing and completely shading the ground. The ET_0 is calculated with the Penmann-Monteith equation. The purpose of calculating ET_0 is to assess irrigation needs and determine irrigation schedule (FAO 1977; FAO–IIASA 2021).

Soil Texture
Soil classes are determined by using soil triangle texture. The most commonly used is the USDA (United States Department of Agriculture) soil triangle texture (Fig. 4.7a), it shows 13 textural classes, these classes are grouped into 3 main soil textural groupings (Fisher et al. 2021) (Fig. 4.7b):

CROPWAT requires introducing soil textures:

- Heavy: for fine texture
- Medium: for medium texture
- Light: for coarse texture

7 Results and Analysis

The results are split into two parts. The first part concerns the prevailing soil texture of the main citrus region in Algeria. The second part shows the values obtained for crop evapotranspiration under standard conditions (ETc), effective rain, and irrigation requirements following CLIMWAT data. We have called these values "actual conditions" and projected them to 2050.

7.1 Soil Texture

Generally, 75% of the soils in Northern Algeria exhibit clay content exceeding 22.2% in the surface horizon and 26% in the sub-surface horizon (Djili 2000). This means that 75% of Algeria's soils have a medium texture. More specifically, and according to our data, 59% of citrus orchards in Algeria are located in soils with medium texture, 18% in soils with a fine texture, and 24% in soils with a coarse texture.

Soil texture is one of the key factors that influence available soil water holding capacity and volume. Considering the climatic context, medium to fine-textured soil would be an advantage for citrus cultivation in the face of drought. The reference available soil water capacity (AWC) typically refers to the amount of water that soil can retain and make available for plants to use. Soil water capacity is a measure of how much water a given soil can hold, and it's an important factor in understanding the water-holding capabilities of different soils and can be helpful to face drought.

In the Mitidja plain, two major soil types are present: the fluvisols with medium to fine texture and vertisols with fine texture. According to FAO–ISRIC (1990), the reference available soil water capacity (AWC) of fluvisols is 250 mm/m, whereas for vertisols, it is 135 mm/m. This suggests that fluvisols are potentially more suitable for citrus cultivation than vertisols.

7.2 Irrigation Requirements Results from CROPWAT 8.0

Irrigation is applied at critical depletion and it refills soil to field capacity. Figure 4.8 displays the CROPWAT sheet results for the CWR of citrus in the specific case of Tipaza.

The results of 4 stations cover principal citrus regions and concern the main parameters: ETC, effective rain, and irrigation requirements for citrus. CLIMWAT for CROPWAT was the main source of climatic data for the actual results (Table 4.3).

Figure 4.9 illustrates the irrigation demands for citrus cultivation in Algeria, calculated using CROPWAT 8.0. The results predict an 8% to 16% increase in irrigation requirements with a mean of 12%.

7.3 Irrigation Efficiency

A comparison has been made between the estimated irrigation requirements and the actual irrigation schedule in the region of Tipaza (Table 4.4).

It appears that the applied irrigation quantity is 37.2% greater than the CROP-WAT estimation. The latter considers only the total amount of water needed for a

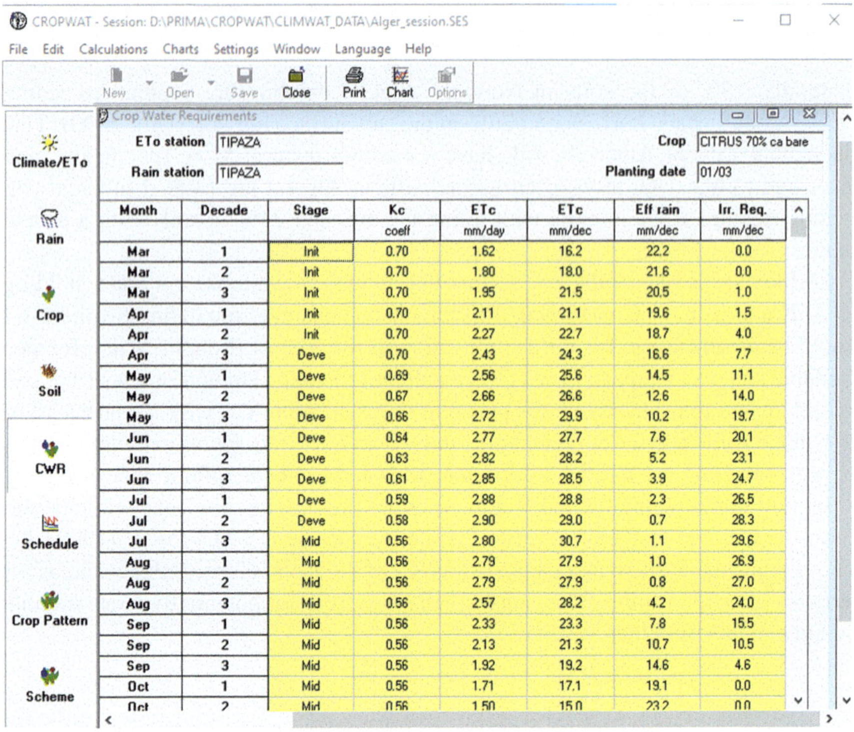

Month	Decade	Stage	Kc	ETc	ETc	Eff rain	Irr. Req.
			coeff	mm/day	mm/dec	mm/dec	mm/dec
Mar	1	Init	0.70	1.62	16.2	22.2	0.0
Mar	2	Init	0.70	1.80	18.0	21.6	0.0
Mar	3	Init	0.70	1.95	21.5	20.5	1.0
Apr	1	Init	0.70	2.11	21.1	19.6	1.5
Apr	2	Init	0.70	2.27	22.7	18.7	4.0
Apr	3	Deve	0.70	2.43	24.3	16.6	7.7
May	1	Deve	0.69	2.56	25.6	14.5	11.1
May	2	Deve	0.67	2.66	26.6	12.6	14.0
May	3	Deve	0.66	2.72	29.9	10.2	19.7
Jun	1	Deve	0.64	2.77	27.7	7.6	20.1
Jun	2	Deve	0.63	2.82	28.2	5.2	23.1
Jun	3	Deve	0.61	2.85	28.5	3.9	24.7
Jul	1	Deve	0.59	2.88	28.8	2.3	26.5
Jul	2	Deve	0.58	2.90	29.0	0.7	28.3
Jul	3	Mid	0.56	2.80	30.7	1.1	29.6
Aug	1	Mid	0.56	2.79	27.9	1.0	26.9
Aug	2	Mid	0.56	2.79	27.9	0.8	27.0
Aug	3	Mid	0.56	2.57	28.2	4.2	24.0
Sep	1	Mid	0.56	2.33	23.3	7.8	15.5
Sep	2	Mid	0.56	2.13	21.3	10.7	10.5
Sep	3	Mid	0.56	1.92	19.2	14.6	4.6
Oct	1	Mid	0.56	1.71	17.1	19.1	0.0
Oct	2	Mid	0.56	1.50	15.0	23.2	0.0

Fig. 4.8 Computation of irrigation requirements for citrus in Tipaza. The main calculated parameters include: Kc, ETC, effective rain, and irrigation requirements. The results are presented by decade, hence each month contain three decades

citrus crop without taking into account potential losses, inefficiencies, or external factors, and can be considered as gross irrigation requirements (GIR) (Eq. 4.4). In contrast, the applied irrigation amount takes into consideration these losses which are actually of tremendous importance and can somehow express the efficiency of the irrigation system. This amount is called net irrigation requirements (NIR):

$$\mathrm{GIR} = \frac{\mathrm{NIR}}{\text{Overall irrigation efficiency}} \tag{4.4}$$

Hence, the overall irrigation efficiency (Eq. 4.5) of Tipaza citrus orchards is:

$$\text{Overall irrigation efficiency} = \frac{460.4}{733.5} = 0,628 \tag{4.5}$$

So, the irrigation system of the citrus orchards in the region of Tipaza is operating at 63% efficiency with about 37% losses. These losses are not solely attributed to

Table 4.3 Results for the current and projected (2050) citrus crop in the some stations in Algeria. **ER**: Effective rain is the part of rainwater stored in the root zone and can be used by the citrus. **IR**.: Irrigation requirement

	Actual			Projection 2050			
Stations	ETC (mm/year)	ER (mm/year)	IR (mm/year)	ETC (mm/year)	ER (mm/year)	IR (mm/year)	% of IR increment
Tlemcen	671.0	77.4	593.6	692.3	52.9	639.4	8
Mohammadia	701.4	269.1	474.2	742.1	219.1	535.5	13
Tipaza	675.7	263.6	460.4	704.5	209.1	514.5	12
Annaba	692.7	545	362.9	733.6	447.4	420.3	16

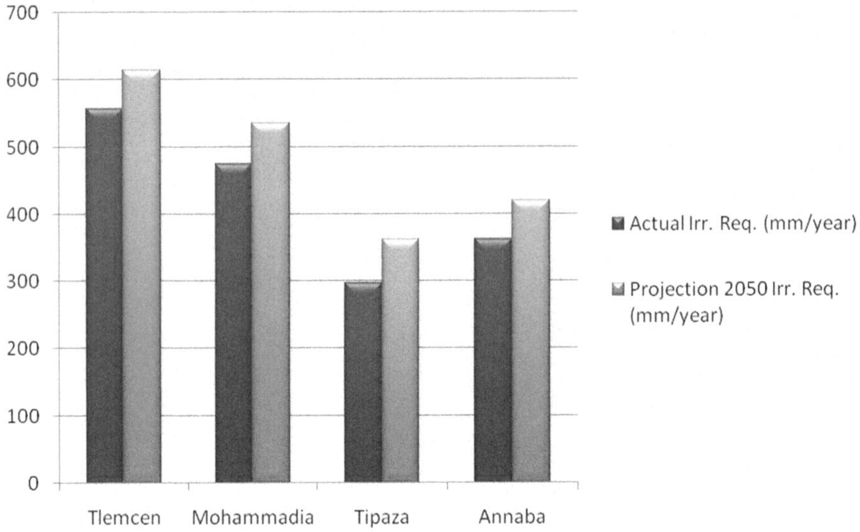

Fig. 4.9 Irrigation requirement results for citrus crop of the different stations in Algeria and the projection in 2050

Table 4.4 CROPWAT estimated irrigation requirements and the actual citrus irrigation schedule in the region of Tipaza (Centre of Algeria)

	Estimated IR (mm)	Actual irrigation (mm)
Jan	0.4	0
Feb	5.5	14
Mar	30.8	46.5
Apr	42.2	75
May	49.5	93
Jun	77.6	90
Jul	89.2	124
Aug	84.8	139.5
Sep	53.2	90
Oct	27.2	46.5
Nov	0	15
Dec	0	0
Total	460.4	733.5

the irrigation system. Differences in cultivation practices between orchards can also impact irrigation requirements (Fig. 4.10). Nevertheless, these results encourage reconsidering the irrigation system to make it more efficient. This discrepancy highlights the need for improved irrigation practices to optimize water usage and enhance water use efficiency in citrus cultivation. Hence, we can assume that the correction of the current losses can effectively address the anticipated water demands arising from climate change estimated to 12% by 2050.

Fig. 4.10 The photo on the left depicts a weed-free soil in a young citrus orchard, while the photo on the right shows untamed orchard alleys filled with weeds (Tipaza, February 2023)

8 Conclusion

The ongoing drought in Algeria threatens agricultural systems, particularly citrus production, and it is expected to worsen in the future. Irrigation of citrus crop is a key factor in its quantitative and qualitative production. This poses additional challenges for farmers, who are also responsible for managing natural resources, especially water.

This study investigated the use of the FAO CROPWAT 8.0 software to monitor citrus crop irrigation in Algeria under present and future climate conditions. The results show that citrus cultivation in Algeria is heavily dependent on irrigation. In addition to the drought, the irrigation system in the most important citrus region of Algeria, Tipaza, is operating at an efficiency of 63%. These losses cannot be solely attributed to the irrigation system, farmer miss skills can also be a source of losses. Nevertheless, these results highlight the need to reconsider the irrigation system and make it more efficient.

Our findings highlight the importance of adopting sustainable and efficient practices in citrus cultivation in North Africa. These practices include improved water management strategies, soil conditions, and irrigation systems. Addressing these factors is essential for ensuring the resilience and productivity of citrus orchards in the region, particularly in the face of evolving challenges posed by climate change and water resource availability. Further research is needed to develop and improve methods and techniques for field-scale implementation to obtain more accurate data and achieve greater irrigation efficiency.

References

Albrigo LG, Stelinski LL, Timmer LW (2019) Citrus, 2nd edn. CABI, Boston

Allen RG, Pereira LS, Raes D, Smith M (1998) Crop Evapotranspiration (guidelines for computing crop water requirements), Drainage and Irrigation Paper N 56. Rome, Italy, Food and Agriculture Organization of the United Nations (FAO).

Balfagón D, Arbona V, Gómez-Cadenas A (2022) The future of citrus fruit: the impact change on citriculture. Mètode Sci Stud J 12:123–129. https://doi.org/10.7203/metode.12.20319

Barnston AG, Livezey RE (1987) Classification, seasonality and persistence of low-frequency atmospheric circulation patterns. Mon Weather Rev 115(6):1083–1126. https://doi.org/10.1175/1520-0493(1987)115<1083:CSAPOL>2.0.CO;2

Blondel L (1974) Influence des porte-greffe sur la qualité des fruits de Citrus. Fruits 29(4):285–290

Brouwer D, Heibloem M (1986) Irrigation water management. Training manual N°3. FAO, Rome

Calabrese F (2002) Soil and cultural practices. In: Digo G, Di Giacomo A (eds) Citrus. The genus *Citrus*. Taylor & Francis, New York, pp 36–48

CNCC (2015) Bulletin des variétés des agrumes. Centre National de Contrôle et de Certification des Semences et Plants, Alger

Cos J, Doblas-Reyes F, Jury M, Marcos R, Bretonnière PA, Samos M (2022) The Mediterranean climate change hotspot in the CMIP5 and CMIP6 projections. Earth Syst Dynam 13:321–340. https://doi.org/10.5194/esd-13-321-2022

Davenport TL (1990) Citrus flowering. Hortic Rev 12:349–408. https://doi.org/10.1002/9781118060858.ch8

Di Prima S, Rodrigo-Comino J, Novara A, Iovino M, Pirastru M, Keesstra S, Cerdà A (2018) Soil Physical Quality of Citrus Orchards Under Tillage, Herbicide, and Organic Managements. Pedosphere 28(3): 463–477. https://doi.org/10.1016/S1002-0160(18)60025-6

Djili K (2000) Contribution à la connaissance des sols du nord de l'Algérie. Création d'une banque de données informatisée et utilisation d'un système d'information géographique pour la spatialisation et la valorisation des données pédologiques. Thèse Doc. d'État, INA, Alger

Drouiche A, Nezzal F, Djema M (2019) Variabilité interannuelle des précipitations dans la plaine de la Mitidja en Algérie du Nord / Interannualvariability of precipitation in the Mitidja plain in Northern Algeria. Rev sciences de l'eau / J Water Sci 32(2):165–177. https://doi.org/10.7202/1065205ar

Duarte A, Fernandes J, Bernardes J, Miguel G (2016) Citrus as a component of the Mediterranean diet. J Spat Organ Dynam 4(4):289–304

FAO (1977) Guidelines for predicting crop water requirements. Irrigation and drainage paper N°24. Food and Agriculture Organization of the United Nations, Rome

FAO (2021) Citrus Fruit Statistical Compendium 2020. Food and Agriculture Organization of the United Nations, Rome

FAO–IIASA (2021) Global agro-ecological zones (GAEZ v4). Model documentation. Food and Agriculture Organization of the United Nations. https://gaez.fao.org/

FAO–ISRIC (1990) Guidelines for profile description, 3rd edn. Food and Agriculture Organization of the United Nations, Rome

Fischer G, Nachtergaele FO, van Velthuizen HT, Chiozza F, Franceschini G, Henry M, Muchoney D, Tramberend S (2021) Global Agro-Ecological Zones v4 - Model documentation. Rome, FAO. https://doi.org/10.4060/cb4744en

Ghada B, Amel O, Aymen M, Aymen A, Amel SH (2019) Phylogenetic patterns and molecular evolution among 'True citrus fruit trees' group (Rutaceae family and Aurantioideae subfamily). Sci Hortic 253:87–98. https://doi.org/10.1016/j.scienta.2019.04.011

Ibáñez JJ, De-Alba S, Lobo A, Zucarello V (1998) Pedodiversity and global soil patterns at coarse scale. Geoderma 83:171–192. https://doi.org/10.1016/S0016-7061(97)00147-X

Ibáñez JJ, Zinck JA, Dazzi C (2013) Soil geography and diversity of the European biogeographical regions. Geoderma 192:142–153. https://doi.org/10.1016/j.geoderma.2012.07.024

Inglese P, Sortino G (2019) Citrus history, taxonomy, breeding, and fruit quality. In: Oxford research Encyclopedia of environmental science. Available via DIALOG. https://oxfordre.com/ . Accessed 8 Dec 2023
INRAA (2006) Rapport national sur l'état des ressources phytogénétiques pour l'alimentation et l'agriculture. Institut National de la Recherche Agronomique d'Algérie, Alger
ITAFV (2012) La culture des agrumes. Institut Technique de l'Arboriculture Fruitière et de la Vigne Tessala El-Merdja—Bir Touta. Alger
Kazi-Tani LM, Gaouar A (2015) Ébauche cartographique des sols dans la région des Traras (NO de l'Algérie). Sketch of soil mapping in the region of Traras (NW of Algeria). Geo-Eco-Trop 39(1):67–86
Kazi-Tani LM, Gaouar A (2016) An application of embedded Markov chain for soil sequences: case study in North Western part of Algeria. Eurasian J Soil Sci 5(3):231–240. https://doi.org/10.18393/ejss.2016.3.231-240
Khan M, Khan I (eds) (2021) Citrus - research, development and biotechnology. Intech Open, London. https://doi.org/10.5772/intechopen.77939
Lagacherie P, Álvaro-Fuentes J, Annabi M, Bernoux M, Bouarfa S, Douaoui A, Grünberger O, Hammani A, Montanarella L, Mrabet R, Sabir M, Raclot D (2017) Managing Mediterranean soil resources under global change: expected trends and mitigation strategies. Reg Environ Chang. https://doi.org/10.1007/s10113-017-1239-9
Luro F, Costantino G, Ollitrault P, Curk F (2022) Origine et évolution de la diversité des agrumes. In: Curk F, Luro F, Minuto G, Nieddu G (eds) Les agrumes du nord de la Méditerranée, edn. Alain Piazzola, Ajaccio, pp 45–75
OEC (2023) The observatory of economic complexity. https://oec.world/en. Accessed 7 Dec 2023
Ortiz JM (2002) Botany: taxonomy, morphology and physiology of fruits, leaves and flowers. In: Dugo G, Di Giacomo A (eds) Citrus—the genus *Citrus*. Taylor & Francis, New York, pp 16–35
Peel MC, Finlayson BL, McMahon TA (2007) Updated world map of the Köppen-Geiger climate classification. Hydrol Earth Syst Sci Discuss 4:439–473. https://doi.org/10.5194/hessd-4-439-2007
Quiroga S, Iglesias A (2009) A comparison of the climate risks of cereals, citrus, grapevine and olive production in Spain. Agric Syst 101:91–100
Rasera JB, da Silva RF, Mourão Filho FAA, Delbem ACB, Saraiva AM, Sentelhas PC, Marques PAA (2023) Climate change and citriculture: a bibliometric analysis. Agronomy 13(3):723. https://doi.org/10.3390/agronomy13030723
Rebour H (1950) Les agrumes. Union des syndicats de producteurs d'Agrumes d'Algérie, Alger
Romanyà J, Povira P (2011) An appraisal of soil organic C content in Mediterranean agricultural soils. Soil Use Manag 27:321–332. https://doi.org/10.1111/j.1475-2743.2011.00346.x
Rosenzweig C, Phillips J, Goldberg R, Carrollh J, Hodges T (1996) Potential impacts of climate change on citrus and potato production in the US. Agric Syst 52:455–479
Smith M (1992) CROPWAT, a computer program for irrigation planning and management. FAO irrigation and drainage paper 26. Food and Agriculture Organization of the United Nations, Rome
Tuel A, Eltahir EAB (2020) Why is the Mediterranean a climate change hot spot? J Clim 33:5829–5843. https://doi.org/10.1175/JCLI-D-19-0910.s1
Verner D, Tréguer D, Redwood J, Christensen J, McDonnell R, Elbert C, Konishi Y, Belghazi S (2018) Climate variability, drought, and drought management in Morocco's agricultural sector. World Bank Group. https://doi.org/10.1596/30603
Vincent C, Morillon R, Arbona V, Gómez-Cadenas A (2020) Citrus in changing environments. In: Talon M, Caruso M, Gmitter FG Jr (eds) The genus Citrus. Woodhead Publishing, pp 271–289. https://doi.org/10.1016/B978-0-12-812163-4.00013-9
Yaalon DH (1997) Soils in the Mediterranean region: what makes them different? Catena 28:157–169
Ziogas V, Tanou G, Marianou G, Kourgialas N (2021) Drought and salinity in citriculture: optimal practices to alleviate salinity and stress. Agronomy 11(7):1283. https://doi.org/10.3390/agronomy11071283

Chapter 5
Survey on the Management of Pesticides in Citrus Farms in Algeria

Fatma Youcefi and Moussa Mokhtari

1 Introduction

Today, the immediate response to growing demand for food production is the intensive use of agrochemicals, which largely include fertilizers and pesticides. Moreover, with the development of resistant populations of pests, each year greater quantities and new chemical compounds are used to protect crops (Carvalho 2006). Every year, around 2.7 million tons of active ingredients are used worldwide in the environment, particularly in agriculture, to prevent crop losses (EPA 2017). Human health and the environment can be directly affected by pesticide residues that accumulate in the environment, contaminating drinking water sources, food sources, or diminishing biodiversity (Syberg et al. 2016). Algeria's population was 42.4 million in 2018 and, according to the national office of statistics ONS extrapolations, will reach 51 million and over 70 million by 2030 and 2050 respectively. It should be noted that the Algerian pesticide market continues to grow, with pesticide imports rising from \$49.5 million in 2007 to \$77 million in 2009 (ONS 2018). Algeria's citrus-growing sector covers 78,000 ha, with a production rate more than 15 million tons in 2021 (Madrp 2022). The Algerian citrus orchard is made up of three age classes: Plantations between 0 and 10 years old, those between 10 and 40 years old (the category that produces the most) and those between 45 and 50 years old, which represent barely 12,000 ha out of the 78,000 ha; these 12,000 ha are part of a rejuvenation program consisting of the gradual uprooting of these plantations and

F. Youcefi (✉)
Laboratoire des Agrosystèmes Fonctionnels et Technologie des Filières Agronomiques, Tlemcen, Algeria

M. Mokhtari
Research Unit in Physicochemical Analysis of Fluid Media and Soil/Scientific and Technical Research Center in Physico-Chemical Analysis (URAPC-MFS/CRAPC), Tipaza, Algeria

A. Melkonyan-Gottschalk, D. Daus (eds.), *Transformation Towards Circular Food Systems*, Greening of Industry Networks Studies 13,
https://doi.org/10.1007/978-3-031-63793-3_5

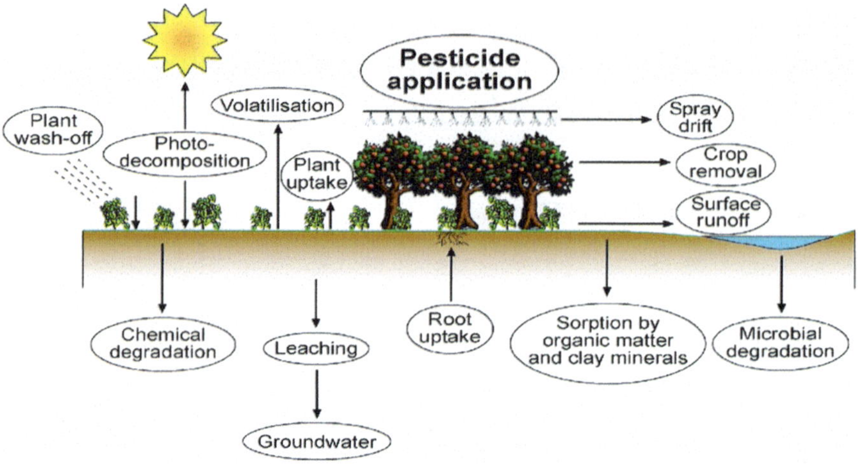

Fig. 5.1 Processes responsible for the fate of applied pesticides in the environment (Sarmah et al. 2004)

their replanting. Algerian citrus growing is plagued by numerous problems, some of them linked to natural climatic factors. As a result, many diseases and animal pests have taken hold, destroying not only orchards but also production, and citrus growers are forced to make excessive use of agrochemicals. In addition, the import, possession, marketing, and use of plant protection products for agricultural use are subject to prior approval issued by the plant health authority, in accordance with the conditions laid down in law no. 87-17 of 1 August 1987 relating to phytosanitary protection. In this regard, we carried out a survey of citrus growers on large farms in order to assess their use of pesticides (herbicides, fungicides, and insecticides). This study provides an overview of citrus-growing practices on large citrus-growing farms located in the regions of: Tipaza, Bejaia, Mostaganem, Temouchent, Tlemcen, Relizane, Oran, Mouhamadia, Blida, and Algiers during the months of December 2022 to March 2023, in order to better understand the influence of citrus growing practices on the environment, the product, and the consumer. The aim of this study is to assess the way in which citrus growers use pesticides, and to assess the knowledge of citrus growers of the possible risks involved (Fig. 5.1).

2 National Framework for the Registration and Control of Pesticides

In Algeria, pesticides are manufactured by autonomous companies like Asmidal and Moubydal. But in today's market economy, several companies have specialized in importing insecticides and various related products. Approximately 400 plant protection products are registered in Algeria and are widely used by farmers (Phytosanitary Index 2017). Algeria ranks among the countries that use between 6000 and 10,000 T/year of pesticides, which corresponds to a rate of 15–20% of the normative needs requirements (Moussaoui and Tchoulak 2005).

The diversity of suppliers makes tighter controls an absolute necessity. Algeria promulgated Law No. 87-17 on plant health protection on 1 August 1987, enabling measures relating to the manufacture, storage, distribution, marketing, and use of plant health products to be detected. Under the law, unauthorized plant protection products may not be sold, imported, or produced. According to the Official Journal of the Algerian Republic N° 9 of 3 February 2010, Algeria has set up a system for approving plant protection products under Executive Decree No. 99–156 of 20 July 1999, on the control of plant protection products for agricultural use. When a plant protection company synthesizes a compound, it is tested on different pest species in vitro or in the greenhouse, and then on small plots of land. If the compound has proved effective, the manufacturer submits an application for marketing authorization to the Ministry of Agriculture, together with full scientific documentation including the following elements:

- Biological documentation providing information on the agronomic effects of the compound formulation and its field of activity.
- Toxicological information on the toxicity of the product for humans (acute and chronic toxicity tests, carcinogenesis, mutagenesis, reproductive toxicity tests, etc.) and for the environment (analysis of effects on fish, birds, insects, microorganisms, and degradation processes in water, air, and soil). After inspection, the experts propose a toxicological classification and preventive recommendations for safe use.

3 Sampling and Conduct of the Survey

The sample surveyed consisted of 25 citrus growers and 25 distributors of phytosanitary products taken at random and concerned different regions of Algeria, taking into account the large farms of central and western Algeria. The farms affected by our survey are those of the cities of: Tipaza, Bejaia, Mostaganem, Temouchent, Tlemcen, Relizane, Oran, Mouhamadia, Blida, and Algiers during the months of December 2022 to March 2023 (Fig. 5.2).

The questionnaire tool was structured in the form of tables containing questions designed to provide information on the various practices and uses of plant protection products. It is made up of questions relating to the identification of the farm, knowledge of the pesticide used during treatment, and the conditions under which it is used.

For more information on the methods used to combat various pests, a list of potential citrus growers in the study regions, pesticide sellers who are in direct contact with citrus growers, we contacted various technical and administrative

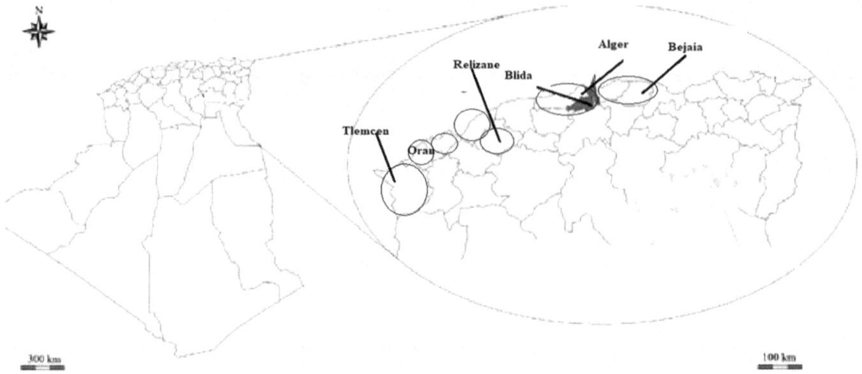

Fig. 5.2 Region study

bodies: the National Plant Protection Institute (INPV), the DSA (agricultural services directorates), and the chambers of agriculture in the regions covered by the survey. It should be noted that when we carried out our survey, we were faced with a difficult situation in which distributors of plant protection products and citrus growers did not cooperate very well, which severely limited our work in the area. In view of the fact that the number of distributors and citrus growers surveyed at the start was 75, this was revised to 25 questionnaires filled in correctly.

We adopted three methods at the same time:

- Surveys, which enabled us to get a closer look at the situation and gain a better understanding of our subject.
- Telephone surveys of a few farmers.
- Surveys completed by volunteer farmers who also agreed to distribute other questionnaires to neighboring farmers.

Each interview lasted 30 minutes, depending on the cooperation of the farmers interviewed. Some answers were verified by direct observation on the farm.

4 Results of the Survey

The survey targeted the conditions of sale of pesticides, with the aim of verifying the application of the regulations on the sale and use of pesticides (Annexes).

After the questionnaires processing, the results revealed:

- Active ingredients
- Pesticides according to the targeted pest
- The chemical families of pesticides used by farmers and distributed by vendors

4.1 Farmers Survey

4.1.1 Age of Farmers

The age of the farmers who took part in the survey ranged from 18 to >60. The age distribution of the farmers is a Gaussian distribution with a normal distribution. More than half of the farmers are aged between 30 and 50 (Fig. 5.3).

The profile of citrus growers is made up of a diversity of players aged between 18 and >50 years, i.e., 90% of the population. Initially, there are more and more young people in the citrus sector, but people over 50 are also represented, accounting for 10% of the population (Fig. 5.3). This could be explained by the fact that older farmers prefer work more than retirement, and young people accept seasonal work because of unemployment. It should also be noted that support for the sector is also provided by the Ministry of Agriculture, which encourages young people to create their own farms, by releasing financial resources through national programs such as the National Agricultural Development Program (P.N.D.A).

4.1.2 Level of Education of Farmers

This part of our study highlights the importance of education and training in the agricultural system. Indeed, education is a key factor in the development of the citrus sector and is likely to have a positive impact on the behavior and performance of economic players in the citrus sector. Figure 5.4 shows the overall configuration of the level of education of the various players, 66.7% of the farmers surveyed have less than secondary education, 14.3% have secondary education, and 19% have university education. This means that most of the farmers do not have the necessary prerequisites for the perception of the dangers and risks associated to the use of plant protection products. Because most of them believe only on what they see or live, they do not or cannot read correctly a document in relation with pesticides risks. As other studies conducted in Algeria have shown, low levels of education and lack of training among citrus growers are the main reasons for pesticide abusage.

Fig. 5.3 Age distribution of citrus growers on citrus farms

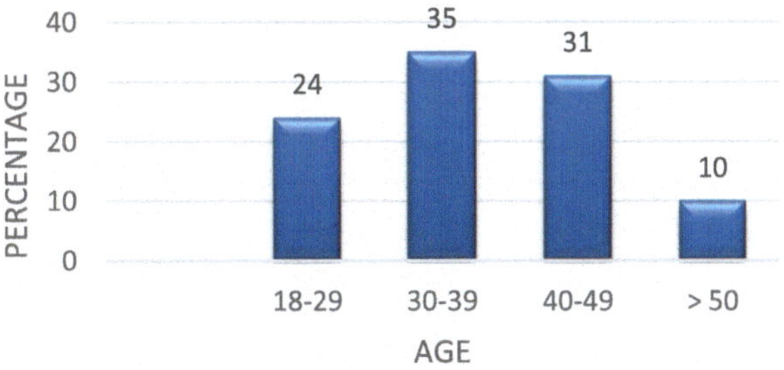

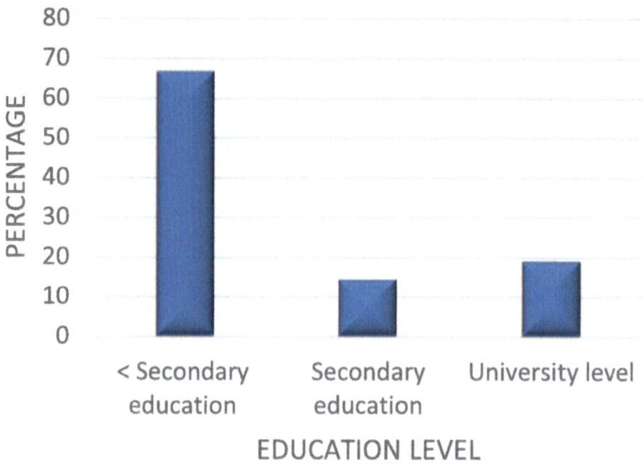

Fig. 5.4 Distribution of citrus growers by level of education

4.1.3 Sources of Information and Risk Related to Pesticides

According to the farmers' responses, most of them (75% of citrus growers) think that treatments are obligatory, and the rest of the farmers think that these treatments are optional. In practice, all the farmers surveyed in the study area use pesticides, although 10% of the farmers surveyed in the study area do not use pesticides. 46% of farmers depend on the sales staff or agronomic technicians from the agricultural services, who have a good level of knowledge and sometimes training in this field, to help them choose which product to use. Relying on their experience, 36% make the choice themselves, and only 18% ask the advice of neighboring farmers. Farmers' general knowledge: 92% of farmers in the study region are aware of the usefulness of pesticides for crop protection, but only 10% are familiar with their effects on the environment and health. When asked about the origin of farmers' knowledge of phytosanitary problems or risks and the use of treatments, the survey revealed that professional experience and recourse to another citrus grower were predominant, while some farmers said that the DSA (Agricultural Services Department) and pesticide sellers were predominant.

4.1.4 Accidents Involving Pesticides

In the area studied, pesticides were applied using motorized knapsack sprayers or prepared manually, which involved direct contact with citrus growers, thus increasing the risk of exposure. In general, the first model covers areas of up to 2 ha, while the second model is used for larger areas. From Fig. 5.5, we can deduce that 52% of the citrus growers we interviewed had suffered accidents linked to the use of pesticides. According to the farmers, these have resulted in poisoning and respiratory problems. Unfortunately, farmers are unaware of the consequences of pesticide misuse. In addition, their low level of education accentuates the problem,

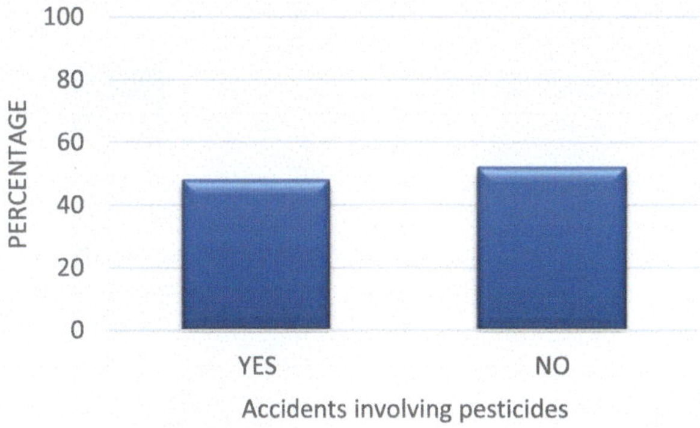

Fig. 5.5 Accidents involving pesticides

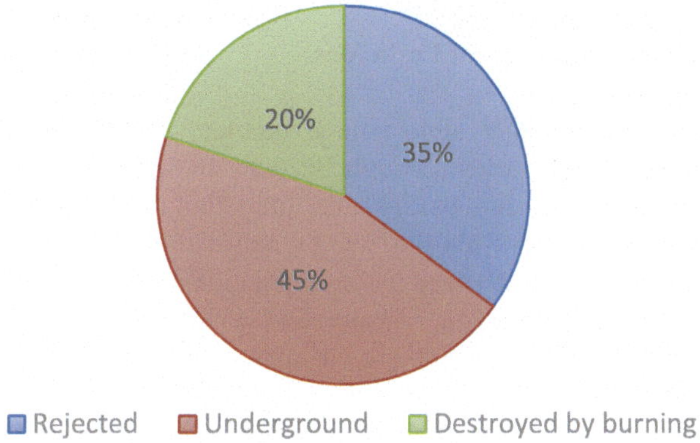

Fig. 5.6 Empty pesticide packaging situation

preventing them from learning about the other risks associated with pesticides, particularly the risk of residues in human food. We also noted that in the majority of cases, the remainder of the tank and the rinsing water from the tanks/sprayers are dumped on the ground on farm.

4.1.5 Empty Pesticide Packaging Situation

Figure 5.6 shows that 45% of farmers bury or incinerate residual pesticide packaging, 20% resort to incineration, and the remaining 35% leave it in the open air in an uncontrolled manner. These unconsidered actions on the part of farmers can contaminate the environment surrounding their crops (soil, air, water) and become a source of undesirable contamination of crop products through their reintroduction into the plant system of cultivated plants.

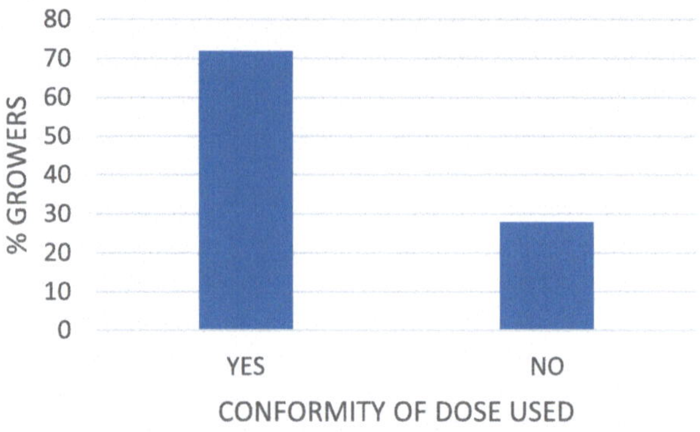

Fig. 5.7 Distribution of farmers according to compliance with approved doses

4.1.6 Compliance in the Use of Pesticides

Figure 5.7 clearly shows that almost three quarters of farmers (72%) do not comply with the pesticide doses recommended by the phytosanitary index. This overdosing of treatments leads to pesticide accumulation in the citrus fruit, resulting in contamination of the harvested produce by active molecules in concentrations exceeding the critical limits required by the regulations in force. The management of pesticides in terms of production, transport and storage can be another problem. It has been found that the handling of these chemicals by farmers is often haphazard. It should also be noted that citrus growers are not sufficiently aware of the risks associated with pesticides, and we observed that citrus growers use cocktails where the dose is not known. According to Ngegba et al. (2022), the overuse of synthetic pesticides has harmful consequences, including environmental degradation, damage to non-target organisms, contamination of food and feed by pesticide residues, resurgence of pests, genetic variation in plants, and negative impacts on biodiversity.

4.1.7 Training and Awareness-Raising for Citrus Growers and Farmers

The survey revealed that 70% of citrus growers have not attended any awareness-raising or training sessions to deal with the risks associated with pesticide management (Fig. 5.8). Hence the importance of training farmers in good agricultural practice and making them aware of the serious risks associated with the incorrect use and handling of plant protection products, awareness-raising, and training campaigns enable citrus growers to:

- Define the health and environmental risks associated with pesticides.
- Describe good practice in preventing health and environmental risks associated with the use of plant protection products.
- Take measures to mitigate the health and environmental risks associated with the use of plant protection products.

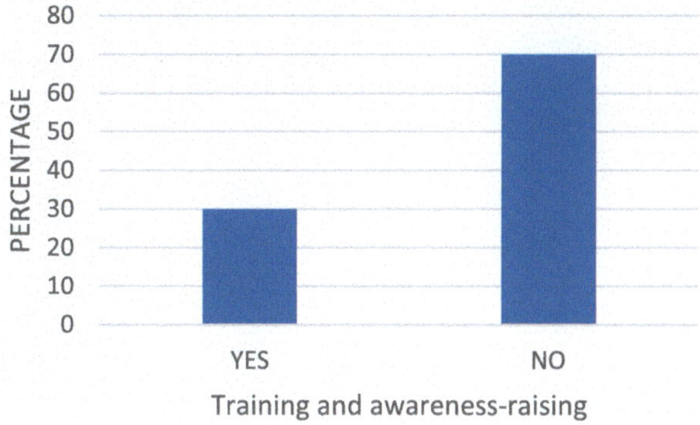

Fig. 5.8 Distribution of farmers according to their training in pesticides

4.1.8 Number of Treatments Carried out by Farmers

Most of the citrus growers surveyed adopted a systematic pesticide application approach and did not use pest and disease surveillance programs or agriculture department alerts. Most of the products used contain multiple active ingredients.

It can be seen from the survey results that phytosanitary treatments are concentrated in two periods: The first period is from February to June, coinciding with the spring sap flow, during which a variety of sap-sucking pests, such as aphids, mealybugs, and whiteflies, targeting sensitive parts of trees, are treated. The second period occurs between August and October and corresponds to the ripening stage of citrus fruits. It is characterized for the treatment of citrus fruit keratitis, which remains a major problem for citrus growers. The results also show that April and June are the months in which we record the most phytosanitary treatments. These treatments were discontinued in July due to very high temperatures during that month.

4.2 A Survey of Pesticide Distributors

4.2.1 Age and Level of Education of Sellers

Our survey revealed that the age of sellers varies between 32 and 40. Analysis shows that 67% of the sellers questioned are between 35 and 40. This is due to the fact that this age group represents the working population. The majority of sellers had a university education, with a rate of 85%. A further 15% had secondary education

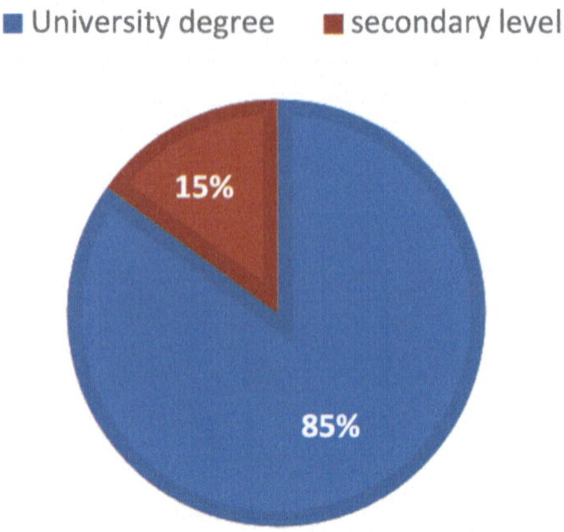

Fig. 5.9 Level of education of sales staff

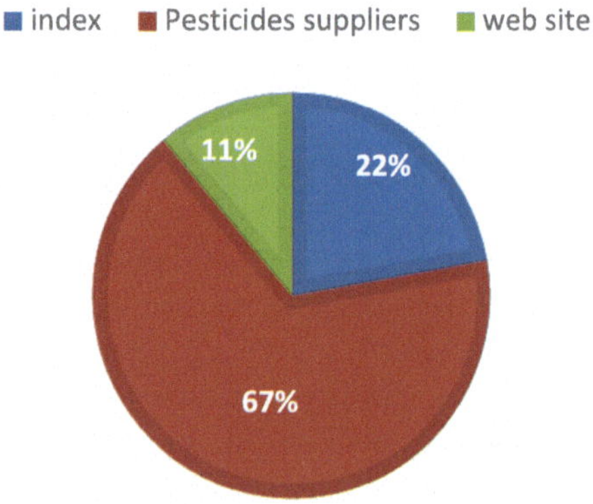

Fig. 5.10 Information sources

(Fig. 5.9). The phytosanitary product sales sector has undergone a reorganization in recent years, and the criterion of competence is a requirement in the recruitment of sales staff. The minimum requirement is a degree in management, but competence in the field of pesticides is not required.

4.2.2 Sources of Information for Choosing Products to Buy

The sources of information on pesticide products vary: The proportion of approved suppliers is 67%, which puts them in first place, the use of the phytosanitary index is 22%, and the number of websites used is 11% (Fig. 5.10). Citrus growers rely mainly on pesticide suppliers represented by technical sales representatives whose main objective is to increase sales. New molecules are constantly arriving on the market. They appear to be more "modern" and more effective, but they are increasingly expensive. Technical sales staff organize demonstration days for farmers and visit them before production to help them choose pesticides and fertilizers. They can also be called out in an emergency.

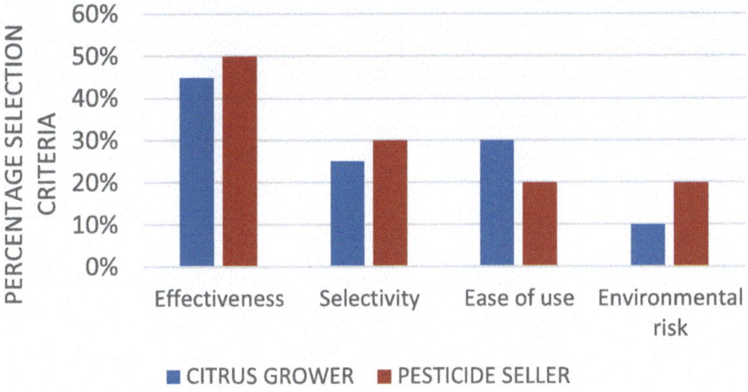

Fig. 5.11 Criteria for selecting pesticides for sale

4.2.3 Criteria for Selecting Pesticides for Sale

Some sellers pointed out that it is sometimes difficult to choose pesticides, especially when there are several products with the same name, the same principle of efficacy, or the same mode of action on the market.

All sellers are aware that plant protection products are sold firstly on the basis of their efficacy and secondly on the basis of their selectivity, and that their harmful effects should be avoided as far as possible (Fig. 5.11). However, most citrus growers do not take into account the environmental risks associated with pesticide toxicity. During the sales process, sellers also offer products that are easier to use. It should be noted that price is an important product selection criterion for farmers. We found that the various players (citrus growers and sellers) did not take environmental factors into account. Pesticides are used in high doses because the advice given on pesticides leads farmers to "chemicalize" agriculture.

This can lead to health and environmental problems. Due to a lack of information and locally available alternatives, farmers demand more effective products, which are often more dangerous and can affect crop resistance to disease.

4.2.4 The Main Pesticide Families Distributed

According to the results presented in Table 5.1, we can see that the distribution of insecticides by vendors is more than 56% of the total distribution followed by fungicides with 27.5%, and acaricides with 11.6% and finally herbicides and nematicides with 2.9% and 1.4% respectively.

Table 5.1 Pesticides distributed by vendors

Type of pesticide	Distribution rate (%)
Insecticide	56.5
Fungicide	27.5
Acaricide	11.6
Herbicide	2.9
Nematicide	1.4

The results show that the 260 registered insecticides represent 100 active substances that are listed. The survey shows that only 15 insecticides are used by citrus growers. As for fungicides, the 111 registered active ingredients are represented by 259 fungicides authorized for sale which are listed. The results show that only 24 fungicides are used by citrus growers. With regard to herbicides, the active ingredients are represented by 120 registered herbicides, of which only 19 are used by farmers. The most commonly used chemical families are Carbamates, Triazoles, Synthetic pyrethroids, and Neonicotinoids.

5 Summary and Conclusion

In this study, we attempted to highlight the risks associated with the use of pesticides on citrus farms in central and western Algeria. We monitored these risks on citrus farms, taking into account the main parameters influencing this chain, particularly the distributors of phytosanitary products. Citrus growers in the study regions use various pesticides to protect their crops and ensure good quality and yield. Pesticides can help reduce crop losses due to pests, but their inappropriate use can harm human health and the environment. The choice of pesticide depends on the type of crop and pest, as well as price, formulation, and availability. The commercial name of the product is not a factor, as there are many products with the same active substance or mode of action (Mokhtari 2011). According to Puel and Grimonprez (2022), farmers fear that if pesticides are not used, their yields will fall, their costs will rise and their revenues will be affected.

Based on the survey results, pesticides are widely used because insects pose a significant threat to crops. In the study area, pesticides were applied using motorized knapsack sprayers or prepared manually, which meant direct contact with citrus growers, increasing the risk of exposure. Typically, the first model covers areas of up to 2 ha, while the second model is used for larger areas. The survey found that farmers did not always follow the recommended dosages, with 72% admitting to using pesticides in excessive amounts. This overuse is often due to low education levels among citrus growers, making it difficult to understand instructions, and a lack of training. In some cases, overdosing is intentional to maximize the products' effectiveness. The data also revealed that most pesticide applicators are unaware of proper protection and hygiene measures, with very few using full protective gear. Many people do not use protective equipment when handling pesticides due to discomfort, and they may not believe that exposure to pesticide residues poses a real threat to their health. It is worth noting that there is currently no legislation in place to regulate the use of pesticides. However, other studies have demonstrated that producers often neglect to use protective equipment, which can increase the risk of poisoning. Van Der Werf (1997) highlights that risks are particularly high during the preparation of the mixture and application to crops if protective equipment is not worn.

As other studies conducted in Algeria have shown, low levels of education and lack of training among citrus growers are the main reasons for pesticide abusage, especially as regards pesticide residues in harvested crops, and some distributors do not help to inform farmers about the risks associated to the pesticide use.

It is crucial to emphasize that quality begins with training and ends with training. Setting up an awareness and information system about these "necessary poisons" has become a national urgency. The systematic control of foodstuffs for pesticide research is almost non-existent due to the high cost of this service, which is not affordable for citrus growers. Phytosanitary protection should start with the implementation of a system for the recovery, storage, and disposal of empty packaging and expired products. Introducing regulations to regulate the permissible minimum concentrations in citrus fruits for consumption or processing is essential. Strengthening prevention systems through mandatory reporting of pesticide-related intoxications is also necessary.

Annexes

Annex 1

QUESTIONNAIRE FOR FARMER		
1. IDENTIFICATION OF THE SURVEY		
1.1. Age	**1.2. Gender** M ☐ F ☐	**1.3. Profession**: ....
1.4. Level of education: None ☐ Primary ☐ Secondary ☐ High school and above ☐		
2. KNOWLEDGE OF THE PESTICIDE		

2.1. What pesticides do you use? (name them and specify the physical aspect: solid, liquid or gas)

....

....

2.2. If the producer does not know the names of the products, ask him why?

....

....

2.3. How do you obtain these products that you use?

at the market ☐ from an authorized dealer ☐ Other ....

2.4. In your opinion, are there any risks involved in being in contact with these products

Yes ☐ No ☐

2.5. If yes, which risks?

....

....

2.6. What happens to empty pesticide containers?

☐ Rejected
☐ Underground
☐ Destroyed by burning

2.7. Do you respect the approved doses when preparing pesticides?

Yes ☐ No ☐

2.8. Have you received training or awareness campaigns on the risks associated with pesticides?

Yes ☐ No ☐

2.9. What are the treatment periods?

Annex 2

DISTRIBUTOR QUESTIONNAIRE

1. IDENTIFICATION OF THE SURVEY

1/. Questionnaire N°

....

2/. Place and Data of survey

....

3/. Age and Level of education of distributor

....

2. KNOWLEDGE OF THE PESTICIDE

4/. Have you had any training on pesticides or plant protection products?

☐ Yes ☐ No

5/. What means do you use to choose plant protection products that meet your customers' requirements?

☐ Index of crop protection products
☐ Potential suppliers
☐ Internet sites

6/. What are the selection criteria when buying pesticides (Distributor)

☐ Effectiveness
☐ Selectivity
☐ Ease of use
☐ Environmental risk

7/. Pesticides potentially used?

☐ Insecticides
☐ Herbicides
☐ Fungicides
☐ Acaricides

8/. What criteria do customers use when choosing a pesticide?

☐ Efficacy
☐ Selectivity
☐ Ease of use
☐ Environmental risk

9/. How do you manage out-of-date products?

☐ Rejected
☐ Returned
☐ Underground
☐ Destroyed by burning

10/. Are you aware of the environmental and crop risks associated with packaging waste?

☐ Yes ☐ No

References

Carvalho FP (2006) Agriculture, pesticides, food security and food safety. Environ Sci Pol 9:685–692

EPA (2017) Pesticides industry sales and usage, 2008–2012 market estimates. United States Environmental Protection Agency, Washington, DC

Madrp (2022) Statistiques-agricoles. http://madrp.gov.dz/agriculture/

Mokhtari M (2011) Recherche de résidus de quelques pesticides par couplage CPG/SM dans quelques fruits et légumes. Thèse de Magister. Ecole Nationale Polytechnique (ENP), Alger, p 103

Moussaoui K.M. Tchoulak Y. (2005). Enquête sur l'utilisation des pesticides en Algérie, Résultats et analyse. Ecole Nationale Polytechnique, Alger, pp 11

Ngegba PM, Cui G, Khalid MZ, Zhong G (2022) Use of botanical pesticides in agriculture as an alternative to synthetic pesticides. Agriculture 12(5):600

ONS (2018) Statistics on population and demography, economic statistics, employment and unemployment, consumer spending

Phytosanitary Index (2017) Ministère de l'Agriculture et du Développement Durable MADR, Algeria

Puel L, Grimonprez B (2022) Gérer les risques inhérents à la sortie des pesticides: état des lieux et perspectives juridiques. Doctoral dissertation. Institut de droit rural-Université de Poitiers

Sarmah AK, Müller K, Ahmad R (2004) Fate and behaviour of pesticides in the agroecosystem—a review with a New Zealand perspective. Soil Res 42(2):125–154

Syberg K, Backhaus T, Banta G, Bruce P, Gustavsson M, Munns WR Jr, Rämö R, Selck H, Gunnarsson J (2016) Vers une approche conceptuelle pour évaluer les risques des mélanges de produits chimiques et d'autres facteurs de stress pour les services écosystémiques côtiers. Intégré Environ Évaluer Gérer 13(2):376–386

Van Der Werf HM (1997) Évaluer l'impact des pesticides sur l'environnement. Le Courrier de l'environnement de l'INRA 31(31):5–22

Part III
Technology Adoption and Advanced Data Analytics

Chapter 6
Technological Circularity Index: Towards a Decision Support Tool

Carlos Pinillos

1 Introduction

As an established trend, the circular economy has won importance and relevance during the last 5 years. The need to reduce material utilization and decarbonize industries has become a worldwide priority, and the circular economy contributes to this goal through its focus on reuse and eco-design. One central aspect is related to the bridge between technological reuse, e-waste generation, and the technology gap closing. These three topics play an important role, when local authorities define an effective e-waste policy by promoting, for example, product remanufacturing procedures inside their territories. In this regard, the circular economy offers a variety of alternatives to close technology gaps, like simple reuse, reparation or remanufacturing strategies, which could relieve investment costs and contribute to decarbonizing supply chains. To better understand the relationship between the above-named topics, it is significant to shed light on how countries adopt and transfer technologies and how those efforts are monitored and measured. In general, a country manages its technological development by enabling and promoting technological advances and their contribution to the well-being of its inhabitants.

From a country's point of view, technological innovation can be seen in two directions, depending on the technological production ownership (Kvedariene 2019). On the one hand, a technology-producing country plays a transfer role, either within its boundaries or across borders. On the other hand, a technology-consuming country adopts technologies, narrowing or widening technological gaps, depending on the local innovation degree and its technological acceptance levels (Gerybadze 2019). Within their autonomous geographical limits, countries regulate and promote

C. Pinillos (✉)
Emporium Partners, Filderstadt, Germany
e-mail: cpinillos@emporiumpartners.com

A. Melkonyan-Gottschalk, D. Daus (eds.), *Transformation Towards Circular Food Systems*, Greening of Industry Networks Studies 13,
https://doi.org/10.1007/978-3-031-63793-3_6

the transfer and adoption of technologies, seeking to efficiently reduce technological gaps while mitigating possible adverse economic, environmental, and social impacts (Romanainen 2021).

In this regard, circularity plays a significant role. As already introduced, the circular economy proposes a technological management based on extending the lifespan of products, either through processes such as reuse, repair, remanufacturing, or the promotion of circular eco-design and high-quality recycling (Bakker et al. 2021). A successful circular technology management indicates the circularity integration level in a territory, depending on a set of decisions that are taken by the authorities and its inhabitants (Ren et al. 2023). To back up this set of decisions, which would impact the local environment but also the level of technology gaps, the decision-makers use measurement tools to support their processes. Those measurement tools are composed of indicators and parameters concerning performance in subjects like, e-waste generation, technology gaps, or a lack of local skills for reparation and remanufacturing. Today, there exists some effort, that concentrates on measuring these subjects but in a separate manner and not complimentarily. Circular indicators are concentrated on parameterizing actions against the generation of e-waste or measuring the threshold of technology gaps (Circular Economy Foundation 2023). However, they don't provide comparative information related to how e-waste can be minimized when technology gaps closing alternatives are conducted through the circular economy. From his perspective, a technological circularity index (TCI) functions as a tool to benchmark countries according to their technology management in frames of the circular economy. It ranks and classifies countries according to an outstanding or deficient circular performance concerning their country's technological management.

The TCI allows its users to interpret the conditions and potentials of a given country by supporting their decision-making processes, whether to transfer or adopt technologies under the circular economy principles and procedures. In the context of the food industry, drawing attention to its technological state-of-the-art, a technological circular strategy would help stakeholders close their technological gaps. In frames of a common goal related to closing technology gaps along the supply chain, suppliers, processors, retailers, and consumers can work collaboratively, improving their margins, delivery times, and quality systems.

In the current article, the TCI's concept and construction process is explained, exposing the complexity and usefulness related to the compilation of information and the role of users. In the first part, a literature review is carried out to understand the TCI's pertinence and functionality in regard of the issues not treated by other indexes or parametrization efforts. After analyzing the existing literature, attention is placed on the construction process, highlighting the methodologies to build a composite index. Afterward, the results and discussions around the TCI applicability are be presented, and a few relevant conclusions are shown, where viable optimization strategies and TCI's usability frameworks are exposed (Utoikamanu 2018).

2 Literature Review

There are some remarkable efforts, which are focused on measuring e-waste management and technology gaps separately. The TCI, as a composite index, built upon the aggregation and unification of other globally well-recognized indexes, seeks to analyze a comparative relation concerning the topic of closing technology gaps throughout technology reuse and circular economy. In this context, the first step to research the existing literature is concentrated on defining the relevance of the sub-indexes, that compose the TCI. They are presented as a review of the existing literature around the mentioned topics by showing its core functionality but also the missing aspects, which the TCI aims to cover.

2.1 A Complimentary Review of the Sub-indexes

Eleven sub-indexes were selected under the criteria of actualization, methodologies, and authors' reputation. The 11 sub-indexes provide information regarding economic development, technology management, and circularity at the level of countries and territories. Separately, these sub-indexes do not provide indicators toward a comparative understanding of territorial technology management and closing technology gaps. By unifying them in a single index, like the TCI, a significant number of correlations can be analyzed and visualized but also many complex scenarios around the topic of technology transfer and closing the gaps can be suggested and supported by figures and professional information sources. The next list is a subindexes' short description with the additions already mentioned:

1. *UNCTAD's Productive Capacities Index*: This index provides information about countries' productive capacities. The PCI covers 194 economies for the period 2000–2022. The set of productive capacities and their specific combinations are mapped across 42 indicators (UNCTAD 2023). It shows issues related to technical skills and economic performance even from a social point of view, but it doesn't deliver information about industrial gaps and how circularity could close those gaps.
2. *Economic Complexity Index*: Economic complexity exposes the diversity and sophistication related to the productive capabilities embedded in the exports of each country (Harvard 2023). This index concentrates on diversification without having a reference around the topic of economic valorization processes throughout the circular economy.
3. *Ease of Doing Business Scores*: The ease of doing business score provides information about the level of regulatory performance over time (The World Bank 2022). This index focuses on defining, how easy it is to settle a business in a certain country, without any attention to regulation, when a circular economy constitutes a relevant part of the operative economy.

4. *S&P Sovereign Index*: The S&P sovereign ratings reflect an analysis of institutional and governance effectiveness, economic structure and growth prospects, external finances, and fiscal and monetary flexibility for each country (S&P Global 2020). The S&P Index handles the subject of countries' financial trust, which is a relevant impulse to implement a circular economy.
5. *SDG Index*: This score measures the total progress toward achieving all 17 SDGs regarding the country's sustainable performance (United Nations 2023). The index shows the country's performance concerning sustainable goals. Within the SDGs, there are some goals, which maintain a connection with the circular economy. The social aspect could be observed to understand the social gaps toward the acceptability of new technologies and procedures.
6. *The Global Innovation Index*: The Global Innovation Index (GII) measures innovation in 132 economies and localizes the top 100 science and technology innovation clusters (INSEAD 2022). With this index, aspects like technology transfer and the derived new procedures threshold might be analyzed, but circularity is not mentioned as a radical innovation, that introduced new ways of doing economy.
7. *Readiness for the Future of Production*: RFP provides a country-level summary and comprehensive view of key levers (factors and institutions) required to effectively transform production systems in the light of rapidly emerging technologies related to the Fourth Industrial Revolution (WEFORUM 2018). In this index, the circular economy is not considered a viable strategy to construct a more environmentally friendly futuristic economy.
8. *Achievement Index (TAI)*: Technology Achievement Index (TAI), which is one of the studies to evaluate the technological performance of countries, classifies countries according to their technological achievements (United Nations-Zenodo 2020). They missed technical skills in frames of reparation and recycling around technology management.
9. *KAPSARC's Circular Carbon Economy (CCE)*: The CCE Performance sub-index measures countries' current performance on the various CCE activities (KAPSAR 2022). This is a remarkable effort to measure comparatively countries' circular performance. Aspects related to technology management or closing the gap are not treated.
10. *The Circularity Gap Report*: Reporting initiative (CGRi) explores how data to inform stakeholder decision-making can best be collected, consolidated, and made available globally (Circular Economy Foundation 2023). The circular gap report delivers data regarding countries but it does not compare country's performance around the topic of technology gap closing.
11. *Circular TAI*: This sub-index is an own production. The circular TAI sub-index aims to complement the normal TAI with circularity figures and indicators to visualize human and social skills according to a circular technology adoption system. In the normalization result Sect. 4.2, this construction is exposed in detail.

After signalizing the focus and missing issues of the sub-indexes that compose the TCI and why the TCI could cover those missing issues, three key aspects are highlighted in the literature review, which help to understand the TCI relevance.

The first key aspect refers to the attention, that the circular economy devotes to EEE (electrical and electronic equipment) management and the resulting WEEE (waste of electrical and electronic equipment). EEE and WEEE policies are the main vehicles through which states manage their technological circularity (ITU 2020). Today, most of the countries have adopted, with greater or lesser commitment, policies on EEE and WEEE. These policies mainly concentrate on the EoL (End of Life) management of equipment and machinery, settling conflicts of interest between users and suppliers.

One of the most significant topics in frames of the EEE and WEEE policies is the debate surrounding the right to repair (European Commission 2023). The right to repair focuses on defending the mechanisms for users to avoid technological obsolescence and loss of value of their technological assets. From the right to repair a second aspect emerges, which is the Extended Producer Responsibility (EPR). The EPR is the key to delimit the threshold in which producers must retain responsibility toward users according to the corresponding right to repair.

The EPR underlines its actions on technological recycling and obliges producers to take care of waste management and recycling. This can mitigate adverse environmental effects for the target country. EPR and the right-to-repair policies are essential TCI components. Unfortunately, those central aspects are not illustrated in any kind of measurement tool, which helps to observe the impacts of circular economy in the territories. One of the main TCI goals is to visualize those impacts in the context of a global policy, which ranks the countries regarding their EPR and the performance of the right-to-repair. This will facilitate the negotiation of better transfer and adoption agreements.

In the food industry, EEE and WEE policies are implemented under the commitment to close technology gaps. In frames of collaboration and circular economy, the supply chain can coordinately act to improve the level of technology adoption and synchronize competitiveness not only from the country perspective but even more from the supply chain perspective, seen as an operative unity that trespasses country boundaries.

2.2 *Circular Economy: An Applied Science to Reach Decarbonization*

The second key aspect to describe the TCI relevance is an extended understanding of what is and what are the functions of the circular economy, seen as a scientific discipline and an instrument for steering the economy eco-efficiently (European Parliament 2023a). The circular economy suggests viable strategies to decarbonize industries by promoting the reduction of raw material use and improving collaboration between stakeholders along supply chains (Oxford Institute 2021). Today, circularity is not considered as an effective strategy to reduce technology gaps. With an extended definition of circular economy, technology gap closing are implemented with scientific and professional approaches, which will valorize more comprehensively the reparation and remanufacturing processes. The TCI bases its rankings on exposing the circularity integration levels related to a certain country's technology management. It allows the visualization and understanding concerning conflicts of interest, barriers, and bridges considering the promotion of a technological circularity. The circular economy, seen through the concept of technological circularity, becomes an essential tool to achieve decarbonization goals and to promote the appropriate stakeholder collaboration in the territories.

2.3 Decision Theory: Individual and Collective Decisions and Their Territorial Impact

A third key aspect in the frame of the literature review is the decision theory. The decision theory adheres to the principle of decision-making in terms of the rational expected utility and the inputs that must be considered to achieve the expected utility (Fishburn 2002). Concerning the expected utility, there is also a differentiation between individual and collective decision-making, which must be highlighted. The implications of individual and collective decision-making condition the final impact on the territory and the community living in a specific territory.

Regarding the circular economy and its bridge to technology management, a collectivity making decision for a collective group can occur, when the environmental authorities regulating e-recycling fees enable a tax reduction for returning discarded technologies according to a specific takeback scheme (Shan et al. 2021). In contrast to collective decisions, an individual makes decisions that only affect his or her environment. For example, a user who imports an overhauled technology from a country that encourages the export of high-quality remanufactured technologies (KIT 2020). The risk and responsibilities lie within his or her boundaries, but it can impact an entire territory if there is a risk of no reparability for that overhauled equipment, which could generate e-waste in the territory. In this sense, two points of view can be underlined:

1. There are stakeholders or decision-makers with collective or individual influence, who can coordinate and promote the technological circular.
2. These decisions in frames of the TCI are taken to improve the territorial adoption capacities.

2.4 *Technology and Circularity: Technical Material Management*

The fourth aspect to study in the literature review is the concept of technological circularity. This concept is introduced in detail through two central aspects. The first aspect is based on a holistic definition of technology, beyond the relationship that users settle with equipment and machinery. A holistic definition of technology facilitates the acknowledgment of a compressive country's technology management (Manceau and Morand 2014). The second aspect describes how countries operationally manage technology, drawing attention to the circular economy management according to their regulations and their closing technology gap initiatives. Based on the statement that technology is much more than a physical linkage with equipment and machinery, other action fields attached to the holistic concept of technology can be elaborated.

Technology:

- Is also the knowledge transfer and adoption inherent to technology management.
- Describes the provider's and users' professionalism level.
- Deals with the end-user's technological acceptance.
- Describes the physical management of equipment and machinery. In the context of the circular economy, this physical management covers manufacturing aspects, the provision of warranty services, maintenance, repairs, overhauling, and recycling operations.

Circularity focuses on the management of materialization and consumption, aiming to avoid and reduce waste (Jansen et al. 2022). Knowledge transfer and other immaterial assets and processes also play an important role in the success of technological circularity, but the circular economy operations core is concentrated on materials and their valorization throughout the diverse life cycles.

Technological circularity is a concept coined by the institution TCO Certified, a certifying body that promotes the reuse and eco-design of electrical and electronic products (TCO Certified 2023). TCO Certified functions as a quality seal for circular design and good practices in the fields of technological repair and remanufacturing. Complementing the conceptualization of TCO Certified concerning the issue of technological circularity, some starting points, where technological circularity demonstrates relevance and sense, could be raised complementarily:

1. *The territorial approach around the issues of technology management, adoption, transfer, commercialization, and disposal*: The territorial approach is fundamental for the TCI's creation and validation. To facilitate an apprehension, countries are taken as comparable geographical units since they contain the most adequate information (reliable information that is easily accessible and updated regularly). If information is available at the level of cities or even smaller geographical regions, the index methodology could be applied to compare these territories.

For the context of the food industry, the citrus market is analyzed to notify the main players in this sector and their technology and circularity integration levels.

2. *The extension of the supplier's responsibility*: It is important to highlight that it is not only the supplier who should extend its obligations towards the products by offering life extension services in EOL or MOL (middle of life) phases. Moreover, it is also important that other stakeholders in the supply chain, such as users, regulators, or retailers are integrated into the circular strategy rooted in the territory.
3. *Local capacity for maintenance, repair, and remanufacturing*: The country adopting the technology must demonstrate a formalized and certified MRO (maintenance, reparation, and overhauling), if possible, granted by the related OEMs. Otherwise, it is very difficult for suppliers to extend their responsibilities in the territory to reduce WEEE significantly and close technology gaps. The installed capacity for maintenance, repair, and remanufacturing must be easily found and cataloged according to their competencies and certifications. A deficient installed capacity brings obsolescence, informality, cost overruns, and pollution to the territories. This happens in emerging producing countries when the attention is shifted to the food supply chains. Local MROs work independently without any kind of certification and recognition of the corresponding OEMs. That is why the food sector in emerging countries suffers obsolescence and expands technology gaps, which worsen the functioning of supply chains.

Concluding with the literature review, where the sub-indexes choice was illustrated throughout their scopes and missing issues, and the critical observation of key aspects like the functionality of the circular economy, the decision theory, and the concept of technological circularity, the methodologies, that back up the construction of the index, were described, starting from the difficulty of standardizing and aggregating the diverse information sources that support the TCI.

3 Methodologies

3.1 Multivariate and Multidimensional Analysis

The core construction methodology implemented is the multivariate analysis. The multivariate analysis facilitates the administration of contexts, realities, and indicators related to the TCI's information environment (Olkin and Sampson 2001). The multivariate approach facilitates an instrument for a suitable integration of the sub-indexes, among which meaningful correlations can be generated to understand and depict complex scenarios. In addition to the multivariate analysis, the TCI represents multidimensional realities, acknowledging not only the variables and their cognitive components but also the variety of layers and action fields that are condensed in the index (De Muro et al. 2009).

The sub-indexes, which comprise the TCI, are grouped into three dimensions. The first dimension describes the sustainable growth dynamic of the territory, contrasting the economic development against the territorial sustainability impacts. The second dimension focuses on highlighting innovation and technological gaps regarding the country's technological management. Finally, the third dimension deals with the circular performance aspects in the context of the country's economic development and its technological gap-closing programs. Each of these dimensions is explained in detail in the results Sects. 4.2 and 4.3.

3.2 *Indexation Methodologies*

The indexation theory introduces three stages for the construction of a composite index: normalization, aggregation, and weighting (Schuschny 2009).

The TCI's normalization process proposes a numerical standardization of the sub-indexes to be able to apply comparations. In other words, the information to be included in the index must maintain a certain parity and unification. The TCI is composed of 11 other indexes (already introduced in Sect. 2). These indexes share multiple elements and can be correlated with each other. The sub-indexes behind the TCI follow two key perspectives, which define the correlated sides of the TCI:

1. The territorial economic development contrasted with the level of sustainability and technological progress.
2. The technological progress observed from the perspective of circularity, optimizing its economic development and minimizing negative impacts.

From that point, the numerical normalization process was carried out by applying two theories, MINMAX and SCALES. MINMAX methodology defines the parameterization based on the highest and the lowest results presented. For the current case, the numerical results of the sub-indexes were singularized in a range from 0 to 1, where 0 means the worst performance and 1 is the best performance. The second methodology is scales. Scales as indexation methodology introduces performance parameters, depending on the ecosystem to be described. Scales are used when the sub-indexes base their ranking on a scale already determined by the authors. For example, the S&P Sovereign Index introduces eight intervals, in which it classifies the country's credit trust. To normalize the S&P Sovereign Index, a scale was proposed within a range between 1 to 100, considering that 100 is 100% of all intervals. After designating intervals and converting them to a numerical representation, the MINMAX indexation methodology was applied again. Below, an example is presented, where the MINMAX and scale parameterizations were concluded based on the normalization exercise (Fig. 6.1, Table 6.1).

Regarding the aggregation process, understood as the process where the numerical normalization is added to a certain interpretative environment, the first aspect to be considered is the periodicity of the sub-indexes' publications, which feed the TCI. As a construction methodology, the TCI implies coordination with the sub-

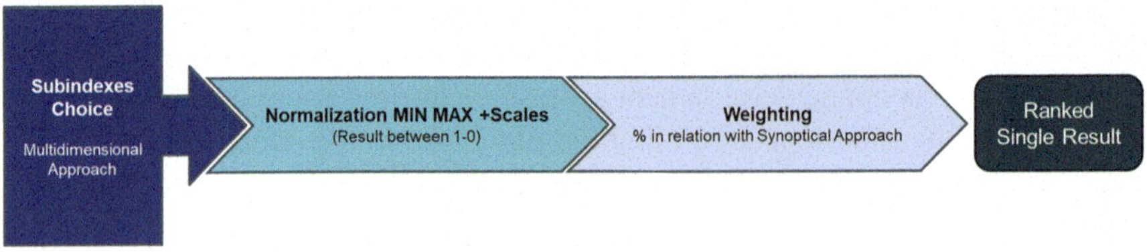

Fig. 6.1 Normalization + weighting process

Table 6.1 Example of a normalization exercise, using MINMAX methodology

Productive capacities index				
	Max	Min		
Country	Dena	Niger	Result	Norm 1
Den	**70.54**	**16.51**	**70.54**	**1**
Niger	**70.54**	**16.51**	**16.51**	**0**
Germany	70.54	16.51	65.8	0.91227096
France	70.54	16.51	62.8	0.85674625
Turkiye	70.54	16.51	54	0.69387377
Tun	70.54	16.51	47.7	0.57727189
Alg	70.54	16.51	43.8	0.50508976
Egy	70.54	16.51	44.1	0.51064224

indexes authors and platform to assure the actualization of figures. This aggregation must be dynamic and must facilitate the future integration of robust sub-indexes. The intention is to condense the information, refining, and building correlations during a year of sub-indexes publication. The TCI could be launched once a year, e.g., during the first week of January, gathering all the information collected and processed during the already finalized year.

Another complexity in dealing with the aggregation process relies on the sub-indexes' actualization and publication policies. In this regard, one example is the TAI publication, which was only officially published until 2018, while the other sub-indexes have been updated constantly. In addition to the updating challenges, there is also a lack of information and data for some countries in some sub-indexes. This was evident during the normalization exercise of the sub-index S&P Sovereign Index. This index presents limited information about North African countries. The lack of data for certain countries may tend to unbalance the aggregation exercise, as it complicates the synchronized updating activities related to the TCI.

Following the aggregation exercise, the weighting exercise corresponds to the index synoptic point of view, introducing a theoretical approach rather than an empirical type of weighting. Certain sub-indexes obtained higher weights to highlight significant correlations. In the weighting process, an index author defines his or her point of view and the need to communicate a certain reality and the interpretation of that reality. The intervals in a weighting process limit its significance in percentage. The results of the weighting process are presented in the next section.

4 Results

In this section, the TCI construction results are shown, beginning with the definition of the potential TCI user and ending with the result of the above-introduced four indexation methodologies.

4.1 Stakeholders as Index Users

To present the result after analyzing the literature review and explaining the basic methodologies behind the TCI, it is appropriate to specify the role and benefits of the stakeholders seen as index users. The following is a brief analysis of their expected utilities, as introduced in the Multivariate and Multidimensional Analysis.

4.1.1 Technology Users

Those are users who adopt technologies under circularity schemes, such as circular new technology products (European Parliament 2023a, b), products transferred or traded on a reuse basis, and repaired or remanufactured products. In the case of the food industry, taking into consideration the relevant technology gaps between producing countries and consuming countries, users could be farmers and processors. But if the entire supply chain is observed, every stakeholder could be considered as a technology user, like retailers, distributors, and even consumers. The following key aspects could be pointed out, which could affect the user's decision process:

- The set of circularity standards regarding the country, where the user lives and the country from where the user imports the technology.
- Technology transfer agreements between countries, which facilitate and promote technological circularity.
- The cost analysis regarding maintenance, reparations, and discarding.

4.1.2 Technology Suppliers

In the framework of the circular economy, suppliers are not only those who manufacture the technology as a product, transfer it within the country or internationalize it, depending on the technological readiness level (NASA 2023). The producers who manufacture the technology are recognized as OEMs (original equipment manufacturers). The OEMs are those who add new materials to the economies, by transforming them into functional products. The circular economy introduces new operations to be granted, such as reuse and life extension. This means that a supplier, in the case of technological reuse, might be the first user or, in a repair case, it might be the repair shop, which commercializes and transfers the technology and not the OEM. There may be cases also in which the OEMs, by operating reverse logistics, recover the discarded product, repair, or remanufacture it, and bring it back into the market (DEKRA 2023). In the food industry, technology suppliers hold a significant market power. Vendor Lock-ins are very common in this market because local farmers and processors mainly depend on one authorized vendor, which significantly impacts the country's technology state-of-the-art (Tatje 2017).

After clarifying the diverse ownerships through the scope of the circular economy, it is important to emphasize that suppliers, depending on their role along the transfer's operations, assume different responsibilities towards the users. These responsibilities range from the provision of ownership documentation to export the products, the management of warranties, and the certified recycling, whether it is mandatory or voluntary according to the target market's legislation. The OEMs must comply with regulations that directly impact the territorial technological management, in which the OEMs are responsible for recycling and final product disposal (OECD 2023). Nowadays, these norms are still unclear if OEMs are not in charge of transferring the technology. These kinds of regulatory gaps distress the dynamics of technological circularity. A global index that measures the conflict resolution between suppliers and users could contribute to the adjustment of public policies and business agreements. This could bring win-win benefits between users and suppliers and significantly enhance their efforts to circularize products and materials. Here are the key aspects, that affect the decision-making process of suppliers.

- The target market's favorable conditions to transfer or commercialize a product portfolio under the circularity principles.
- Technology transfer agreements that promote a technological circularity.
- The risks assumed as a technology supplier when entering products. Regarding warranty compliance, cost overruns in after-sales services, fines, and suspensions of representation licenses.

4.1.3 Collaborative Stakeholders

These are groups that actively support the operations between users and suppliers during adoption or transfer activities. This setup of stakeholders can be identified as academia, business promotion centers (incubators and/or accelerators), third-sector entities, or public and/or governmental entities with market interaction and promotion missions. In the food industry, there are plenty of collaboration cases to study along the supply chain, even at the consumer level. Nowadays, consumers want to dive deep into the supply chain to acknowledge more in detail good practices and optimization strategies toward environmental and social impacts (BWZ 2023). In general, their collaborative activities depend mostly on how they approach their roles during the transfer and/or adoption operations.

4.1.4 Non-collaborative Stakeholders

These are groups that do not deliver any kind of support during the adoption or transfer activities, but rather provide transactional services for users and/or suppliers. This stakeholder group includes regulators, verification, and auditing institutions as well as standardization agencies. Their activities are not collaborative, but they constitute a key piece to dynamizing the technological circularity. They act as observers and qualifiers of the user-provider interactions. Non-collaborative stakeholders affect the dynamic of the food supply chain significantly, promoting their food markets locally and internationally. In the context of closing technology gaps, their decisions conditionate the efficiency of the food industry and impact the competitiveness degree of their local food markets.

4.2 Normalization Results

As mentioned at the beginning of this article, the TCI is divided into three dimensions:

- Dimension 1: Sustainable growth dynamics.
- Dimension 2: Territorial technology management.
- Dimension 3: Circular performance of the territorial technology management.

Those dimensions derived from the multidimensional methodology contain the different sub-indexes' scores normalized between the interval 0 to 1. To assure the normalization interval 0 to 1, all the sub-indexes scores must be transformed into these intervals. By carrying out the normalization some extra exercise must be done to reach the interval 0 to 1. These extra exercises are presented below, to explain, how the author singularized the sub-indexes (Table 6.2).

Sub index "the circularity gap report" special remark This index was also normalized through a combination of minimax and scale methodologies. The Circular Economy Foundation Institute launched this index intending to measure the distance that nations show regarding the circular economy's implementation and integration. The radar infographic positions countries according to the complexity of achieving a sustainable, and circular space for their territorial development.

This space is described as a place where the utilization of materials and space is reduced to a maximum as a goal to ensure sustainability. This is the reason why in the circle closest to success, countries like Uruguay or Liberia are ranked because they are countries that do not require as many resources to achieve circular and eco-efficient well-being as industrial countries. While high-income countries are placed in the circles furthest from the ideal center since they require significant efforts to optimize their economies toward circularity. This could mean a radical change in their production matrices and a considerable reduction in their virgin materials' use, which impact their economic development. In this sense, the parameters were scaled

Table 6.2 Normalization results original intervals with special construction remarks

Dimension	Sub-indices	High value	Low value	Construction remarks
First dimension	UNCTADs Productive Capacities Index (PCI)	Denmark (70.54)	Niger (16.51)	No remark
	Economic Complexity Index	Japan (2.26)	Liber (−2.44)	No remark
	Ease of doing business scores	New Zealand (1)	Niger (−190)	Here, the normalization was carried out considering the first place as number 1 and last place as number 190, where 1 is the country that most facilitates the establishment of businesses. To obtain a correlation, a minus value was integrated. This changes the graduality of the sub-index, where a country with a mark of 1 is the best country and a country with a mark of 0 is the worst, regarding how easy it is to establish businesses in that territory
	S&P Sovereign Index	Norway (100)	Sri Lanka (12.5)	The normalization was made by implementing the scales methodology on the countries' credit credibility, issued by the S&P agency. In the S&P Sovereign Index, countries are qualified between 8 rates, starting from AAA credibility, which was given a range of 100 points for highly confidential countries. While countries with lower positioning ratings, obtain a SD qualification, which is quoted with 12.5 points. The intervals are divided in terms of 8 parameters each with 12.5 points
	SDG index	Finland (86.76)	South Sudan (36.8)	No remark
Second dimension	The Global Innovation Index	Switzerland (67.6)	Angola (10.3)	No remark
	Readiness for the future of production	Structure of production for the future Japan (8.99). Drivers of production	Structure of production for the future Niger (1.66). Drivers of production Cameron (3.24)	No remark
	Technology Achievement Index (TAI)	USA (0.67)	Somalia (0.060)	No remark
Third dimension	KAPSARC's Circular Carbon Economy (CCE)	Norway (69.68)	Uganda (15.56)	No remark
	The Circularity Gap Report	Uruguay (99.6)	Burundi (16.6)	Special remark
	Circular TAI	0	0	Special remark

Table 6.3 TAI indicators complement the construction of the circular TAI

TAI dimension	TAI Zenodo indicators	Circular TAI indicators derived from TAI Zenodo indicators
Development of human competencies	Active labor population/100 people	Number of entities dedicated to the recovery and revaluation of EEE and WEEE
	Schooling, tertiary (% Gross)	Number of professional graduates certified in technological recovery, repair, and remanufacturing
Creation of new technologies	Number of articles from scientific and technical journals	Number of articles from scientific and technical journals specialized in recovery, repair, and technological remanufacturing
	Patent applications, residents + non-residents	Patent applications, residents + non-residents specialized in technological recovery, repair, and remanufacturing
Diffusion of old technology	Access to electricity (% of population)	E-Waste generation per capita
	Telephone + mobile subscribers/100 people	E-Waste recovery per capita
Diffusion of new technologies	People who use the Internet (% of population)	Percentage of authorized providers entering the country with responsibility extension programs
	High-tech exports (% of exports of manufactured products)	High-tech exports (% of exports of manufactured products)

according to the number of concentric circles, which is 6, and each one was assigned a score of 16.6 points per interval from the whole quoted with 100 points. In this sense, Uruguay has the highest score with 99.6 and Burundi the lowest with 16.6.

Sub-index (TAI Circular) special remark This index is an own construction derived from the original TAI developed by United Nations and presented in the normalization of dimension 2. The goal behind a circular TAI, as a supporting index, is precisely to complement the capacities that are denoted in the TAI with circularity figures and indicators, which correspond to the original TAI but also express social aspects (Table 6.3).

4.3 Weighting Results

The weighting results are presented in frame of the synoptic boundaries, which were applied to the TCI. Table 6.4 shows the different percentages for each dimension and correlation with the corresponding sub-index.

Table 6.4 Weighting scheme for the TCI

TCI dimensions	Sub-index	Backed indexes weighting (%)	Dimension weighting 1 (%)	Dimension weighting 2 (%)	Final result weighting (%)	Final result
Sustainable growth dynamics	Capacities index	25	40	40	40	Final weighted result
	Country complexity rankings	25				
	Ease of doing business rank	25				
	S&P sovereign index	25				
	SDG-index	100	60			
Territorial technological management	Global innovation index	50	40	60		
	Readiness for the future of production	50				
	TAI	100	60			
Circular performance of territorial technological management	Circular carbon	50	40	100	60	
	Circular GAP	50				
	TAI circular	100	60			

4.3.1 Synoptic and Theoretical Approach Dimension 1

The intention behind this weighting is to qualify the technological development in the territories higher and measure the dynamism, resilience, and diversification of the observed country. This differential weighting is intended to highlight the importance of sustainability concerning territorial economic development. This condition would facilitate the integration of circularity to optimize gap closings and better the effects on local economic development.

4.3.2 Synoptic and Theoretical Approach Dimension 2

The equivalent weighting conveys the importance of both the technological readiness for the future, as well as examining the territory's skills towards fostering innovation. This is due to the intention of highlighting the technological aspect with the sustainable growth of the observed territory. These two dimensions are also weighted to obtain a complementary result and derive the territorial management impact on the territory's sustainable economic development. For this purpose, dimension 2 is weighted at 60%, while dimension 1 is weighted at 40%. The third dimension is again contrasted with the result obtained from the two dimensions detailed above.

4.3.3 Synoptic and Theoretical Approach Dimension 3

For the circular TAI, the same methodology introduced for the TAI (Dimension 2) was applied, according to which the four dimensions are weighted in equal ranges of 25% for each dimension. These are then summed and divided by four to obtain the cumulative TAI per country.

The TCI intends to singularize the different sub-indexes figures into one single result, which ranks countries based on their circular performance related to their technological management.

Having obtained a single weighted result for the first two dimensions, this result must now be weighted concerning the single result obtained from the weightings of the three sub-indexes that compose dimension 3.

The first two dimensions are given a total weighting of 40%, while the weighting of the third dimension has a weighting of 60%.

This weighting explains the basic overview that was applied for the construction of the TCI. The TCI aims to examine the country's circular performance regarding technology management. For that reason, dimension 3 should be highlighted in the measurement exercise to notice significant changes in the ranking. Table 6.5 shows a singularization exercise, obtained by observing the technological circularity performance of Germany.

Table 6.5 Weighting scheme applied to Germany

TIC dimensions	Sub-indexes	Normalized result of baked	Weighting of baked (%)	Intermidiate results I	Intermidiate results II	Weighting of dimensions (%)	Dimensions results I	Dimensions results II	Weighting of dimensions (%)	Dimensions results III	Dimensions results IV	Weighting singulari-zation (%)	Singulari-zation results	Singulari-zation results II
Sustainable growth dynamics	UNCTAD's Productive Capacities Index (PCI)	0.91	25	0.2275	0.9275	40	0.371	0.929	40	0.3716	0.7724	40	0.30896	0.75
	Economic Complexity Index	0.93	25	0.2325										
	Ease of Doing Business Scores	0.87	25	0.2175										
	S&P Souvereign Index	1	25	0.25										
	SDG index	0.93	100	0.93	0.93	60	0.558							
Territorial technological management	The Global Innovation Index	0.84	50	0.42	0.875	40	0.35	0.668	60	0.4008				
	Readiness for the Future of Production	0.91	50	0.455										
	Technology Achievement Index (TAI)	0.53	100	0.53	0.53	60	0.318							
Circular performance of territorial technological management	KAPSARC's Circular Carbon Economy (CCE)	0.88	50	0.44	0.64	40	0.256	0.736	100	0.74		60	0.44	
	The Circularity Gap Report	0.4	50	0.2										
	Circular TAI	0.8	100	0.8	0.8	60	0.48							

5 Conclusion

After analyzing several ranking exercises, using the TCI and its application around the circular performance of territories and the relationship between countries, this article closes by outlining a set of conclusions and optimizations around the TCI applicability.

The first conclusion focuses on the aggregation exercises concerning the TCI information compilation process. This first choice of indexes was limited to finding indexes that demonstrated professional management in the collection, treatment, and singularization of their data. If the TCI wants to maintain an open aggregation window, depending on new and robust indexes, which could be integrated in the future, those sub-indexes must comply with minimal rules, like steady actualization and transparent information sources. Otherwise, users will lose confidence and interest in using the TCI.

Another important aspect is the implementation of digital extraction technologies for updating, verifying, and contextualizing the TCI. The TCI is developed in frames of a dynamic DSS (decision support system) system (CFI 2020). One of the key components to assure actualization in the future is the integration of computation to expand the information thresholds by implementing digital cloud and artificial intelligence approaches (Elson and Connelly 1995).

In this regard, it is suggested that the index can be fed with information by using web-enabled content and digital extraction technologies. This approach aims to update changes that occur when regulations, standards, or information are modified at the source. When an automated digital content extraction mechanism is introduced, users' trust can be ensured. The sub-indexes provide professional, verified, and immediate information, assisted by extraction transform loads methodologies (IBM 2023). Beyond the upgrading task of digital content extraction, computerized decision support systems will facilitate agile connections through Big Data and artificial intelligence by simulating desired or avoidable scenarios that would be impossible for humans to conceive automatically.

Finally, the TCI as a tool for decision support offers an extended catalogue of solutions, which could be developed in a frame of collaboration and transparency. The TCI could facilitate an immediate apprehension of the country's conditions and achieve the clustering of regions and territories where operations could deeply narrow the technological gaps under coordinated and collaborative eco-efficiency and circularity strategies.

References

Bakker C et al (2021) Understanding and managing product lifetimes in support of a circular economy. J Clean Prod 279:123764. https://doi.org/10.1016/j.jclepro.2020.123764

BWZ (2023) Budesministerium für wirtschafliche Zusammenarbeit und Entwicklung. das Liefer-kettegesetzt

CFI (2020) Decision support system (DSS). https://corporatefinanceinstitute.com/resources/management/decision-support-system-dss/
Circular Economy Foundation (2023) The circularity gap report
De Muro P, Mazziota M, Pareto A (2009) Composite indices for multidimensional development and poverty: an application to MDG indicators
DEKRA (2023) Certified pre-owned (CPO) program design
Elson RB, Connelly DP (1995) Computerized decision support system in primary care. Prim Care 22(2):365–384. https://doi.org/10.1016/S0095-4543(21)00523-6
European Commission (2023) Right to repair: commission introduces new consumer rights for easy and attractive repairs
European Parliament (2023a) Ökodesign: umweltfreundlichere und energieeffizientere Produkte
European Parliament (2023b) Circular economy: definition, importance and benefits
Fishburn P (2002) Utility theory and decision theory
Gerybadze A (2019) Technology and innovation management in a global perspective
Harvard (2023) Economic complexity index
IBM (2023) ETL (extract, transform, load)
INSEAD (2022) The global innovation index
ITU (2020) Global e-waste monitor
Jansen B et al (2022) The technical or biological loop? Economic and environmental performance of circular building components. Sustain Prod Consum 34:476–489. https://doi.org/10.1016/j.spc.2022.10.008
KAPSAR (2022) KAPSARC's circular carbon economy (CCE)
KIT (2020) Decision theory
Kvedariene A (2019) Strategic technology management within global value system. Open Econ 2(1):43–52
Manceau D, Morand P (2014) A few arguments in favor of a holistic approach to innovation in economics and management. J Innov Econ Manag 15(3):101
NASA (2023) Technology readiness levels
OECD (2023) Extended producer responsibility
Olkin I, Sampson A (2001) Multivariate analysis overview. https://doi.org/10.1016/B0-08-043076-7/00472-1
Oxford Institute (2021) Incentivizing decarbonization through the circular economy
Ren Y et al (2023) Technology transfer adoption to achieve a circular economy model under resource-based view: a high-tech firm. Int J Prod Econ 264:108983
Romanainen J (2021) Advanced technologies for industry
S&P Global (2020) S&P sovereign index
Schuschny A (2009) Guia metodológica de desarrollo sostenible
Shan F et al (2021) Comparison of three E-Waste take-back policies. Int J Prod Econ 242:108287. https://doi.org/10.1016/j.ijpe.2021.108287
Tatje J (2017) Farmers face vendor lock-in as agriculture goes digital
TCO Certified (2023) About us. https://tcocertified.com/de/
The World Bank (2022) Ease of doing business scores
UNCTAD (2023) UNCTAD's productive capacities index
United Nations (2023) SDG index
United Nations-Zenodo (2020) Achievement index (TAI)
Utoikamanu F (2018) Closing the technology gap in least developed countries
WEFORUM (2018) Readiness for the future of production

Chapter 7
Debittering of Orange Juice with the Whole Flavour Using Expanded Bed Chromatography in a Vortex Flow Reactor

María J. Ibáñez-González, Tania Mazzuca-Sobczuk, and Elena Bacas-Béjar

1 Introduction

Orange juice production is a consolidated industry in the European Union (EU). However, the EU is a net importer of orange juice. Spain is one of the world's leading citrus producers, with 15% of its citrus production destined for processing (mainly into citrus juice, essential oils and by-products) (USDA Foreign Service Agency 2021).

Oranges, the raw material for orange juice production, are divided into sweet and bitter oranges. The first group is mainly used as edible human food, whole or squeezed (orange juice). Bitter orange juice, on the other hand, is preferred for salads, other fruit preparations and medicinal uses because of its sour taste (Farag et al. 2020).

Oranges are a seasonal fruit. In early and mid-season, oranges are not fully ripe and contain the precursor of the bitter principle. Although the precursor is not bitter, it is responsible for the development of the bitter compound, especially when freshly extracted orange juice is stored. Over time, the orange juice becomes bitter and unpalatable to consumers. The time it takes for orange juice to go from non-bitter to bitter depends on factors such as pH, heating and exposure to air. However, it can be expected to take about 5 hours at room temperature or 24 hours in the refrigerator

M. J. Ibáñez-González (✉)· T. Mazzuca-Sobczuk · E. Bacas-Béjar
Chemical Engineering Department, Almeria University, Almeria, Spain

CEIA3, International Agri-food Campus of Excellence, Cordoba, Spain

CIAMBITAL, Research Center of Intensive Mediterranean Agro Systems and Food Biotechnology, Almeria, Spain
e-mail: mjibanez@ual.es

A. Melkonyan-Gottschalk, D. Daus (eds.), *Transformation Towards Circular Food Systems*, Greening of Industry Networks Studies 13,
https://doi.org/10.1007/978-3-031-63793-3_7

(Swisher 1958). It is, therefore, pointless to package orange juice for delivery to the consumer without a debittering process.

Limonin is the major bitter limonoid constituent of citrus juices and the primary cause of bitterness. The non-bitter precursor of limonin, limonoate A-ring lactone, is found mainly in the albedo and other parts, such as the central bundles, veins and seeds. This substance remains stable in the tissues of the fruit. However, when the fruit tissues are ruptured during orange juice extraction, it is attacked by the acidity of the juice and converted to limonin. Heating and exposure to air also contribute to the conversion.

Late season oranges do not contain this precursor of bitterness. A sustainable solution to produce good quality juice would be to store the early to mid-season orange juice production and blend it with the late season orange juice in the proportion required to produce a juice with a limonin concentration of less than 5 ppm (Fellers 1989). The concentration below this level makes the orange juice palatable to the consumer. Nevertheless, this sustainable process has disadvantages derived from the long storage period, such as refrigeration facilities, the potential growth of microorganisms and the development of off-flavours that affect the quality of the juice.

Using the words "debittered orange juice" as keywords in a Web of Science collection returns 86 publications indexed from 1959, showing the importance of this topic to the orange processing industry. "Limonin debittering" returns 96 scientific papers and 7 reviews. The most cited reviews (Hasegawa and Miyake 1996; Purewal and Sandhu 2021; Puri and Banerjee 2000; Puri et al. 1996) have 170, 90, 47, and 29 citations, respectively and try to summarise the technology and methods used in the debittering process on the one hand, and the role of limonoids in citrus fruits on the other. Orange juice debittering is not a completely solved problem for the industry. It is still looking for improved solutions.

2 Orange Juice Debittering Process by Chromatography

Since 1957, several patented processes for debittering orange juice or concentrated orange juice have appeared. On the one hand, these processes add additives to adjust the pH above seven and avoid the formation of limonin (Swisher 1958) or bitterness modulators by suppressing the perception of bitterness (Guadagni et al. 1977). On the other hand, solvents such as vegetable oils (Swisher 1972) and isopropanol (Pritchett and Calif 1957) are added to separate limonin by liquid-liquid extraction. These processes are not sustainable unless the final orange juice can be considered fresh. Other patented methods, such as enzymes (Hanson 1968, 1969; Hasegawa and Brewster 1975) and microorganisms (enzyme producers) (Hasegawa 1984) are too expensive for industry to develop. They act on the limonin precursor and produce other substances that are not bitter and cannot be converted to limonin by any of the substances present in the citrus juice.

In 1997, the first patent related to adsorbent beads for debittering orange juice was published, proposed by Chandler and Johnson (1977). They patented cellulose ester adsorbents to solve the problems of adsorption of limonin in a batch treatment using different dry polyamide powders (Chandler et al. 1968). Unfortunately, a significant amount of ascorbic acid is removed from orange juices, and clarification to obtain a juice without cloud particles was required. However, these polyamide adsorbents were the beginning of the development of new adsorbents for the food industry.

In order to increase the limonin selectivity, Chandler and Johnson designed different cellulose ester adsorbents that allowed to treat orange juice with its full aroma in batch mode or packed column (Chandler and Johnson 1977). Cellulose ester adsorbents partially adsorb limonin, and ascorbic acid remains practically in the orange juice with pulp in batch treatment where nylon mesh bags hold the gel beads. The bead size ranged from 1.7 to 6.4 mm in diameter to allow all the juice to pass upwards through the packed column (0.502 mm mesh) without removing the pulp, only the coarser particles. Increasing the amount of limonin adsorbed requires increasing the residence time in the packed column. However, the bed was fluidised with a gentle stream of nitrogen, giving the column an extra 20% volume (Johnson and Chandler 1981), to ensure that the whole orange juice did not clog the column.

In 1984, the use of commercially available neutral polymeric adsorbents, chemically defined as a polystyrene adsorbent resin cross-linked with divinylbenzene (Duolite S-861 and Duolite ES-865), was patented (Puri 1984) to reduce the bitterness of various citrus juices or concentrates, including flavonoids and limonoids. These neutral adsorbents adsorb up to 90% of limonin. Due to the small size of the beads, citrus juices are centrifuged or filtered to separate the pulp and added back after packed bed chromatography to minimise the possibility of clogging.

Wilson et al. (1989) again attempted to achieve a stable fluidised bed to reduce the bitterness of clarified orange juices (reconstituted from concentrate) using a variety of adsorbents such as the β-cyclodextrin polymer and two neutral polystyrene divinylbenzene resins from Amberlite approved for food use (XAD-4 and XAD-16). This fluidised bed was of a pilot plant size for industrial implementation. However, in some cases, up to 50% of the polymer adsorbent was lost from the fluidised bed and ended up in the juice, even at lower flow rates. In addition, neutral resins were difficult to fluidise.

2.1 *Current Industrial Orange Juice Debittering Processes*

Since the 1990s, industry debittering processes for orange juice have included a common step of adsorption chromatography in a packed column to remove limonin and off-flavours from clarified orange juice. The neutral adsorbents are widely used in the industry worldwide.

Today, there are two technologies for clarifying orange juice, one by centrifuge and the other by ultrafiltration (Shaw et al. 2000). Figure 7.1 briefly shows the flow

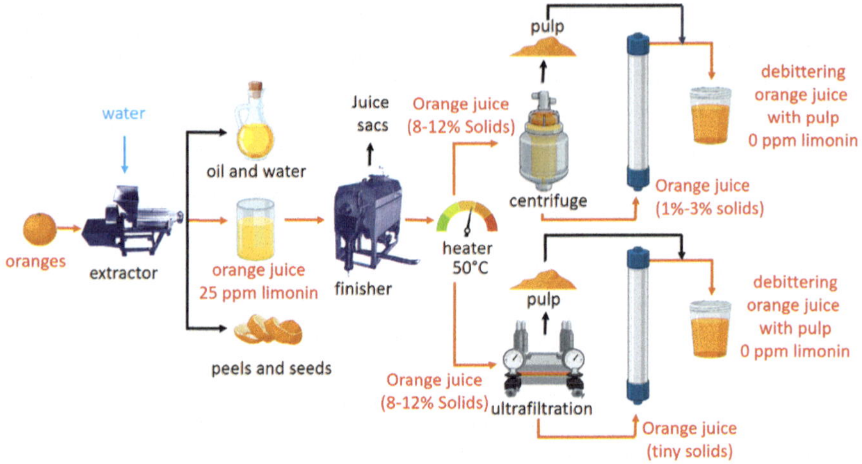

Fig. 7.1 Flowsheets of the two orange juice debittering technologies: centrifugation and packed chromatography and ultrafiltration and packed chromatography

sheet of debittering orange juice from the raw material. In the industry, the orange juice is first extracted in a juice extractor, where the peels and seeds are separated from the orange juice on one side and the orange oil and water on the other side. In the next step, the finisher (screen size ranging from 0.508 to 0.762 mm) removes the juice sacs from the orange juice. The juice is then heated to 50 °C to reduce its viscosity. As the temperature increases, the viscosity decreases, allowing the orange juice to flow more easily through the centrifuge or ultrafiltration membranes and later through the pores of the resin (Hernandez et al. 1992). After centrifugation, the orange juice is left with between 1% and 3% solids. The packed column is often clogged, resulting in shorter runs and more frequent regeneration of the adsorbent. Removal of all solids is essential, so ultrafiltration is the best alternative, although small particles remain in the orange juice (Purewal and Sandhu 2021). As a result, a packed bed is not completely free of clogging. Finally, the pulp is returned to the debittered clarified orange juice.

The cleaning system uses hot water and caustic solution to clean the ultrafiltration membranes and the debittering column. The process takes 4 hours. In 1991, the industrial operating cost of ultrafiltration was less than $0.03 per gallon of orange juice (Wethern 1991). Operating costs, including column and centrifuge or ultrafiltration, are similar (Schofield and Miller 2007).

2.2 *Macroporous Adsorbents*

Packed bed chromatography, which remains the most widely used method for purifying biomolecules from clarified broths, has been improved by the introduction of macroporous resins. They have a porous structure with interconnected channels that facilitate fluid flow and allow high flow rates. The diameter and size of the pore can be adjusted independently along the length of the bed to provide a larger surface area available for adsorption and a lower pressure drop. The pore size varies from 50 to 1000 Å. In this way, an increase in adsorption capacity and flow rate improves productivity (Vega et al. 2017).

The synthetic adsorbents used for the bitterness removal are neutral adsorbents with no functional group. These resins are in the form of spherical beads with particle sizes around 300–1000 μm and are highly porous with large hydrophobic areas. The hydrophobic molecules bind to the adsorbent by van der Waals forces. The ideal pore diameter sizes for the bitterness removal are in the range of 60–200 Å with surface areas in the range of 500–800 $m^2 \cdot g^{-1}$ and pore volumes in the range of 1.2–1.5 mL/g.

The most common matrices that can be used for debittering are (Ghanem 2012):

1. Styrene divinyl benzene resins (HP20, HP21, SP70, SP700, SP710, SP825, SP850, XAD15, FPX66). They have a good affinity for molecules with aromatic rings and conjugated double bonds and a medium affinity for debittering molecules (limonin). These resins are approved under the Federal Food, Drug, and Cosmetic Act and the applicable food additive regulations (21CFR173.65) in the EU.
2. Methacrylic type resin (HP2MG). It has a good affinity for molecules with polar groups (carboxyl, ester, amino, or amide groups) and a low affinity for debittering molecules. This resin is easier to recycle due to its low hydrophobic surface.
3. Brominated styrenic resin (SP207). It has a high affinity for debittering molecules. It has a very high hydrophobic surface which makes it difficult to recycle.

3 Expanded Bed Adsorption

In the 1990s, expanded bed adsorption was developed to address the purification of biomolecules from unclarified culture broths (Chase 1994). The aim is to eliminate clarification and biomolecule concentration steps to increase productivity. To achieve this, adsorbents with higher density than water, and size distribution are designed to create a stable stratified fluidised bed, known as an expanded bed. The largest particles are at the bottom of the column, those of medium size are in the middle of the expanded bed, and the smallest particles are at the top of the expanded bed. The axial flow velocity is between the minimum fluidisation velocity and the terminal velocity. The trend is to use dense solid core materials to increase the terminal velocity of the adsorbents, allowing higher feed rates to achieve higher productivity (Li et al. 2015).

3.1 Steps in Expanded Bed Adsorption

The purification of biomolecules by expanded bed adsorption involves the following steps (Chase 1994):

1. Adsorbent loading in the column in the absence of axial flow.
2. Stabilisation step. A solution is introduced through the bottom of the column and rises to the top, creating the expanded bed. The porous plate at the bottom of the column serves two purposes: to retain the adsorbent in the column and to distribute the inlet flow.
3. Adsorption step. The stabilising solution is changed by the feed. The target biomolecule binds to the adsorbent, while cells, cell debris, and other biomolecules pass through the porosity of the expanded bed without blocking it.
4. Wash step. The feed is replaced with the initial solution to promote the removal of unwanted biomolecules that may have been retained between the adsorbent beads of the expanded bed.
5. Elution step. The eluent is introduced through the top section into the settled bed. During elution, the target biomolecule is purified and concentrated.
6. Regeneration step. The adsorbent is regenerated according to a specific procedure.

3.2 Disadvantages of Expanded Bed Adsorption

The expanded bed has drawbacks, such as partial blockage of the porous plate, resulting in preferential channels or dead zones (Anspach et al. 1999; Hubbuch et al. 2002). To overcome the blockage problem at the column inlet, a perforated blade distributor that rotates and releases the biological extract through the holes has been evaluated (Arpanaei et al. 2008). In this case, this distributor requires very strong adsorbents to prevent them from breaking when the blades rotate. The turbulence generated at the base of the column causes distribution problems in the flow, which are minimised by optimising the height of the expanded bed.

4 Expanded Bed Vortex Flow Reactor

A promising alternative to solve the problems associated with expanded bed chromatography adsorption columns is the vortex flow reactor. In the expanded bed vortex flow reactor (EB-VFR), the axial flow expands the bed into the annular gap. In addition, the rotation of the internal cylinder stabilises the bed and does not create preferential channels (Ibañez-Gonzalez and Cooney 2007). Bed expansion is due to the axial velocity, not the rotational velocity. If there were only rotation, the adsorbent would remain at the bottom of the reactor. The EB-VFR design is as high as an expanded bed column to achieve the required residence time. Ports are located every two centimetres to vary the outlet according to the height of expanded bed (Ibáñez González et al. 2015). The EB-VFR reactor can also operate with unclarified feeds as the feed enters through one side of the reactor without the need to incorporate a support to hold the resin inside the reactor, thus avoiding clogging problems (Ibáñez-González et al. 2016; Redondo Miranda et al. 2021).

4.1 Flow Regime

Taylor (1923) studied fluid motion in the annular zone between two concentric cylinders and demonstrated the existence of vortex flow in the annular gap when the outer cylinder remained stationary and only the inner cylinder moved.

An increase in the rotation rate of the inner cylinder is associated with different flow patterns, which are characterised by a dimensionless number known as the Taylor number, Ta. In addition to the rotation rate of the inner cylinder (ω), the Taylor number depends on the kinematic viscosity of the fluid (ν), the outer radius of the inner cylinder (r_i) and the width of the annular gap (d) according to the following Eq. (7.1):

$$\mathrm{Ta} = \frac{\omega \cdot r_i \cdot d}{\upsilon} \tag{7.1}$$

where $d = r_e - r_i$, being r_e the inner radius of the outer cylinder.

At a low rotational speed, the inner cylinder causes the fluid in the annular gap to move tangentially, a shear flow called laminar Couette flow. As the rotational speed of the inner cylinder increases above a critical level, the fluid motion ceases to be tangential and becomes toroidal, called vortex flow. Toroidal vortices rotating in opposite directions appear alternately. These vortices, called Taylor vortices, are rotationally symmetric and have the shape of "doughnuts" called as vortex pair units with a diameter approximately equal to the gap width.

When the Taylor reactor provides an inlet and outlet for axial flow, the result is a Taylor vortex flow reactor. The Reynolds number characterises the axial flow rate by the following Eq. (7.2):

$$\mathrm{Re} = \frac{U \cdot d}{\upsilon \cdot \varepsilon} \tag{7.2}$$

where U is the axial flow rate, and E is the porosity of the expanded bed.

When the Reynolds number is less than 10 and the Taylor number increases above the critical value, different flow patterns appear, such as laminar Taylor vortex flow when the Ta/Tac ratio is less than 15, wave and spiral vortex flow when Ta/Tac is between 15 and 30, turbulent vortex flow when the Ta/Tac ratio is between 30 and

160 and turbulent flow when the Ta/Tac ratio is higher than 250. The critical Taylor number, Ta_c, is given by the following Eq. (7.3) (Lee and Lueptow 2001).

$$Ta_c = 41.02 \cdot \left(\frac{d}{r_i}\right)^{-0.5} + 25.75 \cdot \left(\frac{d}{r_i}\right)^{0.5} + 1.85 \cdot \left(\frac{d}{r_i}\right)^{1.5} \tag{7.3}$$

Giordano et al. (2000) found that in the laminar Taylor vortex regime when the rotations are small, the vortices move at a speed close to or slightly above than the speed of the axial flow. Under these circumstances, there is little back mixing between the vortex pair units and the vortex reactor behaves as a plug flow similar to a packed column. For this reason, the EB-VFR operates with values of Ta and Re in the laminar Taylor vortex flow regime.

Kataoka et al. (1975) define the laminar vortex regime as an ideal continuous plug flow reactor because the vortices provide intense radial mixing but little back mixing between the vortex pair units. In this way, the vortex pair units act as a series of small stirred tank reactors that form at the bottom of the RFV and ascend through it due to the axial flow. As a result, each vortex pair unit is equivalent to a stirred tank (Sczechowski et al. 1995).

5 Debittering of Orange Juice Using an Expanded Bed Vortex Flow Reactor

The industrial debittering of orange juice (Fig. 7.1) presents a number of problems that need to be solved:

- Adsorbent regeneration is frequent to prevent clogging of the packed bed column with orange juice due to its viscosity and suspended solids.
- Heating degrades the organoleptic properties of the orange juice.
- Ultrafiltration requires high pH cleaning solutions which are treated to release them into the environment.
- Centrifugation requires high maintenance costs.

To solve these problems, it is proposed to debitter orange juice in EB-VFR (Fig. 7.2). The process eliminates heating and clarification steps. This is achieved by changing the packed column to EB-VFR.

The EB-VFR consists of a transparent methacrylate outer stationary cylinder and a stainless steel inner cylinder. Both cylinders are concentric, leaving an annular space between them where the resin is added. The inner cylinder is fixed to the bottom of the reactor by a bearing, and to the top by a motor which controls the speed of rotation of the inner cylinder.

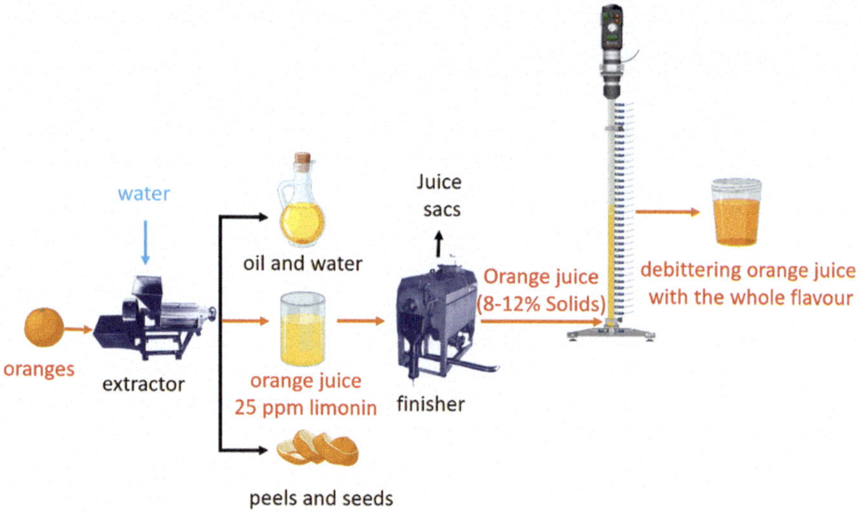

Fig. 7.2 EB-VFR orange juice debittering process setup

5.1 Setting Up an Expanded Bed Vortex Flow Reactor

When using the EB-VFR, the rotation rate of the inner cylinder and the axial flow velocity must be adjusted to obtain a stable expanded bed. Both operating conditions depend on the geometry of the VFR and the physical properties of the orange juice and the adsorbent.

The dimensions of the vortex flow reactor are 1.9 cm outer cylinder diameter, 1 cm inner cylinder diameter and 1 m height. The Tac (Eq. 7.3) depends on the radii, which is 69.2 for this geometry.

The orange juice is taken from the finisher under industrial conditions. The density and viscosity of orange juice are 1045.8 $kg{\cdot}m^{-3}$ and $2.8{\cdot}10^{-3}$ Pa·s, respectively. At a fixed speed of 250 rpm, the Ta number is determined to be 220.1. The ratio of Ta/Tac is 3.2; therefore, the regime of the vortex flow reactor is laminar Taylor vortex flow. Thus, the generated vortices behave as a plug flow to adsorb the limonin from the orange juice.

The axial velocity is determined according to the physical properties of the orange juice and the selected adsorbent. The adsorbent DIAION™ HP2MGL (Mitsubishi Chemical Corporation) fully complies with the Federal Food, Drug, and Cosmetic Act and applicable food additive regulations (21 CFR 1772470). The adsorbent is a hydrophilic resin without functional groups based on a cross-linked polymethacrylate in the form of porous beads of uniform size (355 to 570 μm) and a particle density of 1.09 $kg{\cdot}L^{-1}$. The resin has a specific surface area of 570 $m^2{\cdot}g^{-1}$ and a pore volume of 1.3 $cm^3{\cdot}g^{-1}$. Above 400 $cm{\cdot}h^{-1}$, the top of the expanded bed is unclear, and some particles elutriate from the expanded bed. Therefore, the adsorption process is carried out at 300 $cm{\cdot}h^{-1}$.

5.2 *Stages of the Debittering Process Using an Expanded Bed Vortex Flow Reactor*

7.5 L of orange juice taken from the finisher outlet with a limonin concentration of 10.3 ppm is subjected to a debittering process in an expanded bed vortex flow reactor according to the following steps:

1. Adsorbent loading. 32 mL of DIAION™ HP2MGL adsorbent is added to the annular gap of the vortex flow reactor. The height of the settling bed is fixed at 15 cm to ensure sufficient residence time of the orange juice in the expanded bed for the resin beads to adsorb the limonin.
2. Stabilisation step. Initially, 10.2 mL·min^{-1} of water is passed through the settling bed to initiate resin expansion. The rotation speed is then switched on and fixed at 100 rpm. After stabilisation, the height of the expanded bed is 18 cm (Fig. 7.3).
3. Adsorption step. The water flow is changed by the orange juice. The rotation speed is increased to 250 rpm. The limonin joins the adsorbent, while the rest of the orange juice with its whole flavour passes between the beads of the expanded bed (Fig. 7.4). The height of the expanded bed is initially stabilised at 42 cm and the outlet of the orange juice at 62 cm. The adsorption process took 12.3 hours, demonstrating its robustness. The height of the expanded bed decreases at 32 cm during adsorption, but the orange juice outlet does not change its port. The orange juice is collected in tubes. The 7.5 L of debittered orange juice with the whole flavour contains 2 ppm of limonin, approximately 234 volumes of orange juice per volume of resin. The percentage of solids before and after debittering shows that the concentration is similar, 11.8% for the orange juice and 11.3% for the debittered orange juice. No adsorbent beads were observed in the debittered orange juice.
4. Regeneration step. The DIAIONTM HP2MGL adsorbent is regenerated with a high pH solution (less than 5% in NaOH).

In most industries, limonin is not purified because they do not comply with the organic solvent handling guidelines necessary to purify and concentrate limonin. As a result, washing and elution steps are not performed.

The adsorption time is longer in an expanded bed than in an industrial packed column. The trend in adsorbent design is towards macroporous beads for two reasons: to increase adsorption capacity and to increase axial velocity in packed chromatography. In addition, the beads are large to prevent clogging. The adsorbent design for the expanded bed must increase the size range and density of the beads to increase the terminal velocity and thus the axial flow velocity.

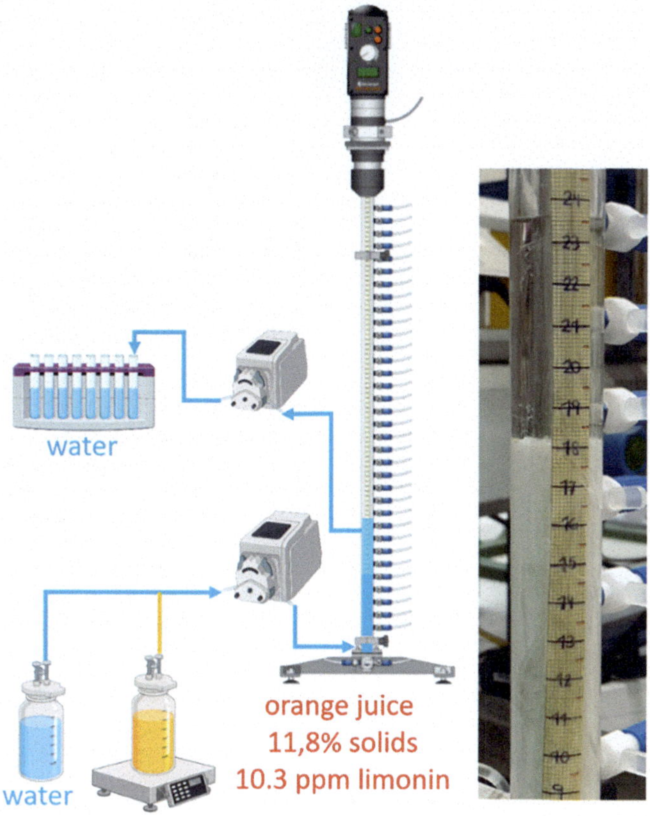

Fig. 7.3 Flow sheet: stabilisation step

5.3 *Process Sustainability*

The new consumer awareness is leading to a shift in the industrial process towards more sustainable and environmentally friendly practices and less treated orange juice.

It is well known that membrane filtration is considered an essential part of the orange juice industry as a clarification tool due to its low temperature treatment (50 °C maximum), high selectivity and low energy consumption. However, membrane fouling compromises these advantages as in-place cleaning of ultrafiltration membranes involves thermal treatments (in the range of 50–80 °C), non-biodegradable chemicals (e.g. alkali solution with a pH of up to 13–14), which require appropriate neutralisation (e.g. acidic solution with a pH of around 1) and significant amounts of water, creating sustainability challenges (Pant et al. 2023).

In addition, fresh water resources are said to be increasingly limited. Efforts are being made to use water more efficiently and the value of reusing water is

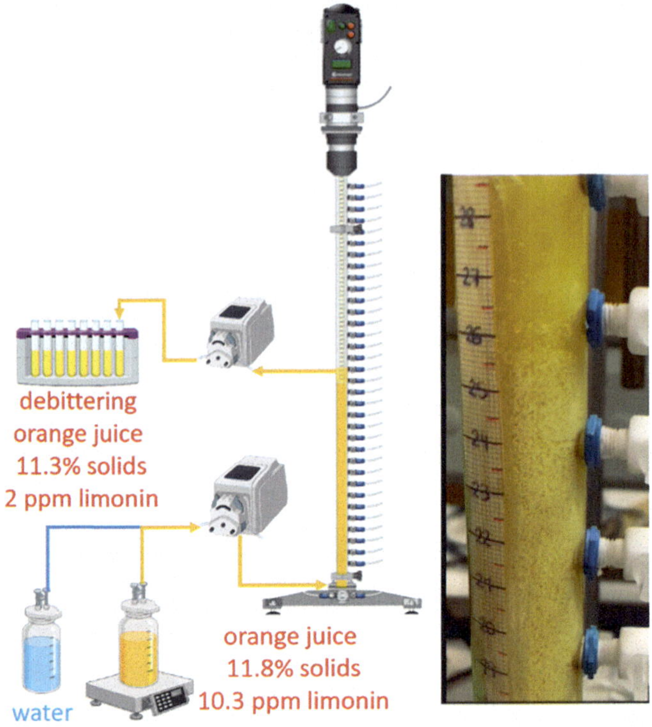

Fig. 7.4 Flow sheet: adsorption step

being recognised, for both environmental and economic reasons. This is expected to become even more critical in the future as the price of water rises and access to good quality water sources close to industries decreases. For example, an orange juice plant processing 800,000 tonnes of oranges per year uses about 756 cubic metres of water to clean ultrafiltration membranes (Bhave et al. 2001). Due to water restrictions, the cleaning water should be reused in other parts of the industry before being discharged.

Membrane filtration also requires frequent downtime due to the unpleasant orange juice flavour identified by the taster, known as ultrafiltration flavour, which means more frequent cleaning of the ultrafiltration membranes.

Eliminating the membrane filtration step would therefore be critical to reducing the operating costs of this process step, such as energy, chemicals and water consumption, and would help to reduce the orange juice carbon footprint and improve the sustainability of the process and the quality of the orange juice required.

6 Conclusions

Debittering of orange juice with the whole flavour is possible with the expanded bed vortex flow reactor.

The expanded bed vortex flow reactor is a sustainable technology that allows the elimination of heating and ultrafiltration or centrifugation. It also eliminates ultrafiltration cleaning solutions and reduces chromatography regeneration solutions, improving the sustainability of the process.

Further research is needed to obtain free limonin orange juice with the whole flavour. On the one hand, increasing the residence time by increasing the height of the settling bed and, on the other hand, reducing the ratio of orange juice to adsorbent. In addition, different resins available in the industry need to be tested. When operating the EB-VFR, the side inlet and the rotation rate provide a stable expanded bed which can operate without clogging for long period of time.

Further research into macroporous adsorbent design, such as size range and density will allow industry to use higher axial flow rates that will improve the productivity using EB-VFR.

References

Anspach FB, Curbelo D, Hartmann R, Garke G, Deckwer WD (1999) Expanded-bed chromatography in primary protein purification. J Chromatogr A 865(1–2):129–144. https://doi.org/10.1016/S0021-9673(99)01119-X

Arpanaei A, Heebøll-Nielsen A, Hubbuch JJ, Thomas ORT, Hobley TJ (2008) Critical evaluation and comparison of fluid distribution systems for industrial scale expanded bed adsorption chromatography columns. J Chromatogr A 1198–1199(1–2):131–139. https://doi.org/10.1016/j.chroma.2008.05.044

Bhave R, Jung G, Sondhi R (2001) Potential savings through ceramic membrane caustic reclaim. Food Process (December):49–50

Chandler BV, Johnson RL (1977) Adsorbent materials. US4024334A

Chandler BV, Kefford JF, Ziemelis G (1968) Removal of limonin from bitter orange juice. J Sci Food Agric 19(2):83–86. https://doi.org/10.1002/jsfa.2740190205

Chase HA (1994) Purification of proteins by adsorption chromatography in expanded beds. Trends Biotechnol 12(8):296–303. https://doi.org/10.1016/0167-7799(94)90046-9

Farag MA, Abib B, Ayad L, Khattab AR (2020) Sweet and bitter oranges: an updated comparative review of their bioactives, nutrition, food quality, therapeutic merits and biowaste valorization practices. Food Chem 331. https://doi.org/10.1016/j.foodchem.2020.127306

Fellers PJ (1989) A review of limonin in grapefruit (Citrus paradisi) juice, its relationship to flavour, and efforts to reduce it. J Sci Food Agric 49(4):389–404. https://doi.org/10.1002/jsfa.2740490402

Ghanem F (2012) Juice debittering: basic science, opitimization, and recent advances. In: ASME 2012 citrus engineering conference, CEC 2012, pp 1–13. https://doi.org/10.1115/CEC2012-5701

Giordano RLC, Giordano RC, Cooney CL (2000) Performance of a continuous Taylor-Couette-Poiseuille vortex flow enzymic reactor with suspended particles. Process Biochem 35(10):1093–1101. https://doi.org/10.1016/S0032-9592(00)00143-6

Guadagni DG, Horowitz RM, Gentili B, Maier VP (1977) Method of reducing bitterness in citrus juices. US4031265A
Hanson J (1968) Navel orange juice concentrate and process for its preparation. GB1134280A
Hanson J (1969) Preparation of serum from liquid containing orange juice solids. GB1171969A
Hasegawa S (1984) Strain of Corynebacterium fascians and use thereof to reduce limonoid bitterness in citrus products. US4447456A
Hasegawa S, Brewster LC (1975) Limonoate dehydrogenase and preparation thereof. US3869345A
Hasegawa S, Miyake M (1996) Biochemistry and biological functions of citrus limonoids. Food Rev Int 12(4):413–435. https://doi.org/10.1080/87559129609541089
Hernandez E, Couture R, Rouseff R, Chen CS, Barros S (1992) Evaluation of ultrafiltration and adsorption to debitter grapefruit juice and grapefruit pulp wash. J Food Sci 57(3):664–666. https://doi.org/10.1111/j.1365-2621.1992.tb08066.x
Hubbuch JJ, Heebøll-Nielsen A, Hobley TJ, Thomas ORT (2002) A new fluid distribution system for scale-flexible expanded bed adsorption. Biotechnol Bioeng 78(1):35–43. https://doi.org/10.1002/bit.10170
Ibáñez González MJ, Mazzuca Sobczuk T, Redondo Miranda RM, Molina Grima E, Cooney C, Ortega Ruíz J, Torres Castañeda J (2015) Reactor de flujo de vórtices para la adsorción y purificación de biomoléculas. ES 2536461 B2
Ibañez-Gonzalez MJ, Cooney CL (2007) Studies on protein adsorption in a vortex flow reactor. Process Biochem 42(12):1592–1601. https://doi.org/10.1016/j.procbio.2007.08.012
Ibáñez-González MJ, Mazzuca-Sobczuk T, Redondo-Miranda RM, Molina-Grima E, Cooney CL (2016) A novel vortex flow reactor for the purification of B-phycoerythrin from Porphyridium cruentum. Chem Eng Res Des 111:24–33. https://doi.org/10.1016/j.cherd.2016.03.032
Johnson RL, Chandler BV (1981) A pilot-plant cellulose acetate gel bead column for the removal of limonin from citrus juices. J Sci Food Agric 32(12):1183–1190. https://doi.org/10.1002/jsfa.2740321208
Kataoka K, Doi H, Hongo T, Futagawa M (1975) Ideal plug-flow properties of Taylor vortex flow. J Chem Eng Jpn 8(6):472–476. https://doi.org/10.1252/jcej.8.472
Lee S, Lueptow RM (2001) Rotating reverse osmosis: a dynamic model for flux and rejection. J Membr Sci 192(1–2):129–143. https://doi.org/10.1016/S0376-7388(01)00493-8
Li P, Gomes PF, Loureiro JM, Rodrigues AE (2015) Proteins separation and purification by expanded bed adsorption and simulated moving bed technology. In: Continuous processing in pharmaceutical manufacturing. Wiley
Pant KJ, Cotter PD, Wilkinson MG, Sheehan JJ (2023) Towards sustainable cleaning-in-place (CIP) in dairy processing: exploring enzyme-based approaches to cleaning in the cheese industry. Compr Rev Food Sci Food Saf 22(5):3602–3619. https://doi.org/10.1111/1541-4337.13206
Pritchett DE, Calif O (1957) Debittering navel orange juice. US2816033A
Purewal SS, Sandhu KS (2021) Debittering of citrus juice by different processing methods: a novel approach for food industry and agro-industrial sector. Sci Hortic (Amsterdam) 276(April 2020):109750. https://doi.org/10.1016/j.scienta.2020.109750
Puri A (1984) Preparation of citrus juices, concentrates and dried powders which are reduced in bitterness. US4439458A
Puri M, Banerjee UC (2000) Production, purification, and characterization of the debittering enzyme naringinase. Biotechnol Adv 18(3):207–217. https://doi.org/10.1016/S0734-9750(00)00034-3
Puri M, Marwaha SS, Kothari RM, Kennedy JF (1996) Biochemical basis of bitterness in citrus fruit juices and biotech approaches for debittering. Crit Rev Biotechnol 16(2):145–155. https://doi.org/10.3109/07388559609147419
Redondo Miranda RM, Ibáñez González MJ, Mazzuca Sobczuk T, Molina Grima E (2021) Vortex flow reactor assessment for the purification of monoclonal antibodies from unclarified broths. J Chromatogr A 1655. https://doi.org/10.1016/j.chroma.2021.462502

Schofield T, Miller C (2007) The development and application of resin systems for the treatment of citrus products containing pulp and cloud
Sczechowski JG, Koval CA, Noble RD (1995) A Taylor vortex reactor for heterogeneous photocatalysis. Chem Eng Sci 50(20):3163–3173. https://doi.org/10.1016/0009-2509(95)00176-6
Shaw PE, Baines L, Milnes BA, Agmon G (2000) Commercial debittering processes to upgrade quality of citrus juice products, vol 758. American Chemical Society Library
Swisher HE (1958) Control of navel bitter in dehydrated juice products. EU2834687A
Swisher HE (1972) Limonin free naval orange juice and drink of milk embodying such orange juice. GB1134280A
Taylor GI (1923) Stability of a viscous liquid contained between two rotating cylinders. Philos Trans R Soc A Math Phys Eng Sci 223(605–615):289–343. https://doi.org/10.1098/rsta.1923.0008
USDA Foreign Service Agency (2021) Citrus annual: European Union, pp 1–38. [Online]. Available: https://www.fas.usda.gov/data/european-union-citrus-annual-0
Vega M, Elviro M, Del Valle E-MM, Cerro R, Galán MÁ (2017) Kinetic and mass transfer model for separation of protein using ceramic monoliths as a stationary phase. Chem Eng Commun 204(7):750–760. https://doi.org/10.1080/00986445.2017.1313243
Wethern M (1991) Citrus debittering with ultrafiltration/adsorption combined technology. In: ASME 1991 citrus engineering conference, CEC 1991, pp 48–66. https://doi.org/10.1115/CEC1991-3704
Wilson CW, Wagner CJ, Shaw PE (1989) Reduction of bitter components in grapefruit and navel orange juices with/3-cyclodextrin polymers or XAD resins in a fluidized bed process. J Agric Food Chem 37(1):14–18. https://doi.org/10.1021/jf00085a004

Chapter 8
A Review of Yield Forecasting Techniques and Their Impact on Sustainable Agriculture

Jorge Celis, Xiangming Xiao, Pradeep Wagle, Paul R. Adler, and Paul White

1 Introduction

Sustainable agriculture is one of the main priorities for the international community and a key indicator for a nation's development (Spânu et al. 2022; Shahmohamadloo et al. 2022). It is one of the priorities on the 2030 agenda and main component of the Sustainable Development Goal to "End hunger, achieve food security and improved nutrition and promote sustainable agriculture" (SDG2), highlighting the need of supporting sustainable agriculture and empowering producers. The population growth projections (10 billion people by 2050) underscore the need to increase food production by more than 50% over the next three decades to meet the projected food demand (Hunter et al. 2017). Sustainable agriculture is a multifaceted concept that includes a range of multiple practices and principles (economic, ecological, and social). Because the SDG goal has the aim of increasing productivity to meet future food demands with social and environmental responsibility, we define sustainable agriculture development for this chapter as optimizing agricultural practices to produce higher yields from a given land area, while minimizing environmental

J. Celis · X. Xiao (✉)
Department of Microbiology and Plant Biology, University of Oklahoma, Norman, OK, USA
e-mail: xiangming.xiao@ou.edu

P. Wagle
USDA-ARS, Oklahoma and Central Plains Agricultural Research Center, El Reno, OK, USA

P. R. Adler
USDA-ARS, Pasture Systems and Watershed Management Research Unit, University Park, PA, USA

P. White
USDA-ARS, Sugarcane Research Unit, Houma, LA, USA

A. Melkonyan-Gottschalk, D. Daus (eds.), *Transformation Towards Circular Food Systems*, Greening of Industry Networks Studies 13,
https://doi.org/10.1007/978-3-031-63793-3_8

harm, avoiding land expansion, and enhancing contributions to natural resources and environmental services (Petersen and Snapp 2015).

Despite the pressing need to optimize productivity in sustainable agriculture, there is a gap between sustainable agriculture and efficient logistics that can be addressed with accurate yield forecasting. Modern industrial food systems prioritize supply-side economics, often compromising food quality and the bond between producers, consumers, and the land, sustainable agriculture emerges as a solution to counteract these challenges (Hardeman and Jochemsen 2012). By accurately predicting crop yields, farmers can make informed decisions that align with sustainable practices, ensuring food quality and environmental preservation. Furthermore, precise yield forecasting can streamline the supply chain, leading to significant logistics cost savings. This not only supports large retailers in optimizing their operations but also has the potential to rejuvenate small farm operations, fostering a more sustainable and connected food system. Forecasting crop yields and providing an expectation of the potential harvest amounts is highly relevant to the entire agricultural production chain. This includes farmers, commodity markets, insurances, seed traders, logistics companies, regional authorities, and food aid programs (Schauberger et al. 2020). These stakeholders require outlooks on expected harvests to adapt their management strategies. Techniques for forecasting crop yields are diverse, both in scientific literature and practical application. This article presents a comprehensive overview of these techniques, explaining some of the fundamental concepts that need to be understood and covering the basic physiological principles that should be comprehended to grasp the yield forecasting concept, and a wide range of methods and scales to estimate crop yield which ranges from traditional methods based on observations to mathematical methods, and remote sensing approaches.

It is important to delve deeper into some of the main concepts involved in crop productivity and forecasting. Crop yield commonly refers to the amount of crop produced per unit area of land, typically measured in fresh weight (FW) with some agreed moisture content. The moisture content is calculated based on FW and it varies between commodities and countries (typically 8–16% moisture), while dry weight (DW) is calculated by subtracting the product of fresh weight (FW) and its moisture content from the FW itself (Fischer 2015). In the scope of global agriculture, it is necessary to highlight that this weight typically is given in three main units (1) tonne/metric ton units (1000 kg or 1 Mg or 2204 lbs), (2) short ton units (907 kg or 2000 lbs), and (3) imperial ton (1017 kg or 2240 lbs).

On the other hand, farm yield (FY), often referred to actual yield (Connor et al. 2011) can encompass different spatial scales (e.g., field, farm, and region). It is typically derived from farmers' yield measurements, often taken from harvesters or total FW data, but can also be taken from surveys or local statistics, and for all cases it provides the average yield per unit area (e.g., kg/ha, Mg/ha, bu/ac, and ton/ac) (Fischer 2015; van Ittersum et al. 2013). However, despite being used as a reference for actual yield its accuracy can sometimes be questionable due to issues like poor data collection, uncertain grain mixtures, moisture content variations, and other data processing complications, resulting inaccuracies and discrepancies between FY sources. For example, Sadras et al. (2014) reported discrepancies between NASS-USDA and NRD yield data in Nebraska ranging from 20% to 35% for individual counties. Moreover, the concept of yield potential is the term that is closest related to yield forecasting but in theory is the yield to be expected with the best adapted plant variety, best crop management practices, ideal weather and nutrient conditions, and lack of stressors (Fischer 2015). This term opens the concept of yield gap in which the factors causing differences between the potential yield (usually higher) and actual yield are studied to improve crop productivity (van Ittersum et al. 2013). For this study, yield forecasting is defined as estimating several weeks in advance the total amount of above ground biomass that will be in the field at harvest time (Schauberger et al. 2020).

Inaccurate yield forecasting can cause economic hardship to farmers, commodity markets, insurance companies, seed traders, and logistics companies (Schauberger et al. 2020; Funk et al. 2019). The information in this chapter aims to support the work of scientists and researchers in government agencies, academia, as well as the private sector, as the variables included in our discussion are critical for sustainable agricultural production. An accurate yield estimate is one of the first and main steps towards reducing the yield gap (Hochman et al. 2013). Our main goal is to provide a comprehensive review of biophysical principles needed to be considered to accurately forecast yield, some of the current methods with their limitations, and the economic impact of the yield gap. These serve as a reference to build, improve, or choose a yield forecasting technique that will result in estimates closer to the actual farm yield, and suggestions to consider into the forecasting technique of the site-specific crop variables that reduce potential yield.

To address these components, this chapter contributes to: (1) *better understanding of the physiological and biophysical components to be considered for an accurate yield forecast*, (2) *providing a clear framework of the current methods and limitations for yield estimates*, and (3) *highlighting the main economic impacts on sustainable agriculture because of inaccurate yield forecast as well as the advantages of some innovative approaches*.

2 Biophysical Elements Involved in Crop Biomass Production

Crop yield, defined as the plant production per unit area, is intrinsically connected with plant growth (Lambers et al. 2008; Sadras and Connor 1991). Biomass production, directly linked with plant growth, is driven and dependents on photosynthesis, respiration, and long distance transport (movement of water, nutrients, and organic compounds throughout the plant) (Lambers et al. 2008). The partition of biomass to grain or fruit is a key component to accurately estimate yield (Sadras and Connor 1991). Plant growth can be analyzed in terms of the allocation of resources for above ground biomass (AGB), which provides positive feedback in the production process with new leaves and below ground biomass (BGB) with plant structures like roots, which contributes to favorably impact the crop water status and nutrient uptake (Schulze et al. 1983). Functional methods aim to distinguish between vegetative and reproductive structures (Sadras and Connor 1991), while mechanistic approaches have focused on the partitioning of biomass among vegetative structures (Stutzel et al. 1988) since total growth includes multiple plant structures (e.g., leaves, steams, and roots) and optimal biomass production takes place if all carbohydrates are partitioned into growth of new leaves unless water uptake is insufficient (Schulze et al. 1983). Multiple studies (Hochman et al. 2013; Sadras and Connor 1991; Schulze et al. 1983, 1994; French and Schultz 1984; Stanhill 1986; Stapper 1986) have highlighted the functional bases that photosynthesis provides to modeling biomass production and the direct connection between transpiration and biomass production.

Plant growth depends on the genotype and the environment as these variables directly affect the resource allocation and the growth of useful biomass in crops (Lambers et al. 1990). Growth rate controls physiological processes through its effect on plant demands for carbon, water, and nutrients (Lambers et al. 2008), while environmental conditions like temperature, radiation, and water conditions influence the rate at which some biophysical process take place and the resources demanded from the plant, directly impacting the pattern of biomass allocation. Stomatal opening and closure control the exchange of CO_2 and H_2O between plants and the atmosphere, influencing photosynthesis directly (Campbell and Norman 1998; Dorman and Sellers 1989; Nobel 1999). Photosynthesis is a widely studied process, where solar radiation is absorbed by chlorophyll to provide the energy needed to transform CO_2 into sugars for plant use and to build plant structures. The role of chlorophyll in the photosynthesis process and the fact that the chlorophyll associated with the photosystem II emits fluorescence (long-wave radiation) allows for the use of chlorophyll fluorescence to quantify the effects of stress and photosynthetic performance (Lambers et al. 2008; Taiz and Zeiger 2002). Stomatal conductance captures tradeoffs between photosynthetic traits and hydraulics, and it can be used to quantify carbon gains and hydraulic risk to determine carboxylation capacity and transpiration (Mencuccini et al. 2019; Prentice et al. 2014; Sperry et al. 2017).

Genotype will directly influence the resource allocation of the photosynthates. For instance, the CO_2 assimilation rate or net assimilation rate (NAR) is linked to the total carbon gain via photosynthesis or gross primary production (GPP) and CO_2 used in plant processes (respiration). The carbon assimilation rate directly depends on the leaf area ratio (LAR). These in turn influence the relative growth rate, where crops with higher LAR will grow faster (Lambers et al. 2008). Previous studies (Lambers et al. 2008; Garnier 1992) suggest that NAR is not the main driver of crop growth rate, but leaf mass ratio has stronger influence on it, and this is correlated with higher yield crops allocating less resources to stems (Chapin et al. 1993). Plants directing more resources towards stem growth divert more resources to storage and consequently have slower-growing rates (Garnier 1992).

At the cellular level, vegetation is composed of organic molecules and phytohormones, that are produced in different plant organs and tissues to regulate growth and development on that plant structure (Fahad et al. 2015). Phytohormones serve as

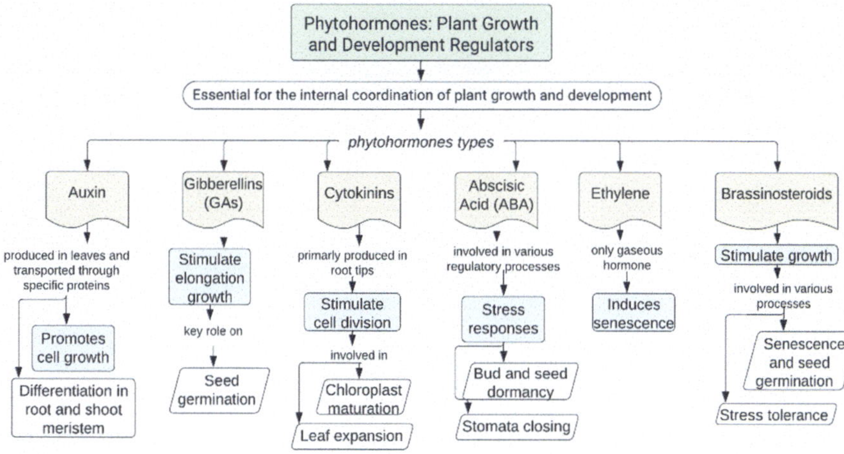

Fig. 8.1 Types of phytohormones and their role in vegetation's response to environmental conditions

a bridge between environmental factors and plant responses, making them crucial to better understand the interactions between crops and environmental conditions (Fahad et al. 2015). Figure 8.1 provides more details into the phytohormones type and their role in the vegetation's response to environmental conditions.

Furthermore, plants have the ability to monitor various aspects of their light environment, adjusting their growth and reproduction accordingly. Phytochrome is a primary system that enables plants to gather information about their light surroundings, helping them adapt and respond accordingly. The primary role of phytochrome is to detect the presence of neighboring plants and mediate a response called shade avoidance. It perceives light's presence, its spectral composition, irradiance level, and direction. Additionally, phytochromes play a pivotal role in various stages of a plant's life cycle, from seed maturation to flowering and senescence. The phytochrome photoconvertible forms are crucial for the high-irradiance response that inhibits germination under prolonged high light exposure (PhyA) and regulating the red to far-red reversibility of seed germination (PhyB) (Lambers et al. 2008).

While plants have inherent growth rates under optimal conditions, natural environments usually present suboptimal conditions. Figure 8.2 displays the plant growth rate responses to specific environmental conditions, given the role of the environment on specific biophysical processes. It is crucial to understand how growth adjusts to these varying environmental factors. Both acclimation (short-term adjustments within an individual) and adaptation (long-term genetic changes across generations) play roles in these responses (Smith 2000). There are many environmental drivers that impact crop rate (Fig. 8.2), including irradiance, temperature, soil water and potential salinity, nutrient supply, soil compaction, wind, and saturated CO_2 environment. The response of vegetation to these conditions and their

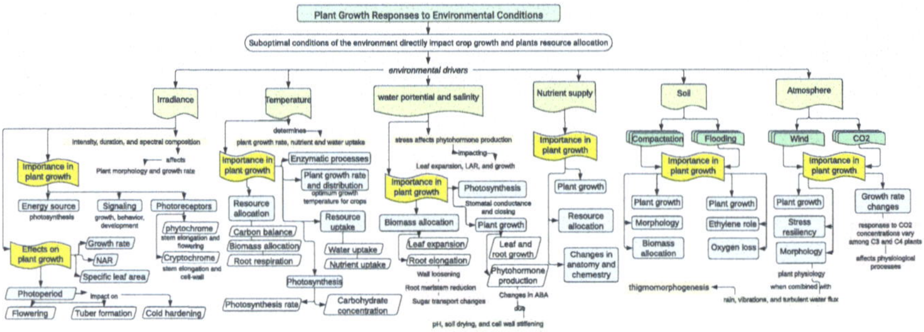

Fig. 8.2 Plant responses to environmental conditions

acclimation will produce different responses and differently affect their growth rate. For example, C3 and C4 crops have different photosynthetic pathways, where some are more efficient under higher CO_2 and temperature conditions. Understanding how plants adjust their growth in response to environmental conditions is essential for predicting plant behavior in changing climates and for optimizing agricultural practices. Both direct resource limitations and anticipatory adjustments play roles in these growth responses, and recognizing the balance between these mechanisms is crucial for accurate modeling and predictions (Poorter et al. 1996; Sharp 2002).

Finally, this happens because crops react to less-than-ideal conditions by slowing their growth rate and reallocating resources to counteract the most limiting factor. For instance, in low light, plants might invest more in leaves, while in nutrient-poor conditions, they might invest more in roots directly impacting the ratios of AGB and BGB production (Brouwer 1963; Garnier 1991). Growth is directly tied to resource availability, and this is known as source-control. For instance, low light would directly reduce photosynthesis, which in turn limits growth (Quail et al. 1995). On the other hand, plants sense unfavorable conditions thanks to phytochrome and adjust growth rates before any resource becomes critically limiting. This proactive approach is known as feedforward control, and it helps maintain a balance to avoid severe resource imbalances (Franklin and Whitelam 2004). Adverse environmental conditions generally lead to reduced growth. Often, these conditions trigger hormonal changes (like shifts in ABA, cytokinins, and gibberellins) that alter growth rates even before the environment directly harms the plant. This feedforward response helps minimize the negative effects on plant growth (Smith 2000). Finally, some yield estimation techniques and computer models in agriculture assume that growth is primarily source-controlled. However, there are situations when feedforward responses are significant, and this directly impacts predictions accuracy about plant responses to changes in environmental conditions and ultimately plant growth (Reski 2006). Figure 8.2 highlights the role of different environmental conditions in plant growth.

3 Main Methods Used and Their Limitations for Yield Forecasting

3.1 *The Concept and Traditional Approaches*

Yield forecasting can be calculated using multiple approaches and the concept is defined in this chapter as estimating crop AGB with time in advance before the harvest. Nonetheless, it is important to clarify that there are semantical distinctions that should be made, when delving deeper into different approaches to estimate yield in advance. This is the case of yield projections that are based on potential scenarios, forecasting typically linked with modeling as a probabilistic statement, and yield prediction that denotates an assertion about a potential yield (Schauberger et al. 2020; Newlands et al. 2014).

Historically, farmers have relied on their experience and observations to predict crop yields, making decisions based on factors such as the upcoming weather (Zuma-Netshiukhwi et al. 2013). Traditional methods of yield forecasting involve expert (agronomist) evaluations of crop status throughout the growing season (Basso et al. 2013). This includes observations and measurements of various parameters like tiller number, spikelet number, stalk number, fertility percentage, and damage from insect and microbial pests. Based on the collected data, yield predictions can be made using regression methods or insights from local experts (Basso et al. 2013; Horie et al. 1992). The earliest approaches to transition into some of the modern methods had a statistical foundation, using rainfall during the growing season in which (Fischer 2015; Fisher 1925) used this method to predict wheat yield and Schultz (1984) (French and Schultz 1984) proposed rainfall-based models to calculate potential yield limits, which are still used in Australia. However, these models can overestimate yields as they do not account for factors like rainfall distribution or access to soil water (Robertson et al. 2005).

In recent times, technological advancements have introduced innovative methods for yield forecasting. Remote sensing, for instance, uses satellite or aerial imagery to monitor crop health and predict yields. Another commonly used approach is the use of crop simulation models (CSMs), which consider various factors, including weather and climate, to provide precise and early yield forecasts. The distinction between forecasts and final estimates lies in their timing; forecasts are made before the crop is fully harvested, while estimates are made post-harvest. The goal of yield forecasting is to provide accurate, scientifically sound, and timely predictions to aid in agricultural decision-making.

3.2 *Statistical Models for Crop Yield Forecasting*

There is a wide range of statistical models available for scientists and enterprises for yield forecasting. Regression models are most often used in the forecasting of yield

Table 8.1 Summary of relevant statistical approaches for yield forecasting

Approach	Advantages	Limitations	Cite
Regression model	Straightforward and easy to understand. Fast identification of general trends in data	May oversimplify complex relationships. Requires a significant amount of historical data	Sridhara et al. (2020)
Meteorological	Can adjust and refine state-level forecasts. Provides a more objective measure	May not capture non-linear trends	Basso et al. (2013); Restrepo-Coupe et al. (2016)
Principal component analysis (PCA)	Aids in reducing over-fitting due to high dimension and interdependency among variables	May not capture all variability with fewer components. Interpretability can be challenging	Sridhara et al. (2020)

(Schauberger et al. 2020). Statistical models leveraging agrometeorological inputs for yield forecasting were some of the earliest approaches (French and Schultz 1984) and to date they maintain high relevance in agricultural research programs (Keong and Keng 2012; Qian et al. 2009). Such models predominantly employ a matrix constructed from historical yield data and various agrometeorological parameters, such as temperature and rainfall (Basso et al. 2013). Through this, a regression equation is formulated, linking yields to one or more agrometeorological factors (Lobell and Burke 2009). The appeal of these models lies in their simplicity, swift computation, and minimal data requirements. However, they are not without limitations. Their predictive scope is confined to the range of values for which they are parameterized, and their outputs, while statistically accurate, might lack agronomic significance. Furthermore, they often overlook the intricate soil-plant-atmosphere continuum relationships among with some of the elements mentioned in Sect. 2, which can be crucial when addressing yield forecast in commercial agroecosystems. Statistical models tend to neglect the timing of water stress during the growth and increased bias occurs during seasons of agricultural drought (Basso and Liu 2019; Rodríguez et al. 2016) which can significantly influence yield outcomes (Basso and Liu 2019; Rodríguez et al. 2016). Efforts are underway to infuse these models with greater depth, such as incorporating crop evapotranspiration or initial soil moisture content via microwave sensing. While these additions enhance predictive accuracy, they still leave certain agronomic queries unresolved. Table 8.1 displays a summary of relevant statistical crop yield forecasting approaches.

3.3 *Process-Based and Machine Learning Models for Crop Yield Forecasting*

Agroecosystems are intricate systems where the resulting crop yield is influenced by numerous interactions such as soil, atmosphere, water, and socio-economic factors. CSMs are designed to consider the continuous interactions between soil, plant, and atmosphere, focusing on the daily changes in biomass and nitrogen accumulation. These models vary in complexity, with some being relatively simple, requiring minimal parameters, while others are intricate, accounting for detailed processes like leaf-level photosynthesis. CSMs have been widely used to assess the agronomic consequences of inter-annual climate variability. They can capture the effects of wet/dry cycles on crop growth, aiding farmers in agronomic planning. Typically used to evaluate agronomic consequences from inter-annual climate variability and to provide farmers with real-time information (Basso and Liu 2019; Asseng et al. 2013; Challinor and Wheeler 2008). However, the adoption of CSMs can be challenging due to the intensive data required for model parameterization. Some of the most relevant process-based models include, but not limited to the Decision Support System for Agrotechnology Transfer (DSSAT) (Jones et al. 2003), the Agricultural Production System Simulator (APSIM) (Hao et al. 2021; McCown et al. 1995), Simulateur multidisciplinaire pour les Cultures Standard (STICS) (Brisson et al. 1998), and The Soil & Water Assessment Tool (SWAT) (Neitsch et al. 2011).

The DSSAT system has been a prominent CSM tool in the field for over 35 years (Jones et al. 2003) and has a modular structure that includes components like weather, soil, plant, soil/plant/atmosphere interface, and management as described in its framework (Jones et al. 2001, 2003; Porter et al. 2000). DSSAT CSM integrates models for 16 diverse crops, and it is designed to stimulate growth, development, and yield on a uniform land area under prescribed or simulated management. Despite the advancements and code integrations that the model has undergone, it still faces several challenges (Jones et al. 2003). These encompass the extensive input data requirements, the intricate complexity within the submodules of each module, the hurdles of integrating new modules as technology progresses and novel data emerges, the substantial computational demands, constraints in its applicability to different crop rotations, challenges in soil representation, and the difficulties in ensuring consistency across different crop models (Jones et al. 2003; Hoogenboom et al. 1999). Nevertheless, the DSSAT system has achieved global success across a broad spectrum of applications. As a result of its impact, subsequent models have been developed using analogous methodologies, but with modifications to address some of the limitations. One notable example is the Land Use Sustainability (Salus) model, which simulates crop yield in rotation, as well as soil, water, and nutrient dynamics (Basso et al. 2012). Other approaches include models based on the ratio of actual to potential evapotranspiration for forecasting (Allen et al. 1998; Condon et al. 2002).

The integration of remote sensing data with CSMs has emerged as a groundbreaking approach in the realm of agricultural research and forecasting. Satellite remote sensing data offers a timely and cost-effective source of information. Several CSMs, such as DSSAT and the World Food Studies (WOFOST) (de Wit et al. 2019), have been used to integrate with remotely sensed data as seen with the integration of the Cropping System Model-CERES-Maize with MODIS leaf area index (LAI) products (Ines et al. 2013; Fang et al. 2008).

However, while these models present numerous advantages, including the ability to simulate core processes like photosynthesis and respiration (Baret et al. 2000), they are not without limitations. The application of crop models is often constrained by uncertainties in their input parameters, such as soil conditions and sowing dates (Ines et al. 2013; Clevers and Van Leeuwen 1996). Moreover, CSMs require extensive parameters and variables as inputs, which often necessitate calibration for specific crop conditions (Ines et al. 2013; Tsuji et al. 1998). Despite these challenges, the fusion of remote sensing and CSMs is increasingly recognized as a promising approach for monitoring growth and estimating yield, offering a comprehensive understanding of crop dynamics, and aiding in effective crop management.

Finally, a common and significant challenge that persists across the diverse range of models. Process-based models necessitate extensive parametrization. This makes their transfer and application in production systems with multiple fields and management practices particularly challenging (Celis 2019). For this reason, they are often considered "point-based" and seen as inadequate for large crop operations and regional/national scale predictions (Bondeau et al. 2007; He et al. 2018) (Table 8.2).

3.4 Remote Sensing-Based Models for Crop Yield Forecasting

Remote sensing (RS) has become an indispensable tool for estimating agricultural yield, offering a panoramic view of vast areas, and facilitating timely evaluations. This technique involves capturing and analyzing data from sensors that detect and measure the energy reflected from the Earth's surface, using the optical data provided like Moderate Resolution Spectroradiometer (MODIS), Landsat, Sentinel-2, and unmanned aerial vehicles (UAV) equipped with spectroradiometers. Once the data is processed, it can provide insights into various crop parameters, including health, growth stage, and potential yield. Several methods have been developed to estimate crop yields using satellite data and some of these methods are summarized in Table 8.3. Hatfield (1983) categorized them into spectral models, albedo models, and thermal models. Horie et al. (1992) identified three primary models: the empirical regression model, the biomass production model based on absorbed solar radiation, and the stress-degree-day model. Meteorological models, which are primarily based on temperature and precipitation data, are also employed since these variables directly influence crop yields (He et al. 2018; White et al. 2011). These

Table 8.2 Summary of relevant process-based approaches for yield forecasting

Approach	Advantages	Limitations	Cite
DSSAT/SALUS	High quality and frequency, widely available, standardized data. Typically have multiple modules and submodules to include different biophysical and physical variables	Requires understanding of various modules within the model and it can be challenging to parametrize over large crop operations	Jones et al. (2003); Hoogenboom et al. (1999); Basso et al. (2012); Loescher et al. (2006)
CSMs	Captures daily changes in biomass and nitrogen accumulation. Can account for effects of wet/dry cycles on crop growth	Intensive data required for model parameterization. Some models may be too complex for practical applications	Asseng et al. (2013); Bondeau et al. (2007)
Combined	Provides real-time information about crops. Combining multiple tools. Offers risk-assessment and monitoring decision support (i.e., yield prophet)	Limited to the capabilities of the agricultural production system simulator (i.e., APSIM)	Hao et al. (2021); McCown et al. (1995)
Remote sensing CSMs	Can update or recalibrate model based on remote sensing data	Added complexity as it requires coupling with a reflectance model	Ines et al. (2013); Clevers and Van Leeuwen (1996)

models often utilize simple regression techniques, with methods based on time-series, changes in space and time, or solely spatial changes (Sridhara et al. 2020; Guan et al. 2016).

The integration of RS with agrometeorological models has become more prevalent given the traditional use of meteorological models to predict yield. The models in programs like Large Area Crop Inventory Experiment (LACIE) (Cheffin and Woolley 1979; Strommen et al. 1979) were statistical, relating yield to air temperature and rainfall (Cheffin and Woolley 1979; MacDonald et al. 1975; MacDonald and Hall 1980). Platforms like NOAA's AVHRR have been instrumental in monitoring large-scale cropping systems and forecasting yield since the 1980s (Tucker et al. 1980). The integration of high spatial resolution data from platforms like Landsat and Sentinel-2 with high-frequency data from MODIS or AVHRR has been a recent trend.

A prevalent approach is the empirical relationship between NDVI (Normalized Difference Vegetation Index) and crop yield. However, a significant limitation of methods relying on statistical relationships between vegetation indices and yield is

Table 8.3 Summary of relevant process-based approaches for yield forecasting

Approach	Description	Cite
Spectral model	Based on the spectral properties of crops	Tucker et al. (1980)
Albedo model	Focus on the reflective properties of crops	Idso et al. (1980b)
Crop models and RS	Integration of RS and CSM for yield forecasting	MacDonald et al. (1975); Strachan et al. (2015)
Physical reflectance models	Simulate interactions between solar beams and canopy	Meek et al. (1984)
Empirical/statistical methods	Build relationships between crop variables and RS or vegetation indices	Clevers and Van Leeuwen (1996); Zurita-Milla et al. (2009)
Integration strategies	Techniques to combine RS and CSM for in-season crop management	Maas (1988)
Temperature-based models	Thermal properties for predictions and models driven temperature measurements like the Stress Degree Day (SDD) and Crop Water Stress Index (CWSI)	Idso et al. (1979, 1980a); Alderfasi and Nielsen (2001); Jackson et al. (1988)
Physiological-based model	Basis on the interactions of physiological principles can have the algorithm background on light use efficiency, surface energy balance principle, or others	Bastiaanssen and Ali (2003)

their empirical nature, often resulting in moderate to low correlation coefficients (Groten 1993; Sharma et al. 1993). These empirical strategies necessitate extensive field measurements to gather yield data, which can be labor-intensive and costly at regional scales. Furthermore, crop yields depend on a multitude of factors beyond just chlorophyll presence and relying on a single vegetation index results in information lacking an agronomic meaning. Rosema et al. (1998) explored an alternative approach for estimating crop yields, leveraging thermal infrared satellite remote sensing and more physically-based models in the South African development community (SADC) region. They found that there is no enough data to apply deterministic models that simulate physiological processes at a regional scale, suggesting a trade-off between physical accuracy and data availability.

The biomass production model proposed by Monteith (1972), which is based on solar radiation and leaf development, has been highlighted in numerous studies for its potential and applicability with satellite data (Celis et al. 2023; Kumar 1982; Zhang et al. 2017). While the physics of this model is simple and direct, its formulation does not demand extensive field data, and it can be integrated with satellite measurements, making it particularly promising for irrigated agricultural conditions. This model, although well-known among global scale ecologists, is seldom tested and applied by the irrigation and water resources community. Despite of using optical and microwave data to estimate biomass through empirical approaches (Blackard et al. 2008; Imhoff 1995; Le Toan et al. 1992), physiological concepts, such as the photosynthetically active radiation (PAR), have been incorporated to make models more reliable. Light use efficiency (LUE) models integrate PAR with other parameters to account for the nonlinear relationship between leaf photosynthesis and absorbed light (Medlyn 1998). They also integrate environmental variables to factor in the crop responses to changes in the temperature, water, and nutrient availability. One relevant component is the maximum light use efficiency (LUE_0), which can be used to estimate biomass and forecast yield through the utilization of a harvest index (Zhang et al. 2017; Xiao et al. 2004; Wu et al. 2021). Other approaches have added surface energy balance models to estimate crop growth and forecast yield in water abundant field conditions, like irrigated fields (Bastiaanssen and Ali 2003).

Finally, other modeling approaches based on canopy temperature measurements, like the Stress Degree Day (SDD) (Idso et al. 1980a; Walker and Hatfield 1979), have been used to forecast crop yield and water status (Hatfield 1983). The increasing availability of high-resolution optical data presents a promising avenue for future yield forecasts.

4 Global Forecasting and Operational Systems

Prominent crop yield forecasting systems worldwide, such as the MARS Crop Yield Forecasting System (MCYFS) of the European Commission and the National Agricultural Statistics Service (NASS) of the United States Department of Agriculture employ various methods to predict crop yields across different spatial scales.

4.1 MARS Crop Yield Forecasting System (MCYFS)

The MCYFS is an intricate tool designed to forecast crop yields across Europe. The system works by breaking down regional crop area statistics across various administrative levels, such as NUTS (Nomenclature of Territorial Units for Statistics) and GAUL (Global Administrative Unit Layers) and mapping them onto a 25 km grid. The primary goal is to consolidate crop indicators from the BioMA-WOFOST crop model resolution to national and sub-national scales for real-time crop yield forecasting, as showcased in MARS Bulletins.

The algorithm underpinning MCYFS integrates crop area statistics from diverse sources like Eurostat and national statistical services. It amalgamates these statistics with land cover data to formulate a hierarchical system of regional weights for each crop annually. These weights represent the proportion of a specific crop area in every spatial unit compared to the national crop area. The foundational model of

MCYFS, the BioMA-WOFOST, operates at the STU (Soil Typological Units) level with 25 km grid meteorological data. The model's indicators are then aggregated based on specific weights, which denote their anticipated contribution to the crop area of a broader administrative unit. This aggregation is a multi-step process, transitioning from the EMU (Elementary Mapping Units) level to the 25 km grid cells and various administrative levels.

However, the report also highlights two primary limitations of the algorithm. Firstly, when statistics are unavailable for a specific NUTS level, land cover data is employed as a substitute. The higher the NUTS level (e.g., NUTS 1), the more the generated weights lean towards land cover distribution, potentially misrepresenting the actual crop distribution. Secondly, the disaggregation of crop area data from NUTS 3 to the grid always relies on land cover data, which remains static year after year, unlike statistical values. This static nature can sometimes lead to inconsistencies between the two data sources. For instance, there are instances where statistical sources indicate area values exceeding the available arable land area at the regional level. This discrepancy can result in the allocation of crop grid areas that surpass the total grid area in some cases. The algorithm addresses these issues through a series of steps, ensuring that the regional weights are coherent and align with country-level statistics. Further details about this approach can be found at Cerrani and López Lozano (2017).

4.2 National Agricultural Statistics Service (NASS)

The USDA NASS prepares early-season forecasts for major crops. To fulfill this mandate, NASS conducts a mix of grower interviews and objective field visits, employing advanced survey sample designs and statistical methodologies. NASS employs the Agricultural Yield Survey (AYS) as a pivotal tool to gather farmer assessments on yield prospects throughout the growing season. This monthly survey targets farmers who have previously reported planting specific crops in parent surveys, such as the March or June Agricultural Surveys. The AYS is designed to cover the yield forecasting needs for most field crops under the NASS program, ensuring comprehensive data collection for all individually published State forecasts. Essentially, the AYS serves as a holistic instrument to guarantee that the entire crop forecasting program is adequately represented.

The AYS adopts a multivariate probability proportionate to size (MPPS) sample design. Unlike the basic probability proportionate to size (PPS) sample design, where units are selected based on the proportion of the commodity of interest compared to other operations, the MPPS design considers multiple commodities or control items to determine a unit's selection probability. This design enhances the targeting of samples to desired commodities, streamlines the sampling process, and is particularly efficient when dealing with multiple commodities. Each commodity of interest with available frame data is assigned a targeted sample size, and a unit's probability of selection is determined by the commodity with the largest proportion of the total and its corresponding sample size.

To draw the AYS samples, respondents from list strata in the March (MAS) and June (JAS) Agricultural Surveys are considered. The small grains (SG) sample, utilized from May to August, is derived from the MAS, targeting respondents who reported having a small grain crop of interest. Conversely, the row crops (RC) sample, used from August to November, is drawn from the JAS, focusing on respondents who reported a row crop of interest. Operations in the largest list strata and non-overlap area tracts are excluded from the AYS sampling. In specific months, such as August for most states, the AYS sample amalgamates operations from both SG and RC samples, necessitating a composite weighting methodology. This approach ensures that all information obtained from AYS responses is maximized, allowing for a more comprehensive and accurate representation of crop yield prospects. Further details can be found at https://www.nass.usda.gov/Education_and_Outreach/Understanding_Statistics/Yield_Forecasting_Program.pdf.

4.3 Limitations and Benefits of These Systems

The MCYFS and NASS are pivotal tools in agricultural forecasting, each with their unique set of challenges. MCYFS faces issues related to data quality and availability, with regional yield statistics not always being as meticulously curated as the national ones (López-Lozano et al. 2015). The system's reliance on land cover data, especially in the absence of specific NUTS level statistics, can lead to potential misrepresentation of actual crop distribution. Additionally, the static nature of land cover data can sometimes result in inconsistencies with statistical values. On the other hand, NASS's forecasting accuracy heavily depends on farmer responses, which can be subjective. The complexity of the multivariate probability is proportionate to the size (MPPS) of sample design and may not always capture every region or crop type's nuances. While national forecasts adeptly capture temporal yield variability, they fall short in detailing spatial variability. For instance, adverse crop conditions in certain regions might be offset by favorable conditions elsewhere (Seguini et al. 2019).

However, the benefits of both systems are undeniable. MCYFS offers granular insights into regional forecasts, capturing spatial variability that national forecasts might overlook. This granularity often leads to more dependable prediction models and less prone to aggregation errors. The system also promotes initiatives to refine data collection and curation protocols, enhancing overall data quality. NASS, with its comprehensive data collection approach through the AYS, ensures a holistic representation of crop yield prospects. Its efficient MPPS sampling design streamlines the process, making it particularly effective for multiple commodities. In essence, despite their challenges, both MCYFS and NASS remain indispensable in the realm of agricultural forecasting.

5 Economic Impact of the Yield Forecasting Accuracy in Sustainable Farming

The Organization for Economic Cooperation and Development (OECD) developed the agri-environmental indicators (AEIs) that were later supported by the European Commission (Spânu et al. 2022). The current 28 AEIs can be used as estimators of the impacts of agriculture and tools to address sustainable agriculture development plans (Briassoulis 2001; Fernandes and Woodhouse 2008; Yli-Viikari et al. 2007). Yield estimates and forecasting play a significant role in evaluating the efficacy of sustainable management practices and in the assessment of practices adopted from the AEIs (van Ittersum et al. 2013; Hochman et al. 2013; Basso et al. 2013). Accurate yield forecast is crucial for commodity investors (Shahmohamadloo et al. 2022; Sadras et al. 2014; Sørensen 2002; Hansen 2005) and in regions where agriculture is a primary economic driver, inaccurate predictions can lead to significant financial losses, food wastage, and market imbalances. Furthermore, inaccurate yield forecasting at regional scales can result in higher crop insurance premiums as some commodities might represent a higher risk under a yield forecast (Goodwin and Ker 1998). These economic repercussions underscore the importance of integrating advanced technology and data analytics to enhance the accuracy of yield predictions. In developing countries, the challenges and economic repercussions of inaccurate yield forecasting are even more pronounced (Jayne and Rashid 2010).

Crop production forecasting is central to making food policy decisions in developing countries (Jayne and Rashid 2010; Barrett 2021). Almost all major food security programs, such as food aid imports, strategic food reserves, granting of licenses for private firms to import or export, local procurement by the government and donors, emergency food assistance, and distribution through social safety net programs, rely on crop forecasts for strategic planning (Jayne and Rashid 2010). These programs are critical to managing food price risks and other humanitarian crises. However, inaccurate crop production estimates can lead to unwarranted or ill-fated policy decisions, causing governments to potentially over-export or import unneeded supplies. Errors in trade decisions lead to accentuated food price fluctuations compared to a system that takes the necessary steps to forecast crop production with precision.

For example, in Africa, there are two primary crop production forecasting systems. The first relies on the large administrative network of Ministry of Agriculture extension workers in rural areas. For instance, in Ethiopia, about 50,000 Development Agents (DA) located throughout the country provide their estimates. These production estimates are then aggregated up to the district, province, and national levels. Countries like Ethiopia, Malawi, Tanzania, Mozambique, and Kenya use this approach annually to produce their official crop production estimates. However, this method has its weaknesses, such as the subjective nature of the estimates and potential biases.

The second approach involves the use of nationally representative annual crop forecast surveys, generally implemented by the National Statistical Office with support from the Ministry of Agriculture. This method, while statistically rigorous, requires careful design and proper weighting, and can be costly. For instance, Zambia's Central Statistical Office spends roughly $380,000 each year to survey about 10,000 households. Discrepancies in national food production estimates can cause great uncertainty in actual production, leading to overshooting or undershooting of actual import, export, and food aid decisions. Such problems exacerbate food price risks and market unpredictability. For example, Malawi, after reportedly having a good harvest, faced challenges with maize prices reaching historic highs, suggesting a potential overestimation of official maize production estimates.

Despite the outgoing efforts to promote sustainable practices in the agriculture supply chain and the pressing need to optimize crop productivity and forecasting, (Schauberger et al. 2020) observed that there is a predominant technical focus on cereals. Figure 8.3 helps to illustrate the amount of efforts displayed as number of studies in yield forecasting for different commodities, where crops used for cereals represent the majority of these efforts. Specifically, wheat, maize, and rice have received the most attention and resources. Following cereals, oilseeds, and certain cash crops used for bioethanol production, such as sugarcane, have been prioritized.

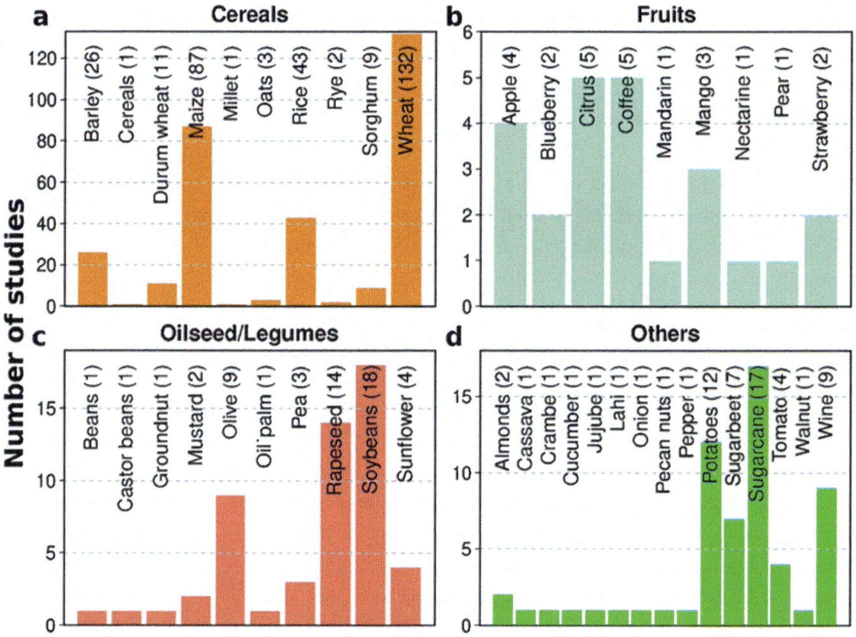

Fig. 8.3 Commodities in which yield forecasting studies (YFS) were conducted. (**a**) Number of YFS per selected cereal crops, (**b**) Number of YFS per selected fruits, (**c**) Number of YFS for oilseed/legumes, and (**d**) Number of YFS in other crops. (Extracted from Schauberger et al. (2020))

Fruits, on the other hand, have seen comparatively less focus, with citrus and coffee being the primary subjects of interest (Fig. 8.3).

Wheat, maize (corn), soybeans, olives, sugarcane, and citrus are essential commodities in the global agricultural market. According to global production data (statista.com) for 2021 and 2022; wheat was largely produced in China, India, Russia, the United States, and France among other nations with a total global production of about 778.6 million metric tons. Maize (corn) is not only consumed directly but is also a crucial ingredient for bioethanol and animal feed. In the 2022/2023 crop year, corn production surpassed other grains with over 1.2 billion metric tons. Soybeans, primarily used for oil extraction and as a protein source in animal feed, have seen consistent demand, especially from countries like China. Olives, predominantly cultivated in the Mediterranean region, are primarily processed into olive oil, a product cherished for its health benefits and culinary applications. Sugarcane is a significant crop in tropical and subtropical regions with Brazil and the USA as some of the largest producers, currently serving as the primary source for sugar production and increasingly for biofuel. Lastly, citrus fruits, including oranges, lemons, limes, and grapefruits, are vital for their nutritional value and widespread consumption both as fresh fruit and juice. Despite that, the current market values for these commodities vary, with maize and wheat leading in terms of production volume. It is essential to note that the yield forecast data provided to different stakeholders trading with these commodities can generate significant economic impact since the yield forecast is used to set the commodity pricing and the number of imports that a nation allows.

Basso and Liu (2019) compare the yield forecasting accuracy for different agricultural commodities through the assessment of average uncertainty levels, as denoted by R^2 values, across three forecasting techniques: statistical-agrometeorology, process-based models, and remote sensing. These techniques were evaluated at three critical growth stages—early, mid, and late season. The findings reveal a discernible pattern; agronomic approaches exhibit lower predictive accuracy early in the season with a gradual increase in R^2 values as the season progresses, suggesting an enhancement in late season forecast precision. Process-based models demonstrate a slight early season advantage over agronomic methods, with a consistent presence of mid-range R^2 values, and present a notable increase in accuracy as the season matures. Remote sensing techniques follow a similar trend, with the most substantial accuracy gains observed in late-season predictions. Collectively, these insights underscore the strategic economic implications of selecting appropriate yield forecasting techniques, which can influence commodity pricing and, by extension, profit margins throughout the agricultural supply chain, thereby affirming the importance of technique selection in relation to the seasonal growth stage for optimizing economic outcomes.

6 Applicability of Improved Yield Forecasting in the Mediterranean's Citrus Industry

Biomass production as mentioned in the previous sections depends on numerous environmental conditions that directly impact crop productivity. Improved yield forecasting allows for more precise planning in the supply chain, and it is critical in sectors with highly perishable products such as citrus farming (Ahumada and Villalobos 2009). Predominantly cultivated in the Mediterranean region for diverse uses, citrus crops benefit greatly from accurate yield predictions (Lezoche et al. 2020). This precision enables producers, distributors, and retailers to adjust their activities based on expected yields, leading to more efficient handling, storage, and distribution of citrus fruits. This reduces the risk of overproduction or underproduction, helping to balance supply with market demand, while ensuring both economic efficiency and sustainability in the citrus industry (Javaid et al. 2022; Nayal et al. 2022).

Citrus fruits in the Mediterranean area, including oranges, lemons, mandarins, and grapefruits, are indispensable for their fresh consumption, juice production, and culinary versatility, enhancing dishes with their zest and juice. They flavor beverages from soft drinks to alcoholic concoctions like Italy's Limoncello and infuse confections and baked goods with zest. Their oils invigorate aromatherapy and cosmetics, while their nutrient profile fortifies health supplements. Improved yield forecasting can bolster the Mediterranean citrus industry, refining supply chain efficiency, reducing waste, stabilizing market prices, and optimizing resources and logistics. Strategic planning for processors and retailers becomes more precise, quality control is enhanced, international trade is better aligned, and small-scale producers gain competitive advantages.

6.1 Enhancing Agricultural Productivity and Efficiency

The improvement of yield forecasting or integration of Artificial Intelligence (AI) in the Mediterranean agriculture is primarily driven by its potential to significantly enhance agricultural productivity and efficiency. AI-driven yield prediction models, such as those utilizing Multilayer Perceptron Neural Networks (MLP-NN) and simple but scalable yield forecasting approaches, are revolutionizing the way farmers approach crop cultivation (European Parliament 2023). These models analyze complex data sets, including indoor climate and microclimate conditions, to predict yields with greater accuracy than traditional methods. This not only leads to increased yields but also improves the quality of produce. For instance, AI and new algorithms have shown promising results in the cultivation of crops like cucumbers and cherry tomatoes, outperforming traditional experience-based methods (Bačiulienė et al. 2023). Nonetheless, stronger efforts should be invested in the modeling of citrus as most of the analysis and yield forecasting tests are performed in oilseed and cereals (Schauberger et al. 2020). The operational decision-making process is also refined through reinforcement learning, optimizing day-to-day climate control and irrigation, thereby enhancing overall crop management. This technological leap contributes to the global competitiveness of the Mediterranean agri-food sector, as higher productivity and quality directly translate into increased revenues for farmers and agribusinesses.

6.2 Cost Reduction and Supply Chain Optimization

Crop growth monitoring and improved yield forecasting have an important role in the supply chain production planning, inventory control, physical distribution; while at the farm level contributes to optimizing resource usage, which are key factor in reducing operational costs in agriculture. By efficiently managing procurement, production (resource management such as such as water, fertilizers, and energy), and distribution, advanced crop growth monitoring and AI technologies help in minimizing expenses, thereby increasing the profitability of agricultural operations (Ahumada and Villalobos 2009; Bačiulienė et al. 2023). This cost-effectiveness is crucial for the economic sustainability of the sector. Accurate yield forecasting allows for better planning and management of the supply chain, reducing waste, and improving the distribution efficiency of agricultural products (Felsberger 2018; Yusuf and Raouf 2013; van der Vorst et al. 2011). This optimization is particularly important in the Mediterranean market, where stable food prices and a reliable supply of products are essential. Crop growth monitoring and AI-driven analytics also aid in managing risks related to climate change, pests, and diseases, which are economically significant for maintaining consistent agricultural productivity (Nayal et al. 2022; Akkaş and Gaur 2022).

Furthermore, more accurate yield forecasts stabilize market prices, allowing producers to anticipate supply needs and market demand more effectively. It also improves resource management, with farmers optimizing labor and input use, which contributes to sustainability. For processors and retailers, reliable yield data informs strategic planning, marketing, and promotions, enhancing overall business operations (Antonaci et al. 2014). Logistics and transportation are optimized, with planning for storage and delivery becoming more efficient, reducing costs and environmental impact (Bačiulienė et al. 2023; van der Vorst et al. 2011).

In the realm of agricultural supply chain (ASC) management, particularly for perishable agri-foods, the integration of accurate yield forecasting into planning models is paramount. The reviewed literature underscores a growing emphasis on perishable agri-foods, driven by their economic significance and food safety concerns (Ahumada and Villalobos 2009). Recent models, ranging from linear programming (LP) to stochastic programming (SP), address various objectives, including cost minimization, revenue maximization, and quality optimization in the context of fresh agricultural products.

The strategic, tactical, and operational planning models incorporate decision variables across different planning horizons, from long-term strategic decisions to immediate operational choices (Ahumada and Villalobos 2009). For instance, Allen and Schuster (2004) and Ten Berge et al. (2000) offer strategic models focusing on capacity and technology selection, while tactical models (e.g., Caixeta-Filho et al. (2002); Ahumada and Villalobos (2011)) optimize crop scheduling and resource allocation. Operational models (e.g., Miller et al. (1997)) delve into the intricacies of short-term planning, crucial for perishable products like those in citrous farming.

Nonetheless, the integration of yield forecasting into these models is critical. Accurate forecasting enables better alignment of production with market demand, thereby enhancing the efficiency of the entire supply chain (Min and Zhou 2002). This alignment is particularly vital for perishable goods, where the shelf life imposes stringent constraints on the supply chain. Models that focus on harvesting decisions (e.g., Ferrer et al. (2008)), exemplify the importance of incorporating yield data to optimize harvest schedules and quality, where models like the vegetation photosynthesis model (VPM) can be incorporated to provide a reliable and adaptable method for crop growth monitoring to support the decision making (Celis et al. 2024).

In summary, the information presented in this chapter highlights the significance of incorporating accurate yield forecasting into ASC planning models for fresh products. This integration not only enhances the efficiency of the supply chain but also aligns production more closely with market demand, a crucial factor for perishable agri-foods like citrus products. The diverse range of models, from LP to SP, and their application across strategic, tactical, and operational planning horizons, underscore the multifaceted approach required to effectively manage the supply chain of perishable agricultural products.

6.3 Market Dynamics, New Business Opportunities, and Regulatory Influence

The implementation of AI and improved remote sensing-based techniques in agriculture are reshaping market dynamics. While it offers substantial economic benefits, there is a potential for larger agribusinesses, which can invest more heavily in these technologies, to gain a competitive edge. This could challenge smaller farmers and enterprises, highlighting the need for policies that support equitable market participation. AI and advanced data analytics open new business opportunities, stimulating economic growth and innovation within the sector. Companies specializing in AI-driven agricultural solutions and services are emerging, offering tailored solutions to enhance agricultural practices. However, the economic impact of these technologies in agriculture is contingent on the regulatory framework set by the European Union (European Parliament 2023). Regulations regarding data ownership, privacy, and AI governance will play a pivotal role in determining the extent of AI adoption, digital agriculture, and overall economic benefits.

Policymakers must therefore carefully consider these aspects to ensure that the benefits of AI and digital agriculture are distributed equitably across the sector, supporting both large and small-scale operations.

Risk management is also refined as producers can better mitigate climate variability and pest-related risks (Brooks and Loevinsohn 2011). Quality control is enhanced, with more precise harvest timings ensuring fruit quality, thus bolstering the Mediterranean citrus reputation globally. Additionally, improved forecasting supports international trade planning, aligning exports with global market demands (Barrett 2021; Lezoche et al. 2020; Ramanathan 2013; Halder and Pati 2011). Small-scale producers particularly benefit from these advancements, as they can plan more effectively and access markets that may have been previously out of reach. In summary, advanced yield forecasting promises more efficient, sustainable, and competitive practices in the Mediterranean citrus industry, with wide-ranging benefits across the supply chain and market dynamics.

7 Summary and Conclusion

This chapter underscores the critical role of accurate yield forecasting in sustainable agriculture, a necessity for achieving the United Nations' Sustainable Development Goal (SDG2) of ending hunger and promoting sustainability. As the global population grows, necessitating a 50% increase in food production by 2050, yield forecasting emerges as a pivotal tool. It bridges sustainable practices with efficient logistics and supply chains, guiding decisions that balance food quality, environmental integrity, and cost-effectiveness.

The chapter delved into the intricacies of farm yield, yield potential, and biomass production mechanisms, essential for understanding regional food productivity and evaluating various crop productivity prediction methods. It highlighted the economic consequences of inaccurate forecasting, which can ripple through the entire agricultural production chain, emphasizing the need for precise estimates to minimize the yield gap and support sustainable production. A comprehensive review of biophysical principles, current forecasting methods, their limitations, and the economic impact of the yield gap is presented. The chapter offers practical guidance on selecting or refining forecasting techniques to align closely with actual farm yields, considering site-specific crop variables that influence potential yield.

Particular attention is given to biomass production and its environmental dependencies, crucial for crop productivity. Accurate forecasting, especially in perishable sectors like citrus farming, is vital for efficient supply chain planning. It allows for the adjustment of activities to anticipated yields, optimizing handling, storage, and distribution, thus reducing overproduction or underproduction, balancing market demand, and ensuring economic efficiency and sustainability.

The chapter concludes by highlighting the integration of AI and precise crop monitoring in Mediterranean agriculture, enhancing productivity and efficiency. The chapter emphasizes the importance of integrating traditional knowledge with innovative approaches, such as remote sensing and CSMs, to achieve precise yield forecasts. However, it notes a gap in research on citrus crops, traditionally overlooked in yield forecasting which has focused on cereals and oilseeds. Accurate forecasting is instrumental in supply chain optimization, cost reduction, and risk management, influencing market dynamics and creating new business opportunities within the agri-food sector. Furthermore, there is a need for a forecasting approach that maintains the agronomic meaning and rigor of CSM, but with the scalability and transferability of remote sensing-based models. AI is shown as an alternative to support this endeavor, but the future of AI adoption and its benefits in agriculture will be shaped by the regulatory environment, underscoring the need for equitable distribution of these advancements. In summary, the chapter calls for a multifaceted approach with agronomic meaning to effectively manage perishable agri-foods' supply chains, integrating accurate yield forecasting for efficiency and market demand alignment.

Acknowledgments This literature review work was supported by research grants from NASA Geostationary Carbon Cycle Observatory mission (GeoCarb Contract #80LARC17C0001), NSF EPSCoR (IIA-1946093), and USDA NIFA (2020-67104-30935).

References

Ahumada O, Villalobos JR (2009) Application of planning models in the agri-food supply chain: a review. Eur J Oper Res 196(1):1–20. https://doi.org/10.1016/j.ejor.2008.02.014

Ahumada O, Villalobos JR (2011) A tactical model for planning the production and distribution of fresh produce. Ann Oper Res 190:339–358

Akkaş A, Gaur V (2022) OM Forum—reducing food waste: an operations management research agenda. Manuf Serv Oper Manag 24(3):1261–1275. https://doi.org/10.1287/msom.2021.1044

Alderfasi AA, Nielsen DC (2001) Use of crop water stress index for monitoring water status and scheduling irrigation in wheat. Agric Water Manag 47(1):69–75

Allen SJ, Schuster EW (2004) Controlling the risk for an agricultural harvest. Manuf Serv Oper Manag 6(3):225–236. https://doi.org/10.1287/msom.1040.0035

Allen RG, Pereira LS, Raes D, Smith M (1998) Crop evapotranspiration-Guidelines for computing crop water requirements-FAO Irrigation and drainage paper 56, vol 300, no. 9. FAO, Rome, p D05109

Antonaci L, Demeke M, Vezzani A (2014) The challenges of managing agricultural price and production risks in sub-Saharan Africa. FAO, Rome. [Online]. Available: https://ageconsearch.umn.edu/record/288979/. Accessed 26 Nov 2023

Asseng S et al (2013) Uncertainty in simulating wheat yields under climate change. Nat Clim Chang 3(9):827–832

Bačiulienė V, Bilan Y, Navickas V, Civín L (2023) The aspects of artificial intelligence in different phases of the food value and supply chain. Foods 12(8):1654

Baret F, Weiss M, Troufleau D, Prevot L, Combal B (2000) Maximum information exploitation for canopy characterization by remote sensing. Asp Appl Biol 60:71–82

Barrett CB (2021) Overcoming global food security challenges through science and solidarity. Am J Agric Econ 103(2):422–447. https://doi.org/10.1111/ajae.12160

Basso B, Liu L (2019) Seasonal crop yield forecast: methods, applications, and accuracies. In: Advances in agronomy, vol 154. Elsevier, pp 201–255. https://doi.org/10.1016/bs.agron.2018.11.002

Basso B, Chou T-Y, Chen C, Yeh M (2012) i-Salus: new web based spatial systems for simulating crop yield and environmental impact. In: Proceedings of international conference on precision agriculture, Indianapolis. ISPAG. [Online]. Available: https://www.ispag.org/abstract_papers/papers/abstract_1283.pdf. Accessed 12 Oct 2023

Basso B, Cammarano D, Carfagna E (2013) Review of crop yield forecasting methods and early warning systems. In: Proceedings of the first meeting of the scientific advisory committee of the global strategy to improve agricultural and rural statistics, FAO Headquarters, Rome, Italy. FAO, pp 1–56. [Online]. Available: https://www.researchgate.net/profile/Mohamed-Mourad-Lafifi/post/Can-anyone-suggest-me-a-dataset-with-multitemporal-images-of-a-Drone-UAV/attachment/59d6432f79197b807799eb17/AS%3A442088039948291%401482413410694/download/Crop_Yield_Forecasting_Methods_and_Early_Warning_Systems_Lit_review.pdf. Accessed 11 Oct 2023

Bastiaanssen WGM, Ali S (2003) A new crop yield forecasting model based on satellite measurements applied across the Indus Basin, Pakistan. Agric Ecosyst Environ 94(3):321–340. https://doi.org/10.1016/S0167-8809(02)00034-8

Blackard JA et al (2008) Mapping US forest biomass using nationwide forest inventory data and moderate resolution information. Remote Sens Environ 112(4):1658–1677

Bondeau A et al (2007) Modelling the role of agriculture for the 20th century global terrestrial carbon balance. Glob Chang Biol 13(3):679–706

Briassoulis H (2001) Sustainable development and its indicators: through a (Planner's) glass darkly. J Environ Plan Manag 44(3):409–427. https://doi.org/10.1080/09640560120046142

Brisson N et al (1998) STICS: a generic model for the simulation of crops and their water and nitrogen balances. I. Theory and parameterization applied to wheat and corn. Agronomie 18(5–6):311–346

Brooks S, Loevinsohn M (2011) Shaping agricultural innovation systems responsive to food insecurity and climate change. Nat Res Forum 35(3):185–200. https://doi.org/10.1111/j.1477-8947.2011.01396.x

Brouwer R (1963) Some aspects of the equilibrium between overground and underground plant parts. Jaarb Van Het Inst Voor Biol En Scheikd Onderz Van Landbouwgewassen 1963:31–39

Caixeta-Filho JV, Van Swaay-Neto JM, Wagemaker ADP (2002) Optimization of the production planning and trade of lily flowers at Jan de Wit company. Interfaces 32(1):35–46. https://doi.org/10.1287/inte.32.1.35.13

Campbell GS, Norman JM (1998) Introduction to environmental biophysics, 2nd edn. Springer, New York

Celis JA (2019) Process based modeling of surface energy fluxes, evapotranspiration, soil moisture, and soil temperature in the US southern plains. Thesis, Department of Geography and Environmental Sustainability

Celis J, Xiao X, Basara J, Wagle P, McCarthy H (2023) Simple and innovative methods to estimate gross primary production and transpiration of crops: a review. In: Digital ecosystem for innovation in agriculture, vol 121. Springer, Singapore, pp 125–156

Celis J, Xiao X, Wagle P, Basara J, McCarthy H, Souza L (2024) A comparison of moderate and high spatial resolution satellite data for modeling gross primary production and transpiration of native prairie, alfalfa, and winter wheat. Agric For Meteorol 344:109797. https://doi.org/10.1016/j.agrformet.2023.109797

Cerrani I, López Lozano R (2017) Algorithm for the dissagregation of crop area statistics in the MARS crop yield forecasting system, vol 21. Joint Research Centre – European Commission, Luxembourg. [Online]. Available: https://agri4cast.jrc.ec.europa.eu/DataPortal/Resource_Files/PDF_Documents/31_rationale.pdf. Accessed 12 Oct 2023

Challinor AJ, Wheeler TR (2008) Crop yield reduction in the tropics under climate change: processes and uncertainties. Agric For Meteorol 148(3):343–356

Chapin FS, Autumn K, Pugnaire F (1993) Evolution of suites of traits in response to environmental stress. Am Nat 142:S78–S92. https://doi.org/10.1086/285524

Cheffin RE, Woolley SK (1979) Large Area Crop Inventory Experiment (LACIE). Evaluation of the LACIE transition year crop calendar model. Lockheed Electronics Co. [Online]. Available: https://ntrs.nasa.gov/citations/19800005335. Accessed 12 Oct 2023

Clevers J, Van Leeuwen HJC (1996) Combined use of optical and microwave remote sensing data for crop growth monitoring. Remote Sens Environ 56(1):42–51

Condon AG, Richards RA, Rebetzke GJ, Farquhar GD (2002) Improving intrinsic water-use efficiency and crop yield. Crop Sci 42(1):122–131

Connor DJ, Loomis RS, Cassman KG (2011) Crop ecology: productivity and management in agricultural systems. Cambridge University Press. [Online]. Available: https://books.google.com/books?hl=en&lr=&id=O2eh7vyvuscC&oi=fnd&pg=PR5&dq=Connor,+D.J.,+Loomis,+R.S.,+Cassman,+K.G.,+2011.+Crop+Ecology:+Productivity+and+Man-+agement+in+Agricultural+Systems.+Cambridge+University+Press,+Cambridge,+UK.&ots=0G1Vnhvgq5&sig=vDxSqOg2PfTdFbNEnaiCLEN3IK8. Accessed 10 Oct 2023

de Wit A et al (2019) 25 years of the WOFOST cropping systems model. Agric Syst 168:154–167. https://doi.org/10.1016/j.agsy.2018.06.018

Dorman JL, Sellers PJ (1989) A global climatology of albedo, roughness length and stomatal resistance for atmospheric general circulation models as represented by the simple biosphere model (SiB). J Appl Meteorol 28:833–855. https://doi.org/10.1175/1520-0450(1989)028<0833:AGCOAR>2.0.CO;2

European Parliament (2023) Directorate general for parliamentary research services. In: Artificial intelligence in the agri-food sector: applications, risks and impacts. European Union, Brussels. [Online]. Available: https://data.europa.eu/doi/10.2861/516636. Accessed 26 Nov 2023

Fahad S et al (2015) Phytohormones and plant responses to salinity stress: a review. Plant Growth Regul 75(2):391–404. https://doi.org/10.1007/s10725-014-0013-y

Fang H, Liang S, Hoogenboom G, Teasdale J, Cavigelli M (2008) Corn-yield estimation through assimilation of remotely sensed data into the CSM-CERES-Maize model. Int J Remote Sens 29(10):3011–3032. https://doi.org/10.1080/01431160701408386

Felsberger A (2018) An overview of measurement approaches for evaluating the impact of the digital transformation on sustainable supply chain management. PhD thesis, Alpen-Adria-Universität Klagenfurt. [Online]. Available: https://netlibrary.aau.at/obvuklhs/content/titleinfo/5464891/full.pdf. Accessed 26 Nov 2023

Fernandes LA d O, Woodhouse PJ (2008) Family farm sustainability in southern Brazil: an application of agri-environmental indicators. Ecol Econ 66(2):243–257. https://doi.org/10.1016/j.ecolecon.2008.01.027

Ferrer J-C, Mac Cawley A, Maturana S, Toloza S, Vera J (2008) An optimization approach for scheduling wine grape harvest operations. Int J Prod Econ 112(2):985–999

Fischer RA (2015) Definitions and determination of crop yield, yield gaps, and of rates of change. Field Crop Res 182:9–18. https://doi.org/10.1016/j.fcr.2014.12.006

Fisher RA (1925) III. The influence of rainfall on the yield of wheat at Rothamsted. Philos Trans R Soc Lond B 213(402–410):89–142. https://doi.org/10.1098/rstb.1925.0003

Franklin KA, Whitelam GC (2004) Light signals, phytochromes and cross-talk with other environmental cues. J Exp Bot 55(395):271–276

French RJ, Schultz JE (1984) Water use efficiency of wheat in a Mediterranean-type environment. I. The relation between yield, water use and climate. Aust J Agric Res 35(6):743–764. https://doi.org/10.1071/ar9840743

Funk C et al (2019) Recognizing the famine early warning systems network: over 30 years of drought early warning science advances and partnerships promoting global food security. Bull Am Meteorol Soc 100(6):1011–1027. https://doi.org/10.1175/BAMS-D-17-0233.1

Garnier E (1991) Resource capture, biomass allocation and growth in herbaceous plants. Trends Ecol Evol 6(4):126–131

Garnier E (1992) Growth analysis of congeneric annual and perennial grass species. J Ecol 80(4):665–675. https://doi.org/10.2307/2260858

Goodwin BK, Ker AP (1998) Nonparametric estimation of crop yield distributions: implications for rating group-risk crop insurance contracts. Am J Agric Econ 80(1):139–153. https://doi.org/10.2307/3180276

Groten SME (1993) NDVI—crop monitoring and early yield assessment of Burkina Faso. Int J Remote Sens 14(8):1495–1515. https://doi.org/10.1080/01431169308953983

Guan K et al (2016) Improving the monitoring of crop productivity using spaceborne solar-induced fluorescence. Glob Chang Biol 22(2):716–726. https://doi.org/10.1111/gcb.13136

Halder P, Pati S (2011) A need for paradigm shift to improve supply chain management of fruits & vegetables in India. Asian J Agric Rural Dev 1(1):1–20

Hansen JW (2005) Integrating seasonal climate prediction and agricultural models for insights into agricultural practice. Philos Trans R Soc Lond B Biol Sci 360(1463):2037–2047. https://doi.org/10.1098/rstb.2005.1747

Hao S, Ryu D, Western A, Perry E, Bogena H, Franssen HJH (2021) Performance of a wheat yield prediction model and factors influencing the performance: a review and meta-analysis. Agric Syst 194:103278. https://doi.org/10.1016/j.agsy.2021.103278

Hardeman E, Jochemsen H (2012) Are there ideological aspects to the modernization of agriculture? J Agric Environ Ethics 25(5):657–674. https://doi.org/10.1007/s10806-011-9331-5

Hatfield JL (1983) Remote sensing estimators of potential and actual crop yield. Remote Sens Environ 13(4):301–311

He M et al (2018) Regional crop gross primary productivity and yield estimation using fused Landsat-MODIS data. Remote Sens 10(3):3. https://doi.org/10.3390/rs10030372

Hochman Z et al (2013) Reprint of 'Quantifying yield gaps in rainfed cropping systems: a case study of wheat in Australia'. Field Crop Res 143:65–75. https://doi.org/10.1016/j.fcr.2013.02.001

Hoogenboom G, Wilkens PW, Thornton PK, Jones JW, Hunt LA, Imamura DT (1999) Decision support system for agrotechnology transfer v3.5. DSSAT Version 3:1–36

Horie T, Yajima M, Nakagawa H (1992) Yield forecasting. Agric Syst 40(1):211–236. https://doi.org/10.1016/0308-521X(92)90022-G

Hunter MC, Smith RG, Schipanski ME, Atwood LW, Mortensen DA (2017) Agriculture in 2050: recalibrating targets for sustainable intensification. Bioscience 67(4):386–391

Idso SB, Hatfield JL, Jackson RD, Reginato RJ (1979) Grain yield prediction: extending the stress-degree-day approach to accommodate climatic variability. Remote Sens Environ 8(3):267–272

Idso SB, Reginato RJ, Hatfield JL, Walker GK, Jackson RD, Pinter PJ Jr (1980a) A generalization of the stress-degree-day concept of yield prediction to accommodate a diversity of crops. Agric Meteorol 21(3):205–211

Idso SB, Pinter PJ Jr, Jackson RD, Reginato RJ (1980b) Estimation of grain yields by remote sensing of crop senescence rates. Remote Sens Environ 9(1):87–91

Imhoff ML (1995) Radar backscatter and biomass saturation: ramifications for global biomass inventory. IEEE Trans Geosci Remote Sens 33(2):511–518

Ines AVM, Das NN, Hansen JW, Njoku EG (2013) Assimilation of remotely sensed soil moisture and vegetation with a crop simulation model for maize yield prediction. Remote Sens Environ 138:149–164. https://doi.org/10.1016/j.rse.2013.07.018

Jackson RD, Kustas WP, Choudhury BJ (1988) A reexamination of the crop water stress index. Irrig Sci 9:309–317

Javaid M, Haleem A, Singh RP, Suman R (2022) Enhancing smart farming through the applications of Agriculture 4.0 technologies. Int J Intell Netw 3:150–164

Jayne TS, Rashid S (2010) The value of accurate crop production forecasts. Michigan State University. [Online]. Available: https://ageconsearch.umn.edu/record/97032/. Accessed 12 Oct 2023

Jones JW, Keating BA, Porter CH (2001) Approaches to modular model development. Agric Syst 70(2):421–443. https://doi.org/10.1016/S0308-521X(01)00054-3

Jones JW et al (2003) The DSSAT cropping system model. Eur J Agron 18(3):235–265. https://doi.org/10.1016/S1161-0301(02)00107-7

Keong YK, Keng WM (2012) Statistical modeling of weather-based yield forecasting for young mature oil palm. APCBEE Procedia 4:58–65

Kumar M (1982) Remote sensing of crop growth. In: Plants and daylight spectrum. Academic Press, London, pp 133–144

Lambers H, Freijsen AHJ, Poorter H, Hirose T, Van der Werf AK (1990) Analyses of growth based on net assimilation rate and nitrogen productivity: their physiological background. In: Lambers H, Cambridge ML, Konings H, Pons TI (eds) Causes and consequences of variation in growth rate and productivity of higher plants. SPB Academic Publishing, The Hague, pp 1–17

Lambers H, Chapin FS III, Pons TL (2008) Plant physiological ecology. Springer Science & Business Media

Le Toan T, Beaudoin A, Riom J, Guyon D (1992) Relating forest biomass to SAR data. IEEE Trans Geosci Remote Sens 30(2):403–411

Lezoche M, Hernandez JE, Díaz M d MEA, Panetto H, Kacprzyk J (2020) Agri-food 4.0: a survey of the supply chains and technologies for the future agriculture. Comput Ind 117:103187

Lobell DB, Burke M (2009) Climate change and food security: adapting agriculture to a warmer world. Springer Science & Business Media

Loescher HW, Law BE, Mahrt L, Hollinger DY, Campbell J, Wofsy SC (2006) Uncertainties in, and interpretation of, carbon flux estimates using the eddy covariance technique. J Geophys Res Atmos 111:D21S90. https://doi.org/10.1029/2005JD006932

López-Lozano R et al (2015) Towards regional grain yield forecasting with 1 km-resolution EO biophysical products: strengths and limitations at pan-European level. Agric For Meteorol 206:12–32

Maas SJ (1988) Use of remotely-sensed information in agricultural crop growth models. Ecol Model 41(3–4):247–268

MacDonald RB, Hall FG (1980) Global crop forecasting. Science 208(4445):670–679. https://doi.org/10.1126/science.208.4445.670

MacDonald RB, Hall FG, Erb RB (1975) The use of Landsat data in a large area crop inventory experiment (LACIE). In: LARS Symposia. IEEE, p 46. [Online]. Available: https://docs.lib.purdue.edu/cgi/viewcontent.cgi?article=1047&context=lars_symp. Accessed 12 Oct 2023

McCown RL, Hammer GL, Hargreaves JNG, Holzworth D, Huth NI (1995) APSIM: an agricultural production system simulation model for operational research. Math Comput Simul 39(3–4):225–231

Medlyn BE (1998) Physiological basis of the light use efficiency model. Tree Physiol 18(3):167–176. https://doi.org/10.1093/treephys/18.3.167

Meek DW, Hatfield JL, Howell TA, Idso SB, Reginato RJ (1984) A generalized relationship between photosynthetically active radiation and solar radiation. Agron J 76(6):939–945

Mencuccini M, Manzoni S, Christoffersen B (2019) Modelling water fluxes in plants: from tissues to biosphere. New Phytol 222(3):1207–1222. https://doi.org/10.1111/nph.15681

Miller WA, Leung LC, Azhar TM, Sargent S (1997) Fuzzy production planning model for fresh tomato packing. Int J Prod Econ 53(3):227–238

Min H, Zhou G (2002) Supply chain modeling: past, present and future. Comput Ind Eng 43(1–2):231–249

Monteith JL (1972) Solar radiation and productivity in tropical ecosystems. J Appl Ecol 9(3):747–766

Nayal K, Raut R, Priyadarshinee P, Narkhede BE, Kazancoglu Y, Narwane V (2022) Exploring the role of artificial intelligence in managing agricultural supply chain risk to counter the impacts of the COVID-19 pandemic. Int J Logist Manag 33(3):744–772

Neitsch SL, Arnold JG, Kiniry JR, Williams JR (2011) Soil and water assessment tool theoretical documentation version 2009. Texas Water Resources Institute. [Online]. Available: https://oaktrust.library.tamu.edu/bitstream/handle/1969.1/128050/TR-406_SoilandWaterAssessmentToolTheoreticalDocumentation.pdf?sequence=1. Accessed 15 Oct 2023

Newlands NK et al (2014) An integrated, probabilistic model for improved seasonal forecasting of agricultural crop yield under environmental uncertainty. Front Environ Sci 2:17. [Online]. Available: https://www.frontiersin.org/articles/10.3389/fenvs.2014.00017. Accessed 11 Oct 2023

Nobel PS (1999) Physicochemical & environmental plant physiology. Academic Press

Petersen B, Snapp S (2015) What is sustainable intensification? Views from experts. Land Use Policy 46:1–10. https://doi.org/10.1016/j.landusepol.2015.02.002

Poorter H, Roumet C, Campbell BD (1996) Interspecific variation in the growth response of plants to elevated CO2: a search for functional types. In: Carbon dioxide, populations, and communities. Academic Press, San Diego, pp 375–412

Porter C, Jones JW, Braga R (2000) An approach for modular crop model development. Int Consort Agric Syst Appl 2440:13

Prentice IC, Dong N, Gleason SM, Maire V, Wright IJ (2014) Balancing the costs of carbon gain and water transport: testing a new theoretical framework for plant functional ecology. Ecol Lett 17(1):82–91

Qian B, De Jong R, Warren R, Chipanshi A, Hill H (2009) Statistical spring wheat yield forecasting for the Canadian prairie provinces. Agric For Meteorol 149(6–7):1022–1031

Quail PH, Boylan MT, Parks BM, Short TW, Xu Y, Wagner D (1995) Phytochromes: photosensory perception and signal transduction. Science 268(5211):675–680. https://doi.org/10.1126/science.7732376

Ramanathan U (2013) Aligning supply chain collaboration using Analytic Hierarchy Process. Omega 41(2):431–440

Reski R (2006) Small molecules on the move: homeostasis, crosstalk, and molecular action of phytohormones. Plant Biol 8(3):277–280. https://doi.org/10.1055/s-2006-924190

Restrepo-Coupe N et al (2016) MODIS vegetation products as proxies of photosynthetic potential along a gradient of meteorologically and biologically driven ecosystem productivity. Biogeosciences 13(19):5587–5608

Robertson MJ, Kirkegaard JA, Robertson MJ, Kirkegaard JA (2005) Water-use efficiency of dryland canola in an equi-seasonal rainfall environment. Aust J Agric Res 56(12):1373–1386. https://doi.org/10.1071/AR05030

Rodríguez JAC, Vergara H, Vanegas DX (2016) Evaluación del nivel de mejora en el monitoreo de sequías a escala regional en Colombia, a partir de indicadores ecohidrologicos basados en teledetección. Rev Tecnol 15(2):15–42

Rosema A, Roebeling RA, Van Dijk A, Nieuwenhuis GJA, Huygen J, Kashasha DA (1998) ACMP agromet and crop monitoring project in the SADC region. BCRS. [Online]. Available: https://research.wur.nl/en/publications/acmp-agromet-and-crop-monitoring-project-in-the-sadc-region. Accessed 12 Oct 2023

Sadras VO, Connor DJ (1991) Physiological basis of the response of harvest index to the fraction of water transpired after anthesis: a simple model to estimate harvest index for determinate species. Field Crop Res 26(3):227–239. https://doi.org/10.1016/0378-4290(91)90001-C

Sadras VO, Grassini P, Costa R, Cohan L, Hall AJ (2014) How reliable are crop production data? Case studies in USA and Argentina. Food Secur 6(4):447–459. https://doi.org/10.1007/s12571-014-0361-5

Schauberger B, Jägermeyr J, Gornott C (2020) A systematic review of local to regional yield forecasting approaches and frequently used data resources. Eur J Agron 120:126153. https://doi.org/10.1016/j.eja.2020.126153

Schulze ED, Schilling K, Nagarajah S (1983) Carbohydrate partitioning in relation to whole plant production and water use of Vigna unguiculata (L.) Walp. Oecologia 58(2):169–177. https://doi.org/10.1007/BF00399213

Schulze E-D, Kelliher FM, Körner C, Lloyd J, Leuning R (1994) Relationships among maximum stomatal conductance, ecosystem surface conductance, carbon assimilation rate, and plant nitrogen nutrition: a global ecology scaling exercise. Annu Rev Ecol Syst 25(1):629–662

Seguini L, Bussay A, Baruth B (2019) From extreme weather to impacts: the role of the areas of concern maps in the JRC MARS bulletin. Agric Syst 168:213–223

Shahmohamadloo RS, Febria CM, Fraser EDG, Sibley PK (2022) The sustainable agriculture imperative: a perspective on the need for an agrosystem approach to meet the United Nations Sustainable Development Goals by 2030. Integr Environ Assess Manag 18(5):1199–1205. https://doi.org/10.1002/ieam.4558

Sharma T et al (1993) Procedures for wheat yield prediction using Landsat MSS and IRS-1 A data. Int J Remote Sens 14(13):2509–2518. https://doi.org/10.1080/01431169308904289

Sharp RE (2002) Interaction with ethylene: changing views on the role of abscisic acid in root and shoot growth responses to water stress: ABA, ethylene and root and shoot growth. Plant Cell Environ 25(2):211–222. https://doi.org/10.1046/j.1365-3040.2002.00798.x

Smith H (2000) Phytochromes and light signal perception by plants—an emerging synthesis. Nature 407(6804):585–591

Sørensen C (2002) Modeling seasonality in agricultural commodity futures: modeling seasonality. J Futur Mark 22(5):393–426. https://doi.org/10.1002/fut.10017

Spânu I-A, Ozunu A, Petrescu DC, Petrescu-Mag RM (2022) A comparative view of agri-environmental indicators and stakeholders' assessment of their quality. Agriculture 12(4):4. https://doi.org/10.3390/agriculture12040490

Sperry JS et al (2017) Predicting stomatal responses to the environment from the optimization of photosynthetic gain and hydraulic cost. Plant Cell Environ 40(6):816–830

Sridhara S et al (2020) Weather-based neural network, stepwise linear and sparse regression approach for rabi sorghum yield forecasting of Karnataka, India. Agronomy 10(11):11. https://doi.org/10.3390/agronomy10111645

Stanhill G (1986) Water use efficiency. In: Brady C (ed) Advances in agronomy, vol 39. Academic Press, pp 53–85. https://doi.org/10.1016/S0065-2113(08)60465-4

Stapper M (1986) Modelling plant growth and development. [Online]. Available: https://publications.csiro.au/rpr/pub?list=BRO&pid=procite:7458e129-d71c-46f5-bace-f9dcbc2710dd. Accessed 10 Oct 2023

Strachan IB, Stewart DW, Pattey E (2015) Determination of leaf area index in agricultural systems. In: Hatfield JL, Baker JM (eds) Agronomy monographs. American Society of Agronomy, Crop Science Society of America, and Soil Science Society of America, Madison, pp 179–198. https://doi.org/10.2134/agronmonogr47.c9

Strommen ND, Sakamoto CM, LeDuc SK, Umberger DE (1979) Development of LACIE CCEA-1 weather/wheat yield models. In: Proceedings of technical sessions, Volumes 1 and 2. NASA. Johnson Space Center. [Online]. Available: https://ntrs.nasa.gov/citations/19800007197. Accessed 15 Oct 2023

Stutzel H, Charles-Edwards DA, Beech DF (1988) A model of the partitioning of new above-ground dry matter. Ann Bot 61(4):481–487. https://doi.org/10.1093/oxfordjournals.aob.a087579

Taiz L, Zeiger E (2002) Plant physiology, 3rd edn. Sinauer Associates, Sunderland

Ten Berge HFM, Van Ittersum MK, Rossing WAH, Van de Ven GWJ, Schans J (2000) Farming options for The Netherlands explored by multi-objective modelling. Eur J Agron 13(2–3):263–277

Tsuji GY, Hoogenboom G, Thornton PK (1998) Understanding options for agricultural production. Springer Science & Business Media

Tucker CJ, Holben BN, Elgin JH Jr, Mcmurtrey JE III (1980) Relationship of spectral data to grain yield variation. Photogramm Eng Remote Sens 46:657–666. [Online]. Available: https://ntrs.nasa.gov/citations/19800051130. Accessed 12 Oct 2023

van der Vorst JG, van Kooten O, Luning PA (2011) Towards a diagnostic instrument to identify improvement opportunities for quality controlled logistics in agrifood supply chain networks. Int J Food Syst Dyn 2(1):94–105

van Ittersum MK, Cassman KG, Grassini P, Wolf J, Tittonell P, Hochman Z (2013) Yield gap analysis with local to global relevance—a review. Field Crop Res 143:4–17. https://doi.org/10.1016/j.fcr.2012.09.009

Walker GK, Hatfield JL (1979) Test of the stress-degree-day concept using multiple planting dates of red kidney beans. Agron J 71(6):967–971. https://doi.org/10.2134/agronj1979.00021962007100060019x

White JW, Hoogenboom G, Wilkens PW, Stackhouse PW, Hoel JM (2011) Evaluation of satellite-based, modeled-derived daily solar radiation data for the continental United States. Agron J 103(4):1242–1251. https://doi.org/10.2134/agronj2011.0038

Wu X, Xiao X, Yang Z, Wang J, Steiner J, Bajgain R (2021) Spatial-temporal dynamics of maize and soybean planted area, harvested area, gross primary production, and grain production in the Contiguous United States during 2008-2018. Agric For Meteorol 297:108240. https://doi.org/10.1016/j.agrformet.2020.108240

Xiao X et al (2004) Satellite-based modeling of gross primary production in an evergreen needleleaf forest. Remote Sens Environ 89(4):519–534

Yli-Viikari A, Hietala-Koivu R, Huusela-Veistola E, Hyvönen T, Perälä P, Turtola E (2007) Evaluating agri-environmental indicators (AEIs)—use and limitations of international indicators at national level. Ecol Indic 7(1):150–163. https://doi.org/10.1016/j.ecolind.2005.11.005

Yusuf I, Raouf A (2013) Reverse logistics: an empirical study for operational framework. Proc Pak Acad Sci 50(3):201–210

Zhang Y et al (2017) A global moderate resolution dataset of gross primary production of vegetation for 2000–2016. Sci Data 4:170165

Zuma-Netshiukhwi G, Stigter K, Walker S (2013) Use of traditional weather/climate knowledge by farmers in the South-Western Free State of South Africa: agrometeorological learning by scientists. Atmosphere 4(4):4. https://doi.org/10.3390/atmos4040383

Zurita-Milla R, Kaiser G, Clevers J, Schneider W, Schaepman ME (2009) Downscaling time series of MERIS full resolution data to monitor vegetation seasonal dynamics. Remote Sens Environ 113(9):1874–1885

Chapter 9
Advanced Data Analytical Methods for Citrus Crop Yield Forecasting

Kiran Anwar, Lotfi Mustapha Kazi-Tani, and Metin Türkay

1 Introduction

In supply chain planning, accurate forecasting of future demands is an important factor. Efficient use of resources and prevention of catastrophic events beforehand are aided by accurate forecasting. Food is a fundamental necessity of life, and it has a long journey from farm-to-fork. Food supply chains aim to manage the process of planning and executing all of the actions that span a collection of the food at farms to final consumption by people at their tables. The citrus supply chain involves many interlinked stages where the ripe fruit is collected at the farm, then transferred to food processing facilities, the products and by-products are processed, and the delivery from food processing facility to markets and finally sales to customer and the consumption by the customer. The efficiency of citrus supply chain is described as the ability to meet the customer demand at the right location, the right time, the right quantity, and the right price. There are critical uncertainties at the supply side (farm) and also the demand side (consumer). In this chapter, we consider the forecasting of citrus yield at the farm side using a variety of statistical learning and machine learning methods. The accurate forecast of the crop yield will allow to achieve great efficiency in citrus supply chains.

Time series analysis serves as a valuable tool for forecasting future values by using historical data and associated underlying trends identification. It is a sequential arrangement of quantitative random values observed at uniformly distributed time intervals. Thus a time series is generally modeled via a stochastic procedure $\mathbf{X(t)}$.

K. Anwar · M. Türkay (✉)
Koç University, Istanbul, Turkey
e-mail: kanwar16@ku.edu.tr; mturkay@ku.edu.tr

L. M. Kazi-Tani
University of Tlemcen, Tlemcen, Algeria

A. Melkonyan-Gottschalk, D. Daus (eds.), *Transformation Towards Circular Food Systems*, Greening of Industry Networks Studies 13,
https://doi.org/10.1007/978-3-031-63793-3_9

In forecasting, our objective is to predict the value of X at time **X(t+h)** by utilizing the information available at time **t** (Davide Burba 2019). Sufficient historical data is required to estimate the parameters of a forecasting model. In the case of a cyclic or seasonal series, an ample number of cycles or seasons is essential to accurately model and capture the variations associated with these patterns. Hence, seasonal series requires a greater quantity of data points compared to nonseasonal ones (Yaffee and McGee 2000). To address missing data in a series, various algorithms are utilized, such as replacing missing values with the average of the preceding values. However, in case of a substantial amount of missing data in economic or social factors, the dataset might not be appropriate for conducting time series analysis; it can compromise the accuracy of the forecasts.

Long-term forecast tends to be less accurate as compared to short-term forecast, and there is always some inaccuracy in forecast (Chopra and Meindl 2016). Decision-makers can make crucial decisions regarding growth strategies and prices when demand or price forecast is made ahead of time. For example, if a business has a reasonably accurate estimate of future product sales, the relevant department can proactively place orders for a quantity of units that closely aligns with the expected demand. Precisely predicting the annual demand for a product is of importance to business professionals, facilitating improved decision-making. A precise forecast enhances the likelihood of the business generating profits when the actual demand aligns with the forecast values. However, if the actual demand falls short of the initial prediction, the business may incur losses. Forecasting future demands serves the purpose of minimizing inventory holding costs while simultaneously enhancing customer satisfaction and overall profitability. Hence, the accuracy of future demand forecasts is crucial for effective decision-making and the optimization of a supply chain system. Our emphasis is on developing an automated forecasting system for time series analysis in this chapter.

We use the quantitative forecasting method and considered Holt's–Winters' model, exponential smoothing, simple moving average, and Box–Jenkins ARIMA and SARIMA models. The objective is to develop an automated and precise forecasting system for enhancing the use of forecasting for future demand planning. This implies that all essential steps for generating a forecast are automatically executed by the system without requiring human judgment or modification. The system is designed to deliver precise predictions within reasonable time frame. The system is user-friendly, requires input data from user only, automatically processes input data, and generates future forecasts as its output. The general data processing involves the following steps:

(1) Verifying data to ensure the absence of negative values
(2) Applying transformations as needed
(3) Testing the performance of all forecasting models on the dataset
(4) Analyzing and comparing the outcomes of all models
(5) Choosing the best-fitting model based on performance evaluation
(6) Generating predictions for future periods

The forecasting system comprises basic and advanced models. Holt's and Winters' model, simple moving average, and exponential smoothing are the basic models, whereas Box–Jenkins ARIMA (Auto-Regressive Integrated Moving Average) and SARIMA (Seasonal Auto-Regressive Integrated Moving Average) models are advanced models for this system. The basic models consider data as yield comprised of systematic and random components. Level, Trend, and Seasonality are the systematic components. The "Level" represents the current de-seasonalized demand in the series. It indicates the baseline or average demand without the influence of seasonal variations (Chopra and Meindl 2016). "Trend" denotes the rate at which the demand for the next period is increasing or decreasing, reflecting a sustained upward or downward movement over time. On the other hand, "seasonality" reveals predictable fluctuations in demand that occur at regular intervals, often associated with specific seasons, months, or other recurring patterns. The random component represents unpredictable variations or noise in the data. In forecasting, the objective is to avoid incorporating the direction of the random component into the forecast. Instead, the objective is to filter out this noise to generate more accurate and reliable predictions.

ARIMA and SARIMA models yield inaccurate future forecast when the data is nonstationary. There is no strict rule about a minimum number of observations for time series data, and a common guideline is to have minimum 30 data points for Box–Jenkins ARIMA and SARIMA. These models are complex often requiring a significant amount of expert judgment and involvement for proper implementation. The long implementation time is attributed to the Box–Jenkins methodology, as it involves numerous iterative procedures to select the most fitting model. These considerations highlight demand for Box–Jenkins methodology to be automated.

The research conducted by Seide, as referenced in (Anvari et al. 2016), primarily emphasizes the automatic conversion of nonstationary data into a stationary form. We draw inspiration from her work and develop our own algorithm for identifying data stationary and conducting stationary testing. Furthermore, within the literature, the selection of the model order is determined through the examination of Auto-Correlation Function (ACF) and Partial Auto-Correlation Function (PACF) plots. Specifically, ACF plots are used for selecting the order of Auto-Regressive (AR) and Seasonal Auto-Regressive (SAR) models, whereas PACF and time plots are utilized for determining the order of Moving Average (MA) and Seasonal Moving Average (SMA) models. The human interpretation of these plots may sometimes result in the selection of incorrect model orders. To eliminate the reliance on human judgment, we opt for an enumeration technique, utilizing computer programs to carry out this step.

Our integrated algorithm comprises the following steps:

(1) Utilizing mathematical modifications on the data and a sequence of statistical tests to create a stationary dataset
(2) Using enumeration process to determine model's order parameters and estimation of coefficient

(3) Bases on diagnostic checks choosing best-fitting model
(4) Concluding the model selection by evaluating accuracy parameters for performance, and
(5) Predicting outcomes and implementing necessary transformations

The rest of this chapter is divided as follows: Related literature review is discussed in Sect. 2, detailed methodology in Sects. 3 and 4 system's platform, and datasets and results are discussed. It is important to note that our system is generic and capable of generating forecasts for any time series dataset.

2 Related Literature

Forecasting time series data is a subject of numerous research work, mainly focusing on forecast for road transport, tourism demand, air traffic, predicting inflation, crop yield, crop products and by-products, electricity use, and gasoline prices. A review of the time series forecasting models for various demands types is discussed in this chapter. In particular, we focus on identifying the relations between demand and other variables. There are three subsections for discussion on univariate time, forecast combination techniques, and automated forecasting systems.

2.1 Univariate Time Series Models

Forecasting methods are categorized into qualitative and quantitative approaches. The former is utilized in case there is a lack of authentic or limited data availability concerning the events to be anticipated. For instance, this approach might be applied in scenarios such as the introduction of a new product. For quantitative forecast, different methods are utilized for the future demand prediction. These methods are time series models, multimode models, neural networks, market surveys, and multiple regression (Anvari et al. 2016). Furthermore, in selecting the optimal model, one must take into account the specific forecasting plan, such as whether it pertains to tactical or strategic decisions, and if the estimates are for the medium to long term or short to medium term (Krasic and Gatti 2009). For example, univariate time series analysis proves to be beneficial in the short-term prediction of the number of passengers (Uysal and Crompton 1985).

Box–Jenking models (ARIMA and SARIMA) have been extensively utilized for the forecasting of air passenger numbers. To forecast outbound air passengers form the United Kingdom, Coshall (1996) used Holt's, Winters', and ARIMA models along with Naive 1 and Naive 2 models. The articles's conclusion was that, with the exception of one scenario, the ARIMA model outperformed all other models. Oh and Morzuch (2005b) studied the models for inbound Singapore tourists arrival proposed by Fong-Lin (1998) and Chan (1993), alongside two of their own. The authors determined that ARIMA models yield accurate results compared to Winters' model in post-sample forecasts. Conversely, Winters' model outperforms ARIMA when utilized for within-sample forecasts.

Emiray and Rodríguez (2003) utilized six different time series models for analyzing data on Canadian air passenger numbers. The models used were SARIMA model, Periodic Auto-Regressive model (PAR), linear model with seasonal unit roots, Auto-Regressive AR(p) and Auto-Regressive AR(p) with seasonal unit roots, and Structural Time Series Model (STSM). Models such as exponential smoothing were not considered. The authors concluded that the accuracy of forecast is significantly influenced by the forecasting horizon and performance parameters.

The studies have explored the combination of both quantitative statistical and nonstatistical methods to predict future horizons. Mwakalonge and Badoe (2014) concluded that relying solely on data acquired for recent time intervals for forecasting can lead to variations in low- and high-frequency values beyond what is anticipated, due to instability of travel behavior and economic conditions. Their focus was on an urban region, and they gathered multiple cross-sectional datasets for this area at various time periods. The forecast for it was made in two categories: In the first category, they aggregated from different periods all the model parameters, whereas in the other category, the data from various points is merged. The research work's conclusion was that utilizing multiple datasets provides an opportunity to develop improved models. State space modeling framework is proposed by De Livera et al. (2011) to deal with dual calendar effect, complex seasonal patterns, and multiple seasonal periods. In addition to Exponential Smoothing, the framework includes ARMA error correction, Box–Cox transformations, and the incorporation of time-varying coefficients through Fourier representations. The authors claim that their framework is proficient in modeling both linear and nonlinear time series effectively. The framework is capable of generating accurate out-of-sample forecasts with minimal parameter estimations, in contrast to Holt–Winters' exponential smoothing model as well.

Omane-Adjepong et al. (2013) acquired monthly data spanning nearly 40 years from the Ghana Statistical Service. Holt–Winters' and SARIMA models were utilized to generate out-of-sample forecast. It was concluded that the SARIMA model is best-fitting model as compared to other models and is the most suitable for short-term out-of-sample forecasts for Ghana's inflation. Nguyen and Hansen (2017) produced short-term forecasts for load demand utilizing ARIMA and SARIMA models. It was determined that SARIMA models outperform ARIMA model. However, it was noted that as the future horizon increases, SARIMA model's accuracy also reduces.

2.2 *Forecast Combinations*

"Forecast Combinations" is an alternative approach to single model's forecast. In case of non-reliable past data, forecast combination is utilized. In this scenario, the number of observations is limited, and/or individual forecasts are not specified appropriately (Timmermann 2013). This concept by Bates and Granger (1969) increases the accuracy of forecasts. The concept is expected to enhance forecasting performance, particularly when there is no correlations in combined forecasting models (Brown and Murphy 1996).

Bougas (2013) utilized variance–covariance methodologies and simple averaging as combination forecasts for air travellers data in Canada. The methods involve the use of a weighted average of single model forecasts, but the computation procedure for the weights varies. In simple averaging, predictions obtained from a single model combination are assigned equal weight. In contrast, different weights are assigned to each individual model and then based on its past performance combined in the variance–covariance methodology. The author claims utilizing combination models enhances forecast accuracy compared to using single models. While ARIMA and SARIMA may outperform in certain cases, the combination model consistently outperformed its corresponding single model, even when the single model exhibited worse performance.

2.3 Automated Forecasting

In this section, several research works centered around automatic forecasting systems are reviewed. Box–Jenkins methodology is compared with neural network by Sharda and Patil (1992). A commercial program AUTOBOX (Inc 1998) is used for Box–Jenkins methodology. The software can receive input data, go through the phases of the Box–Jenkins methodology, and determine the most suitable model. The experimental work concluded that the simple training neural network methodology is generating better outcomes for 50 series, whereas AUTOBOX outperformed it in 22 series. For long-term forecasts, the Box–Jenkins methodology exhibited slightly less accuracy compared to short-term future horizons. However, it is worth noting that the AUTOBOX program is not open source, thus preventing a direct academic comparison.

The best-fitted model through automated procedures for model selection and the Box–Jenkins approach is utilized by Hwang et al. (2012), thus developing an Automated system for Time series Material cost Forecasting, known as ATMF. Their proposed system, ATMF, proves valuable in the construction industry for decision-makers, providing accurate out-of-sample forecasts and trend predictions specifically for construction material costs. This system is specifically beneficial for accurately predicting material costs; however, no testing was conducted for other time series within its scope.

Multiple time series models are compared by Burkom et al. (2007) for bio-surveillance forecasting. These models are Box–Jenkins ARIMA model, nonadaptive regression model, adaptive regression model, and generalized exponential smoothing. The research indicates that the model comparison is automated, but for ARIMA, automation involves a preselected model. Additionally, regarding smoothing constants in the Holt–Winters' approach, specific values were set through a grid search using a clustering approach.

By utilizing R-trees and the delay coordinate embedding technique, Chakrabarti and Faloutsos (2002) constructed an automated forecasting system for predicting future time series data. The authors utilized fractals in the development of their system, introducing a novel approach involving the use of lag plots. They claim that their method is characterized by speed, reliability, and accuracy. There is a built-in package in R by Hyndman and Khandakar (2007) namely "forecast package for R." The package has the capability to forecast in both state space models and ARIMA models. However, it lacks ability to autonomously select and test all models. The user is required to manually code the testing of each model and then comparing results.

3 Methodology

In this section, we outline the strategies employed in developing our automated forecasting system, including implementation details of various time series forecasting models. A brief description of the preprocessing for error estimation methods and data modification is provided as well. This section primarily includes our approach for implementing and optimizing model parameters. For a comprehensive understanding of the basic models usage, accompanied by relevant examples, we recommend referring to Chopra and Meindl (2016) book. Figure 9.1 illustrates the flow of the forecasting system.

3.1 Null or Negative values

Prior to feeding the data into forecasting models, the initial step involves checking for the presence of null or negative values and assessing the length of series. We find the null and negative values and length of the series using built-in commands in Python. In the presence of any negative or missing values, they are substituted with 0. Recognizing that not all null values necessarily imply missing data, we replace these null values with a number within the range of 0–1.

The length of the series plays a crucial role in partitioning it into training and testing data. For the forecasting models, the data values are segmented into training and testing datasets. As previously noted, the minimum requirement for the training data is to have at least 30 observations. In cases where the data consists of fewer than 30 observations, our system undergoes testing exclusively with and Holt's and Winters' model, Simple Moving Average, and Exponential Smoothing. The training data is supplied to the models within the system, where optimized parameters for each model are estimated. Subsequently, the performance of these models is compared using the test data. The forecasting models produce predictions with the

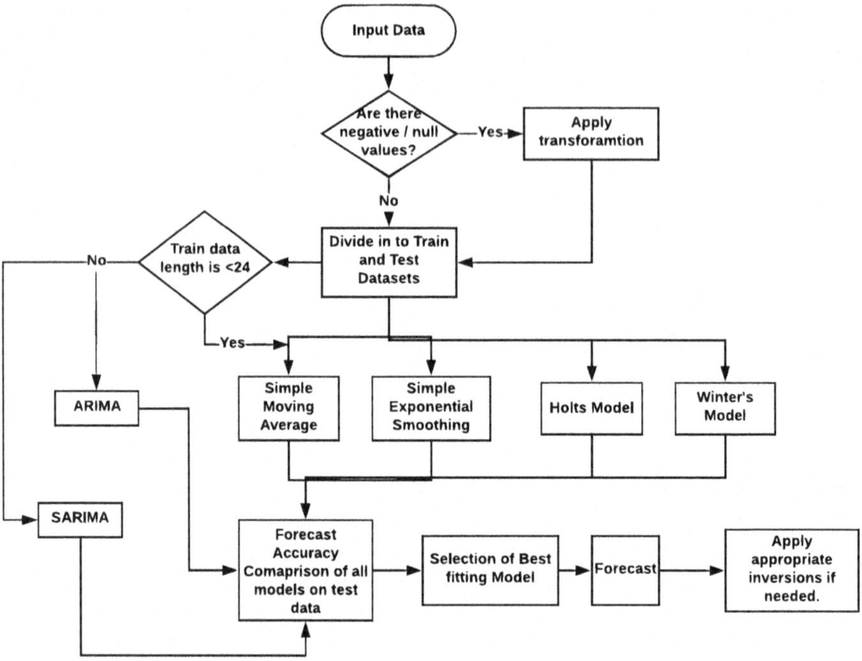

Fig. 9.1 Flowchart of automated multi-model forecasting system

same length as the test data. Subsequently, the predictions from each model are compared with the test data, and a forecasting error is computed. The model with the minimum error is considered the best-fitting model. This selected model is then used to forecast future yield values.

3.2 Error Estimation Methods

Various measures of forecasting errors exist in the literature to assess the accuracy of models, with no universally favored metric. The choice of models varies depending on the selection criteria used (Oh and Morzuch 2005a). We use RMSE for ARIMA and SARIMA in order to choose the best estimated model's order. For reporting and comparison purposes and comparisons of the models we use MAPE error. The MAPE error expressed in percentage terms facilitates a more intuitive understanding of the magnitude of the error. AIC (Akaike Information Criterion) and SBC (Schwarz Bayesian Criterion) are both minimum information criteria utilized for evaluating a parsimonious fit (Yaffee and McGee 2000). These accuracy measures have a tendency to over-parameterize and penalize the number of terms in the model. Hence, we prefer to utilize RMSE for model selection, while using MAPE for reporting purposes. Equation 9.1 shows RMSE calculation:

$$RMSE = \sum_{t=1}^{i} \sqrt{\bar{e}_t} \tag{9.1}$$

3.3 *Basic Time Series Forecasting Models*

The subsequent notation is used to explain the components of basic models: The components are updated, after each yield observation:

1. L_t: *Level estimate at the end of period t*
2. T_t: *Trend estimate at the end of period t*
3. S_t: *Seasonal factor estimate for period t*
4. F_t: *Forecast for period t*
5. D_t: *Actual yield in period t*
6. E_t: *Forecast error in period t*

The following subsections provide details of times series forecasting models.

3.3.1 Simple Moving Average

In instances where the yield lacks observable seasonality or trend, we opt for the use of a simple moving average. **Level** is the only systematic component of yield in this case. The average yield over most recent **N** periods is computed for estimating the systematic component level at period t. Equation 9.2 is used for evaluating it:

$$L_t = (D_t + D_{t-1} + - - - + D_{t-N+1})/N \tag{9.2}$$

The forecast at period t corresponds to the level calculated for period t. Further forecasts are calculated by revising the estimate for level $t + 1$, $t + 2$, etc. (Eq. 9.4).

$$F_{t+1} = L_t \tag{9.3}$$

$$L_{t+1} = (D_{t+1} + D_t + - - - + D_{t-N+2})/N \tag{9.4}$$

$$F_{t+2} = L_{t+1} \tag{9.5}$$

The parameter "N" is optimized for the training data by assessing the model's performance within the range $N = [2, 8]$. The value of "N" for which the RMSE values are minimized is considered the optimal choice. The results are saved, and train data is sent to Simple Exponential Smoothing model.

3.3.2 Simple Exponential Smoothing

Level is the only systematic component of yield in this case as well. The initial value for the level is determined by computing the average of all observations in the training data. The first forecast for period **t+1** will be equal to the first level estimated (Eq. 9.6. All the future forecasts are computed by revising the estimates of level at period **t+1** by using Eq. 9.7.

$$F_{t+1} = L_t \tag{9.6}$$

$$L_{t+1} = \alpha D_{t+1} + (1 - \alpha) L_t \tag{9.7}$$

The smoothing constant for level is α in Eq. 9.7. The value ranges between 0 and 1 for α. The optimization process entails training our training data for all potential values of α. The optimal value for α is the one that yields the minimum RMSE and is considered best-fitting model for simple exponential smoothing. The forecast for period >**t+1** is revised in Eq. 9.8:

$$F_{t+2} = L_{t+1} \tag{9.8}$$

3.3.3 Holt's Model

Trend and level are the systematic components of yield in this model. A linear regression between yield D_t and the time period t is run to calculate the initial estimate of trend T and level L. The linear regression is of the form (9.9):

$$D_t = at + b \tag{9.9}$$

In Eq. 9.9 the initial trend T_0 is the slope a. At Period $t = 0$, the initial estimate of Level L_0 is equal to the constant b. For further periods the forecast is calculated using Eq. 9.10

$$F_{t+1} = L_0 + T_0,\ F_{t+n} = L_t + nT_t \tag{9.10}$$

The revised level and trend are calculated using Eq. 9.11 and 9.12. The forecast is then calculated using Eq. 9.13

$$L_{t+1} = \alpha D_{t+1} + (1 - \alpha)(L_t + T_t) \tag{9.11}$$

$$T_{t+1} = \beta(L_{t+1} - L_t) + (1 - \beta)T_t \tag{9.12}$$

$$F_{t+2} = L_{t+1} + T_{t+1} \tag{9.13}$$

The smoothing constants α and β in Holt's model range between 0 and 1. The optimization of our Holt's model involves training the training data for all possible values of α and β. The model that results in the minimum RMSE values is considered the best-fitting Holt's model.

3.3.4 Winters' Model

In this model, all three systematic components of yield are considered. Additional parameter periodicity **p** is utilized in this model. It signifies the number of periods after which a repetition of seasonal cycles occurs. For instance, if the yield is measured on a yearly basis, then $p = 12$. To estimate the level and trend in the Winters' model, the initial step involves de-seasonalizing our training data. Using Eq. 9.14 we determine if periodicity is even or odd.

$$\bar{D}_t = \begin{cases} \left[D_{t-(p/2)} + D_{t+(p/2)} + \sum_{k=t+1-(p/2)}^{t-1+(p/2)} 2D_k\right]/(2p), & \text{for} p \text{ is even} \\ \sum_{k=t-[(p-1)/2]}^{t+[(p-1)/2]} D_k/p, & \text{for} p \text{ is odd} \end{cases} \tag{9.14}$$

The initial values for the level and trend are estimated by conducting a linear regression between the de-seasonalized yield and the time period t. Since the original training data is nonlinear and encompasses seasonality, the linear regression is performed on the de-seasonalized training data. The relationship is expressed in the form of Eq. 9.15:

$$\bar{D}_t = L + Tt \tag{9.15}$$

From the de-seasonalized train data, the seasonal factors are calculated using Eq. 9.16. Using periodicity and seasonal factors of de-seasonalized train data, the final seasonal factors are calculated by Eq. 9.17. These seasonal factors are then used for forecast calculation.

$$\bar{S}_t = D_t/\bar{D}_t \tag{9.16}$$

$$S_k = \sum_{j=0}^{r-1} \bar{S}_{(j+p+k)} \text{ where } 1 \le \text{k} \le \text{p} \tag{9.17}$$

The r in Eq. 9.17 refers to seasonal cycles. Using Eqs. 9.15 and 9.17 we calculate the forecast in period t as follows:

$$L_{t+1} = \alpha(D_{t+1}/S_{t+1}) + (1 - \alpha)(L_t + T_t) \tag{9.18}$$

$$T_{t+1} = \beta(L_{t+1} - L_t) + (1 - \beta)T \tag{9.19}$$

$$S_{t+p+1} = \gamma(D_{t+1}/L_{t+1}) + (1 - \gamma)S_{t+1} \tag{9.20}$$

The smoothing constants for level, trend, and seasonal factors are α, β, and γ, respectively. These constants, ranging between 0 and 1, are crucial parameters in Winters' model. The optimization process for Winters' model involves finding the best-fitting values for these constants. The results for which RMSE is the least are optimal values for α, β, and γ.

3.4 Advanced Time Series Models

We classify Box–Jenkins ARIMA and SARIMA models as advanced time series forecasting models.

3.4.1 Box–Jenkins ARIMA

The Box–Jenkins approach is valuable for forecasting future values in a time series. These time series values are linked or statistically dependent (Bowerman and Connell 1979). Box–Jenkins ARIMA methodology works best on stationary data. ARIMA is a combination of AR, MA models and a parameter **d** called differencing. There are four fundamental steps in applying Box–Jenkins ARIMA models: **(1)** Identification, **(2)** Estimation, **(3)** Diagnostic checking, and **(4)** Forecasting. Figure 9.2 represents the flowchart of automated Box–Jenkins ARIMA forecasting model.

In the initial step of 'Identification', the process examines if stationary data is provided. If the data is nonstationary, it undergoes conversion into stationary data. Alternatively, in the Identification step, the model determines Moving Average (MA) and Auto-Regressive (AR) order. The second step "Estimation" involves estimating the coefficients for the AR and MA components. The third step "Diagnostic checking" verifies whether the model chosen in the earlier steps is accurate and the best fit. If the model is deemed the best fit, the Box–Jenkins ARIMA model is then used to forecast future values.

The mathematical equation for ARIMA model is 9.21

$$X_t = \sigma + \phi_1 X_{t-1} + \phi_2 X_{t-2} + \ldots + \phi_p X_{t-p} + \in_t - \theta_1 \in_{t-1} - \theta_2 \in_{t-2} - \theta_q \in_{t-q} \tag{9.21}$$

X_t is the value to be forecast-ed at time **t**. The AR model defines the term **p**. $X_{t-1}, X_{t-2}, \ldots, X_{t-p}$ are the previous observed values at time $t-1$, $t-2$ till $t-p$. The AR coefficients ϕ_1,ϕ_2,...ϕ_p define the term σ. The MA coefficients are the terms θ_1, θ_2,... The q term is defined by the MA model. $\in_t$, $\in_{t-1}$,... are the residuals terms. The original Box–Jenkins methodology is shown in Fig. 9.3.

In our proposed automated forecasting system, we employ the Box–Jenkins ARIMA methodology. The primary objective is to eliminate the dependency on human logical reasoning by using computer algorithms. Instead of manually

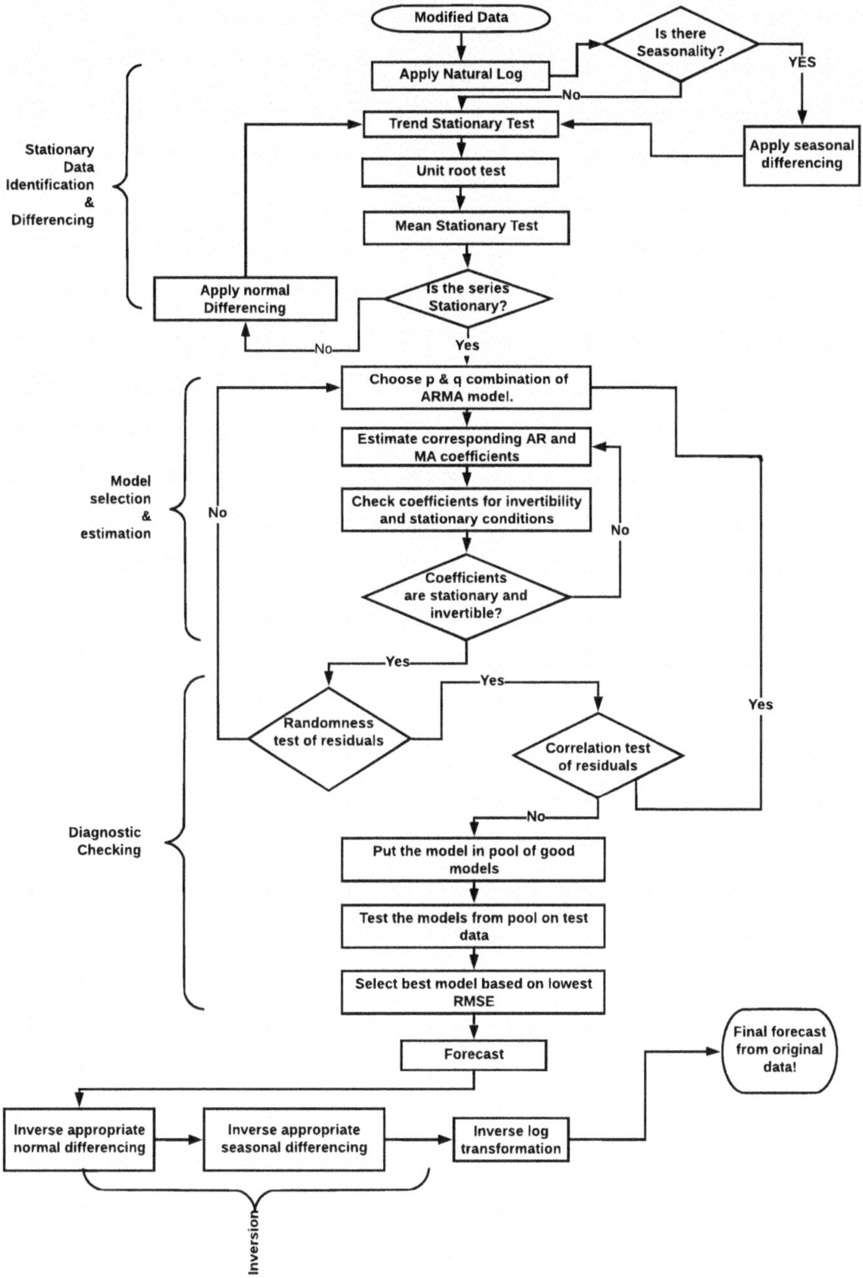

Fig. 9.2 Flowchart of automated Box–Jenkins ARIMA forecasting model

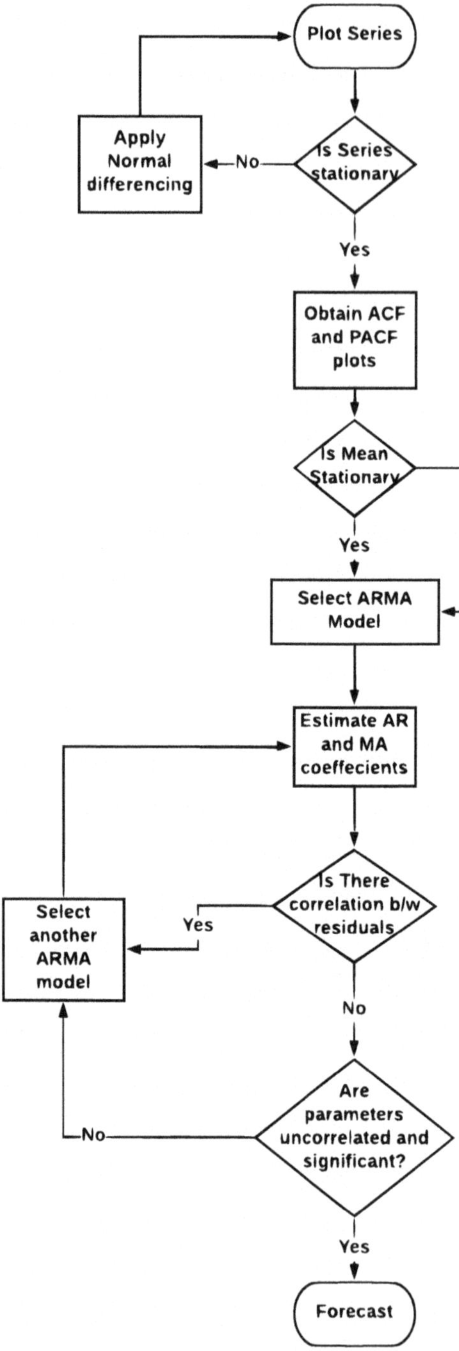

Fig. 9.3 The conventional Box–Jenkins ARIMA model

selecting model parameters **p, d, and q** based on ACF and PACF plots, we enhance accuracy by systematically exploring all possible models. Furthermore, the best-fitting model is chosen based on its performance accuracy. The algorithm steps details are followed next.

3.4.2 Stationary Data Identification and Differencing

The original data has been preprocessed to handle missing and null values. The assessment of nonstationary data often involves human graphical judgment through Auto-Correlation Function (ACF) and Partial Auto-Correlation Function (PACF) plots. For instance, if the ACF plot exhibits a significant spike or decays slowly, reaching unity, it is nonstationary data (Hamed 1999). Our automated process assesses data stationarity through the application of the following tests and transformations:

- Any outlier/variance from the training data is mitigated by applying the natural logarithm. This aids in reducing high variance, and even in cases of low variance, we have observed that the forecast remains accurate.
- Seasonal differencing is required, and it is determined using the OSCB test (Bougas 2013). This test produces an output of either 0 or 1. A result of 1 indicates that the changes in seasonal patterns are significant enough to warrant appropriate seasonal differencing. Conversely, an output of 0 suggests that no seasonal differencing is necessary.
- A stationary data does not exhibit any trend. To check if the train data has any trend, we use Kwiatkowski, Phillips, Schmidt, and Shin test (Kwiatkowski et al. 1992). We examine the significance of the p-value. If the p-value from the test result is greater than 0.05, then normal differencing is applied. Otherwise, we proceed to the next test.
- The mean stationary test is implemented to assess if the mean is time-dependent, utilizing the T-test for this purpose. If dependence is identified from the T-test result, normal differencing is applied.
- A unit root test is conducted to examine if the data is stochastically nonstationary. The Augmented Dickey Fuller test is utilized for this purpose. If the p-value is found to be insignificant, normal differencing is applied.

All tests are automated and do not require human judgment. The result is a stationary training dataset. The progression of applied differencing transformations is tracked, and over-differencing is prevented by imposing a threshold limit, typically set at 3.

3.4.3 Identification and Estimation

Many studies highlight that the identification and estimation of ARIMA models should be an iterative process involving selection based on human judgment, coefficient estimation, reassessment, and re-selection (Anvari et al. 2016). However, this method is susceptible to errors and challenging to implement as a computer program (Lu and Abourizk 2009).

To mitigate uncertainties and errors associated with human judgment, we utilize enumeration approach to select the orders of Auto-Regressive (AR) and Moving Average (MA) components (denoted as p- and q-values). We utilize the observation that the order is typically less than or equal to two, as suggested by Lu and Abourizk (2009). An order exceeding 2 can lead to over-fitting of the series. In our investigation, we explore up to order 3 and examine all possible 15 combinations of MA and AR models.

3.4.4 Estimation of AR and MA Coefficients

After determining the order, we proceed to estimate the Auto-Regressive (AR) and Moving Average (MA) coefficients using the Yule–Walker equation. In Python, we utilize the statsmodels library and its built-in functions for this purpose. It is crucial to verify whether the estimated coefficients of the AR and MA models are stationary and invertible, respectively. This ensures:

- To be stationary: The sum of AR coefficients should always be less than one (Lu and Abourizk 2009).
- To be invertible: The sum of MA coefficients should always be less than one (Lu and Abourizk 2009).

If, in a particular case, the sum is not less than 1, we reestimate the parameters using the lagged version of the training data. The constant term σ in Eq. 9.21 is calculate as $1 - \phi_1 - \phi_2 \ldots$

3.4.5 Diagnostic Checking

During diagnostic checking, various tests are conducted to assess how well the selected ARIMA model fits the data. Each ARIMA model undergoes tests for randomness and correlation of residuals. If a model successfully passes these tests, it is added to the pool of considered good models:

- The residuals of the best model are anticipated to be randomly distributed (Hamed 1999). To verify this, we use the Runs test, utilizing the statsmodels library in Python (Perktold et al 2009).
- For a well-fitted model, there should be no auto-correlation among the residuals (Hamed 1999; Krasic and Gatti 2009). To examine auto-correlation, we utilize the Ljung–Box test (lbq test), implemented in Python.

If the tests are not rejected, the model is added to the pool of good models. Otherwise, another combination of p and q for the model is chosen, and this process continues until all 15 combinations are checked. Next, all the models from the pool are evaluated on our test data. The final model is the one that exhibits the minimum RMSE value, and this selected model is utilized to forecast future periods.

The forecast undergoes an inversion process before reaching the final forecast. In cases where normal and seasonal differencing has been applied, inverse differencing is performed on the forecast. Finally, inverse log transformation is applied to the forecast to complete the process.

3.4.6 Box–Jenkins SARIMA

A seasonal ARIMA model is referred to as SARIMA. In SARIMA, additional seasonal terms $(P, D, Q)_s$ are incorporated into the ARIMA model. The variable s denotes the number of observations per year in the seasonal terms of the model. The SARIMA model follows the same steps: **(1)** Identification, **(2)** Estimation, **(3)** Diagnostic checking, and **(4)** Forecasting.

SARIMA $(p, d, q, P, D, Q)_s$ is expressed in the following form (Eq. 9.22):

$$\Phi_P(B^s)\phi_p(B)(1 - B^s)^D(1 - B)^d X_t = \Theta_Q(B)^s\theta_q(B) \in_t \tag{9.22}$$

where

$$\theta_q(B) = 1 + \theta_1(B) + \cdots + \theta_q(B)^q \tag{9.23}$$

$$\Theta_Q(B)^s = 1 + \Theta_1(B)^s + \Theta_1(B)^{2s} + ... + \Theta_Q(B)^{Qs} \tag{9.24}$$

$$\phi_q(B) = 1 - \phi_1(B) - \phi_2(B)^2 - ...... - \phi_p(B)^p \tag{9.25}$$

$$\Phi_Q(B)^s = 1 - \Phi_1(B)^s - \Phi_1(B)^{2s} - .... - \Phi_p(B)^{ps} \tag{9.26}$$

SARIMA $(p, d, q, P, D, Q)_s$ has two parts: nonseasonal part (p, d, q) and seasonal parts (P, D, Q, s):

- p-order of nonseasonal AR terms
- d-order of nonseasonal differencing
- q-order of nonseasonal MA terms
- P-order of seasonal AR terms (i.e., SAR)
- D-order of seasonal differencing
- Q-order of seasonal MA terms (i.e., SMA)

The steps for the SARIMA model are also automated. The primary distinction lies in the inclusion of seasonal terms and the absence of a constant term. We expand the back-shift operator in Eq. 9.22 automatically. After implementing the appropriate differencing, we enumerate the selection of p, q, P, and Q up to order three. In the automated SARIMA model, this results in a total of 81 combinations. While testing this many combinations might appear resource-consuming, our experimental work has demonstrated that this approach enhances accuracy and reduces errors associated with human judgment. The flowchart of automated SARIMA algorithm is shown in Fig. 9.4.

3.4.7 Nonstationary Data Identification and Differencing

To identify nonstationary data, we apply the statistical test mentioned in Sect. 3.4.2. The difference here is the addition of seasonal differencing order of SARIMA model. We initially apply our seasonal differencing order **D** and then check for nonstationary data. If the data remains nonstationary, we subsequently apply normal differencing **d**.

3.4.8 Estimation of AR, SAR, MA, and SMA Coefficients

Once the order has been finalized, we estimate the AR, SAR, MA, and SMA coefficients using Yule–Walker equation. Using statsmodels library in python, we use built-in python function for this purpose. We keep in check if the AR and MA model estimated coefficients are stationary and invertible, respectively, this suggests:

- To be stationary: The sum of AR and SAR coefficients should always be less than one (Lu and Abourizk 2009).
- To be invertible: The sum of MA and SMA coefficients should always be less than one (Lu and Abourizk 2009).

3.4.9 Diagnostic Checking

For SARIMA, diagnostic checking involves running the same tests as mentioned in Sect. 3.4.5, namely the Run test for verifying the random distribution of residuals and the Ljung–Box test for detecting auto-correlation among residuals. The models that pass these tests without rejection are included in the pool of considered good models. If the tests are rejected, an alternative combination of p, P, q, and Q models is chosen. The model with the minimum RMSE is regarded as the best-fitting model, and this model is used to predict the SARIMA forecast. Subsequent to the forecast, appropriate inversions are applied. The forecast after inversion constitutes the final forecast, preserving all the trends and seasonality present in the original data.

3.5 Machine Learning Models

Combining historical yield, atmospheric, biophysical, and meteorological data to forecast seasonal crop yield is widely used in the literature (Basso and Liu 2019).

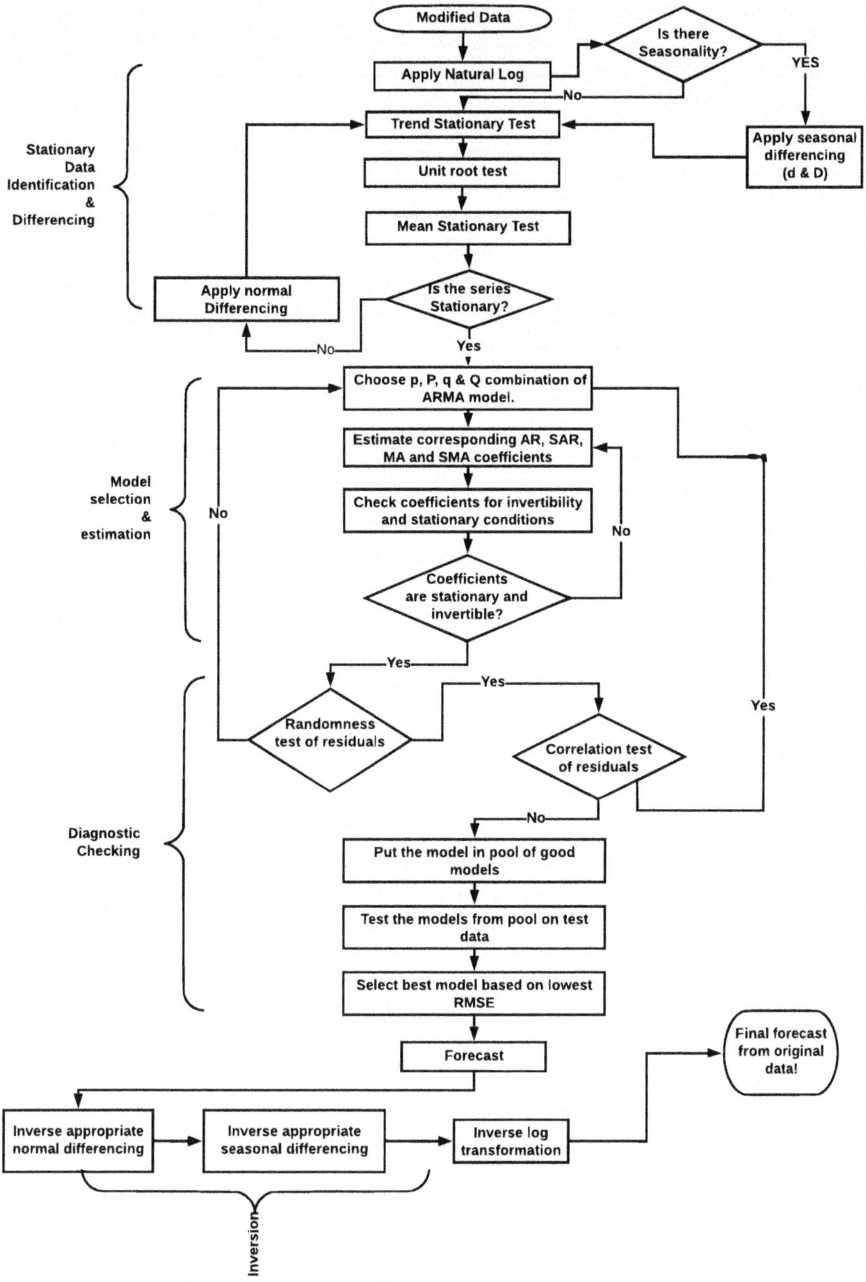

Fig. 9.4 Flowchart of automated Box–Jenkins SARIMA forecasting model

We use four supervised machine learning algorithms to test the accuracy of this method on real data. The algorithms used are:

- AdaBoost Regressor
- Support Vector Machine 0
- Bagging Regressor
- Random Forest Regression

In each case, the target variable is crop yield, and the predictor variables are atmospheric, biophysical, and meteorological data.

4 Computational Results

In this section, the results of all the models are compared. The software and hardware details for developing the system are in Sect. 4.1, whereas the dataset used for testing the forecasting system is described in Sect. 4.2. Section 4.3 describes all the results gathered from testing our forecasting system.

4.1 Computational Platform

We develop our forecasting system in Python 3.9 Spyder IDE. Utilizing some of the statistical tests already available in Python, all the models are integrated into a single file. The user is only required to provide the data file, and the system performs all computations automatically. There is no need for human judgment at any step of the computation. Our workstation is equipped with 3.0 GHz processor speed, 64-bit operating system, ×64-based processor, and 192 GB RAM.

4.2 Dataset

We tested our time series forecasting models on real dataset of citrus yield in Algeria. The dataset comprises of four citrus crops namely oranges, combined citrons and limes, combined grapefruit and pomelo, and combined Tangerines, Mandarins, and clementines yield. The yield data is from 1961–2021.

For our machine learning models, the satellite data was acquired from MODIS (Moderate Resolution Imaging Spectrometer) and downloaded from LP DAAC (2023). MODIS data provides consistent spatial and temporal time series and is split into two types. The first one (MOD15A2H) includes:

- Fraction of photo-synthetically active radiation (FPAR), which is a biophysical variable directly linked to the amount of incoming solar radiation, mainly in the blue and red bands, that is absorbed by chloroplasts to activate photosynthesis.

- Leaf area index (LAI) is also a biophysical variable linked to biomass, because leaf surfaces are an important interface between the atmosphere, vegetation, and soil, where energy and mass exchange occurs, such as gas exchange, photosynthesis, and evapotranspiration (Fang and Liang 2014).

The MOD15A2H data is an 8-day composite dataset with 500 m spatial resolution, hence permitting phenologic analysis. In this chapter 505 images were downloaded to cover a period from 2010 to 2020, with three to four images per month. The second type (MOD13Q1) data is a 16-day composite dataset with 250 m spatial resolution, permitting also phenologic analysis. MOD13Q1 provides one of the most used vegetation indices the NDVI (normalized difference vegetation index). It is commonly expressed as Kazi-Tani and Bannari (2022)

$$NDVI = (\rho NIR - \rho Red)/(\rho NIR + \rho Red) \tag{9.27}$$

where in Eq. 9.27 ρNIR is the reflectance in the near-infrared band (760–900 nm), which is strongly reflected by leaf cellular structures and ρRed is the reflectance in the red band (630–690 nm) which is characterized by chlorophyll content absorption at the canopy. Total 132 MOD13Q1 images were used corresponding to one image per month during 11 years (2010–2020). The biophysical data (NDVI, LAI, and FPAR) is combined with mean temperature, precipitation, and citrus yield data for testing machine learning models forecast accuracy. Using interpolation methods we increase the dataset range from 2010–2020 to 2005–2020 years.

4.3 Computational Results

We judge the performance of our forecasting system on the basis of following criteria:

- Accuracy of time series forecasting models on the basis of MAPE
- Testing for the conversion of nonstationary data to stationary data in case of automated ARIMA and SARIMA models
- Accuracy of machine learning models on the basis of MAPE

4.3.1 Forecast Accuracy

Changing the forecasting horizon can affect the accuracy of forecast (Fong-Lin 2009). We test our methods with different forecasting horizons. We mentioned forecasting error comparison criteria in Sect. 3. We use RMSE to select best-fitting model and compare the accuracy results using MAPE in Eq. 9.28

$$MAPE = \sum_{t=1}^{i} |e_t|/y_t * (100/i) \tag{9.28}$$

Table 9.1 1-, 2-, and 3-period-ahead MAPE values of forecasting models. (Simple Moving Average (S.M.A), Exponential Smoothing (Ex. S.))

Yield	S.M.A.	Ex.S.	Holt's	Winters'	ARIMA	SARIMA
One-period-ahead forecast						
Oranges	04.71	54.18	07.84	36.50	01.42	02.55
Citrons & Limes	03.09	44.22	01.51	15.81	01.19	08.80
Grapefruit & Pomelo	03.77	49.28	04.61	15.13	01.83	05.80
Tangerine, Mandarin, & Clementines	08.43	46.17	09.55	09.83	05.03	01.73
Two-period-ahead forecast						
Yield	S.M.A.	Ex.S.	Holt's	Winters'	ARIMA	SARIMA
Oranges	04.62	36.01	03.63	19.96	02.81	08.80
Citrons & Limes	06.20	30.82	07.03	07.99	01.25	10.39
Grapefruit & Pomelo	04.26	35.71	03.63	12.32	01.47	07.35
Tangerine, Mandarin, & Clementines	08.63	28.07	04.34	15.25	01.17	09.08
Three-period-ahead forecast						
Yield	S.M.A.	Ex.S.	Holt's	Winters'	ARIMA	SARIMA
Oranges	09.69	30.72	12.23	09.92	11.84	18.84
Citrons & Limes	10.32	30.98	12.08	16.24	10.06	14.84
Grapefruit & Pomelo	06.41	31.22	03.54	24.09	05.07	08.20
Tangerine, Mandarin, & Clementines	10.48	25.77	13.50	08.53	05.72	13.85

We test accuracy of models for up to six periods ahead for different citrus series. Tables 9.1 and 9.2 show MAPE values for 1–3- and 4–6-period-ahead forecast values of all our models available in our forecasting system. We also calculated the total MAPE of all the time series in our dataset for 1–6-period-ahead forecast in Table 9.3.

By comparing the results in Tables 9.1, 9.2, and 9.3, it is evident that ARIMA outperforms all other models for up to four-period-ahead forecasting horizon. Figures 9.5, 9.6, 9.7, 9.8, and 9.9 show best (with the lowest MAPE) 2- and 4-period-ahead forecast for all the series. As we move toward 6-period-ahead forecast, the accuracy of models reduces.

4.3.2 Accuracy of Nonstationary Series Identification and Transformation

It is crucial for ARIMA and SARIMA models to identify the nonstationary data and convert it into stationary data. This is achieved using statistical tests and differencing on the data. By comparing the graphical plots of nonstationary and stationary data after conversion, we tested our model's accuracy. Figures 9.10a and 9.11a show our input data from two different time series. From the plots, trend, level changing, and increasing variance are observed, thus confirming nonstationary nature of the input

Table 9.2 4-, 5-, and 6-period-ahead MAPE values of forecasting models. (Simple Moving Average (S.M.A), Exponential Smoothing (Ex. S.))

Series Name	S.M.A.	Ex.S.	Holt's	Winters'	ARIMA	SARIMA
Four-period-ahead forecast						
Oranges	21.90	37.96	17.05	15.75	12.69	17.46
Citrons & Limes	14.53	31.62	10.70	17.31	10.43	09.52
Grapefruit & Pomelo	07.65	31.16	01.28	12.10	04.03	04.01
Tangerine, Mandarin, & Clementines	15.67	30.09	10.99	15.83	08.01	09.06
Five-period-ahead forecast						
Series Name	S.M.A.	Ex.S.	Holt's	Winters'	ARIMA	SARIMA
Oranges	18.84	38.81	09.44	20.88	15.56	24.59
Citrons & Limes	12.60	31.74	04.64	18.87	07.38	15.18
Grapefruit & Pomelo	05.75	28.53	05.60	10.48	05.87	04.96
Tangerine, Mandarin, & Clementines	12.44	27.13	06.52	14.18	09.24	15.21
Six-period-ahead forecast						
Oranges	11.57	19.90	08.30	13.46	25.04	25.64
Citrons & Limes	08.56	24.73	07.37	12.55	21.58	22.76
Grapefruit & Pomelo	06.56	26.50	11.99	09.19	09.09	15.88
Tangerine, Mandarin, & Clementines	07.29	12.76	07.35	13.03	16.01	19.94

Table 9.3 1–6-period-ahead forecast average MAPE of the dataset. (Simple Moving Average (S.M.A), Exponential Smoothing (Ex. S.))

Total average MAPE of dataset						
Periods Ahead	S.M.A.	Ex.S.	Holt's	Winters'	ARIMA	SARIMA
One	5.01	48.4	5.78	19.2	2.45	4.72
Two	5.92	32.6	4.66	13.9	1.67	8.90
Three	9.31	29.7	10.3	14.7	8.15	13.9
Four	14.9	32.6	10.1	15.3	8.79	10.2
Five	12.4	33.4	6.54	16.1	9.51	14.9
Six	8.49	20.9	8.75	12.1	17.9	21.3

data. In our automated ARIMA and SARIMA the identification of nonstationary data is carried out automatically without human intervention. Figures 9.10b and 9.11b show the transformed stationary data plots. This is also confirmed by the ACF plots in Figs. 9.12 and 9.13. The train data ACF plots decay slowly, hence confirming nonstationary data, while the transformed plots in Figs. 9.12b and 9.13b show that plots reach 0 quickly, thus confirming stationary data.

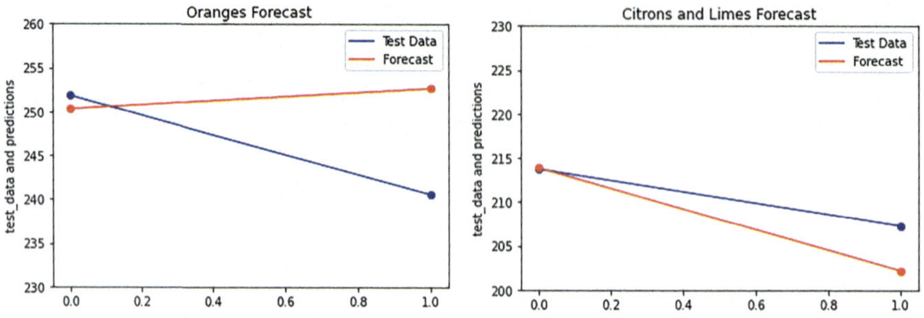

(a) (b)

Fig. 9.5 Two-period-ahead test data and predictions for (**a**) Oranges and (**b**) Citrons and Limes series

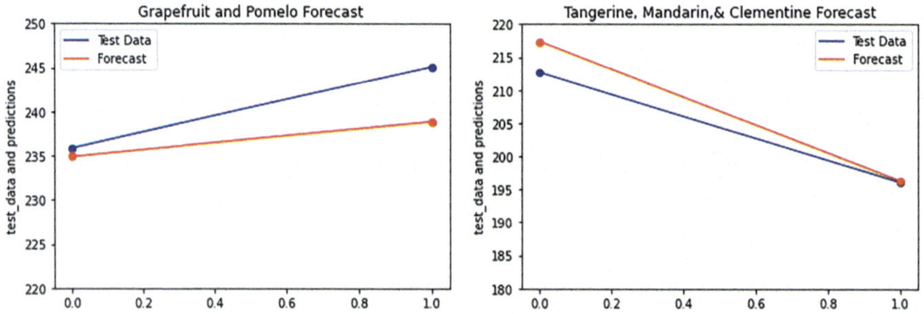

(a) (b)

Fig. 9.6 2-period-ahead test data and predictions for (**a**) Grapefruit and Pomelo and (**b**) Tangerine, Mandarin, and Clementine series

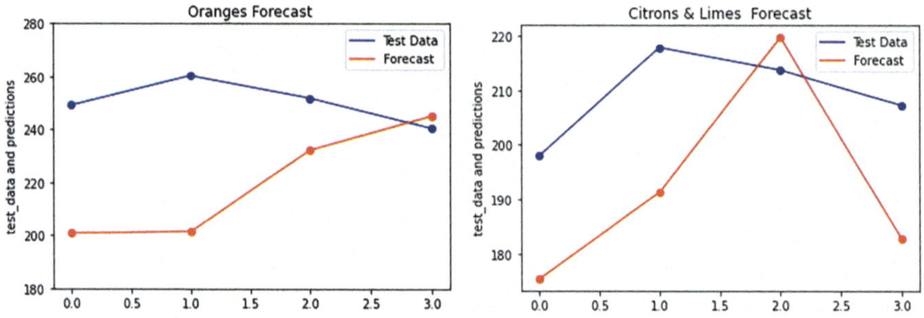

(a) (b)

Fig. 9.7 4-period-ahead test data and predictions for (**a**) Oranges and (**b**) Citrons and Limes series

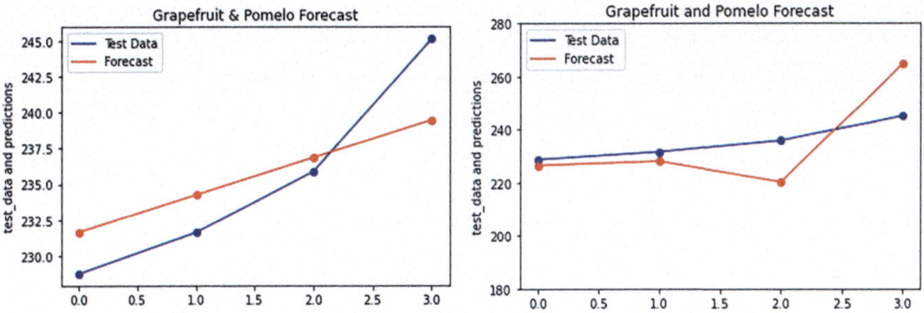

(a) (b)

Fig. 9.8 4-period-ahead test data and predictions for Grapefruit and Pomelo (**a**) using Holt's model and (**b**) using Arima model

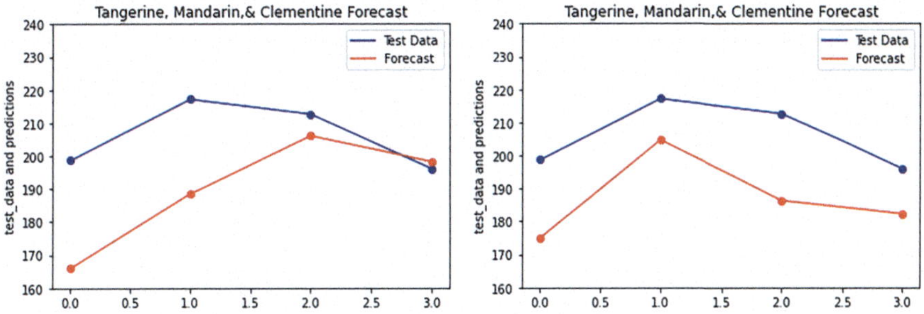

(a) (b)

Fig. 9.9 4-period-ahead test data and predictions for Tangerine, Mandarin, and Clementine (**a**) using ARIMA and (**b**) using SARIMA

4.3.3 Machine Learning Models Accuracy

This subsection provides tabular and graphical results of machine learning models accuracy (Table 9.4). Overall, the Adaboost regressor and Bagging regressor MAPE values are comparable and lowest frequently (Figs. 9.14 and 9.15).

5 Enhancing Sustainability Through Advanced Forecasting Algorithms

Ensuring sustainability and environmental care is essential in today's agriculture. Our work on forecasting algorithms, which utilizes both statistical and machine learning techniques to predict citrus crop yields, is intrinsically linked to these goals. By accurately forecasting citrus crop yields, we can significantly contribute to more sustainable agricultural practices in several key ways:

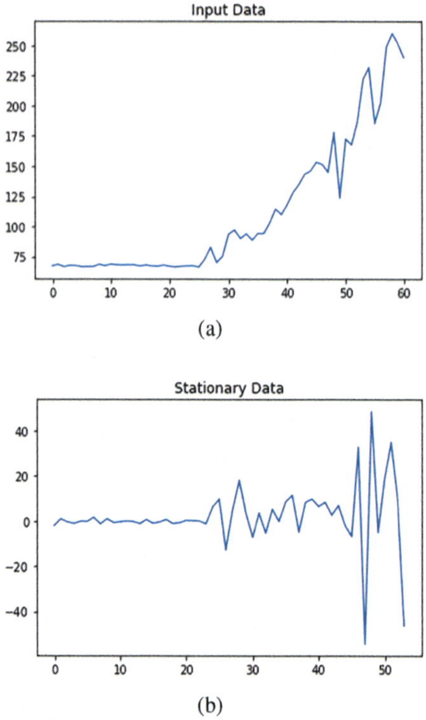

Fig. 9.10 Oranges data: (**a**) nonstationary input data plot and (**b**) stationary data after transformation plot

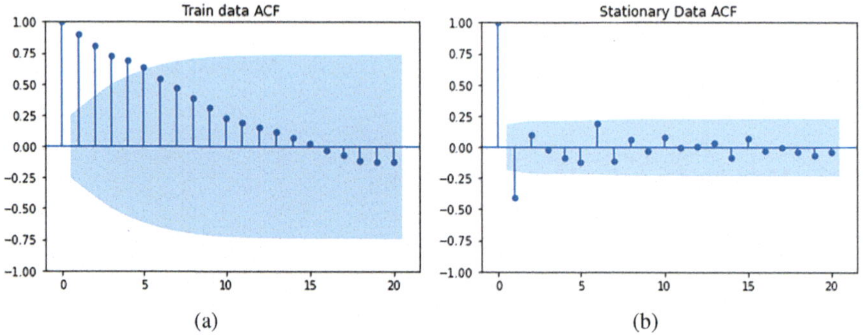

Fig. 9.11 Tangerine, Mandarin, and Clementine data: (**a**) nonstationary input data and (**b**) stationary data after transformation

- Efficient Resource Utilization:
  Accurate crop forecasts allow farmers to optimize their use of resources like water, fertilizers, and pesticides. By predicting yields more precisely, farmers can adjust resource applications to match the crops' actual needs, thus reducing waste and minimizing environmental impact. This conserves valuable resources and helps reduce runoff and soil degradation.

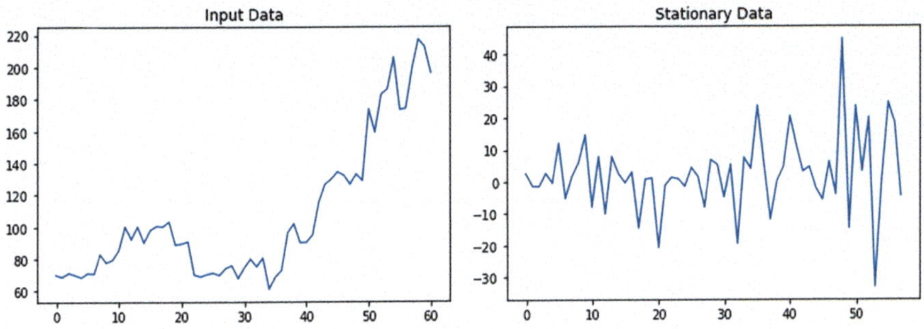

(a) (b)

Fig. 9.12 Oranges data: (**a**) nonstationary input data ACF plot and (**b**) stationary data after transformation ACF plot

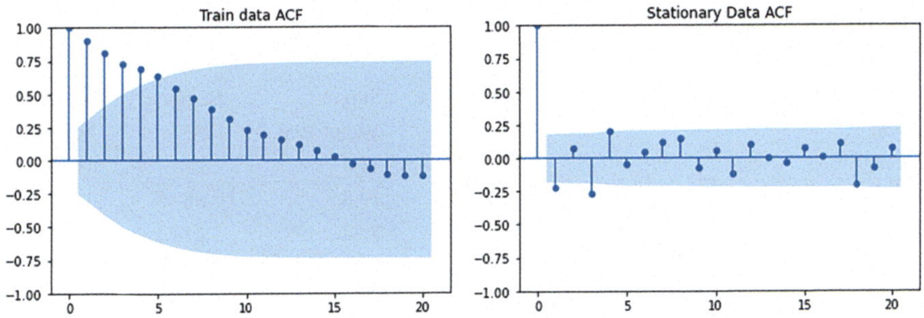

(a) (b)

Fig. 9.13 Tangerine, Mandarin, and Clementine data: (**a**) nonstationary input data ACF plot and (**b**) stationary data after transformation ACF plot

- Reduced Environmental Footprint:
  Our forecasting models assist in planning and implementing more efficient agricultural practices. For example, predicting high-yield periods allows farmers to prepare in advance to handle larger quantities of produce and reduce post-harvest losses. This results in a more efficient supply chain with less waste, contributing to a lower overall carbon footprint for citrus farming operations.
- Supporting Climate Adaptation:
  Climate change poses significant challenges to agriculture. Our machine learning models incorporate climate and biophysical data with historical yield to predict citrus crop yields. This proactive approach allows farmers to adapt their practices to mitigate the impacts of climate variability, such as adjusting planting schedules or selecting more resilient crop varieties.

Table 9.4 1-, 2-, 3-, and 4-period-ahead MAPE values of machine learning regression models

Yield	Adaboost Reg.	Support Vector Reg.	Bagging Reg.	Random Forest
One-period-ahead forecast				
Oranges	1.05	29.1	6.83	10.5
Citrons & Limes	1.93	20.6	2.04	6.10
Grapefruit & Pomelo	7.12	7.14	1.80	6.66
Tangerine, Mandarin, & Clementines	3.13	18.5	4.95	7.40
Two-period-ahead forecast				
Yield	Adaboost Reg.	Support Vector Reg.	Bagging Reg.	Random Forest
Oranges	16.3	31.3	13.1	12.7
Citrons & Limes	12.1	21.8	10.8	10.8
Grapefruit & Pomelo	4.69	7.12	5.67	4.68
Tangerine, Mandarin, & Clementines	7.50	22.7	12.6	10.3
Three-period-ahead forecast				
Yield	Adaboost Reg.	Support Vector Reg.	Bagging Reg.	Random Forest
Oranges	17.3	31.3	15.1	16.7
Citrons & Limes	9.12	19.8	7.88	10.3
Grapefruit & Pomelo	4.15	7.32	5.21	4.42
Tangerine, Mandarin, & Clementines	6.50	24.7	10.6	10.3
Four-period-ahead forecast				
Yield	Adaboost Reg.	Support Vector Reg.	Bagging Reg.	Random Forest
Oranges	14.7	28.1	14.12	12.5
Citrons & Limes	18.1	17.2	7.51	9.81
Grapefruit & Pomelo	3.71	5.28	2.61	3.80
Tangerine, Mandarin, & Clementines	4.71	21.1	8.44	8.50

- Sustainable Economic Growth:
  Accurate yield predictions enable better market planning and pricing strategies, which can stabilize the economic viability of citrus farming. By reducing the uncertainties in crop production, farmers can achieve more consistent revenues, supporting the long-term sustainability of the agricultural economy and contributing to rural development.

By aligning our work on forecasting algorithms with the principles of sustainability and greening, we contribute to the development of a more resilient, efficient, and environmentally friendly agricultural sector. Our research not only supports the immediate needs of citrus farmers but also paves the way for a sustainable future in agriculture.

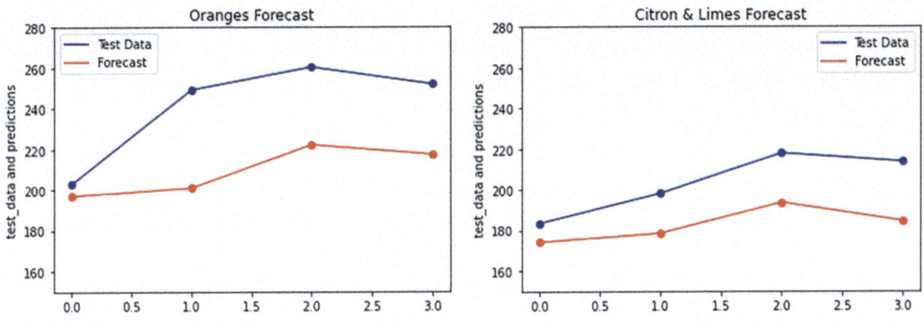

(a) (b)

Fig. 9.14 4-period-ahead test data and predictions for (**a**) Oranges and (**b**) Citrons and Limes series using machine learning algorithms

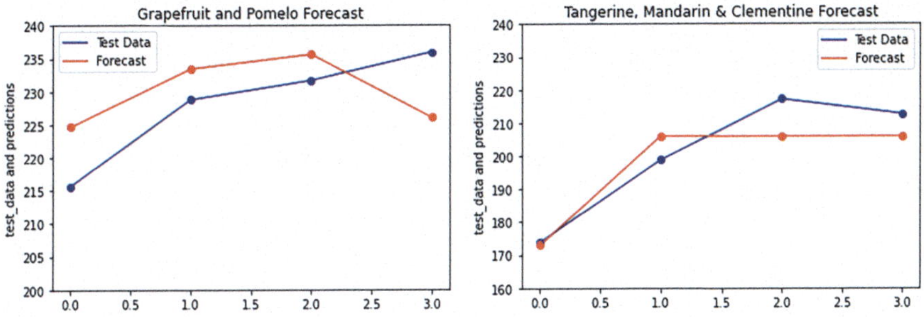

(a) (b)

Fig. 9.15 4-period-ahead test data and predictions for (**a**) Grapefruit and Pomelo and (**b**) Tangerine, Mandarin, and Clementine using machine learning algorithms

6 Conclusion and Discussion

Forecasting plays a crucial role in supply chain planning. In this chapter, we have developed an automated multi-model time series forecasting system. Although we have tested the system using citrus crop yield data from Algeria, it is not confined solely to this geographic region. This system can predict the yields of different crops in various places that share similar traits. We also test accuracy of the machine learning algorithms for multivariate citrus dataset. Our main objective is to select the best-fitting forecasting method for citrus yield to provide accurate future crop yield predictions. We applied our approach to two distinct datasets: four citrus crops from Algeria and the spatial and temporal time series data with satellite data acquired from MODIS. The key observations from this research work include:

- As the forecasting horizons increase, the forecasting deviations also increase. This means that 1- or 2-year-ahead forecasts are more reliable than long-term forecasts.

- Among the time series forecasting methods, ARIMA outperforms other forecasting models for the given dataset. This may be due to the fact that ARIMA method was able to capture the dynamic changes in the crop yield more accurately.
- For up to 2-period-ahead forecast, ARIMA model is able to forecast with $\leq 2\%$ MAPE for 3 out of 4 series. This suggests that for Algeria's citrus crop yield dataset, ARIMA is the best-fitting forecasting model.
- The Holt's model MAPE values are lower, but it is not able to catch the trend in the forecast.
- The machine learning algorithms can perform better than time series forecasting models if more data covering more years/seasons are available. Since the satellite imaging systems are started to be used only recently in the agriculture sector, data is limited to a few recent years. If the machine learning algorithms were tested after more data is accumulated, then the accuracy values with the machine learning methods would increase.
- To increase the accuracy of the models, more information needs to be included about historical data. In general, the historical data lacks annotation about the events affecting the crop yield: drought, hurricanes, social unrest, pandemic conditions, economic recession, etc. Certainly, there is need to learn the external factors affecting the yield and thus remove the anomalies and outliers.

The accurate future yield can help in estimating the potential revenues and planning for harvesting season. In the event of a forecast predicting lower yields, decision-makers can proactively devise innovative approaches to mitigate the risks associated with reduced production. The supply chain decisions that are directly affected by the crop yield include:

- Product prices
- Pesticide and fertilizer usage: timing and amount
- Workforce management at the farm
- Product and by-product allocations
- Logistical planning: number of vehicles to move fruit from farms to processing centers
- Capacity planning

Acknowledgments We are grateful to TÜBİTAK Akdeniz Narenciye Urunleri Tedarik Zincirinde Inovasyon project (Project code: 121N260), Partnership for Research and Innovation in the Mediterranean Area (PRIMA), and ImPUlSe (Innovation in the by-Product Supply Chain of Citrus in the Mediterranean area) for their support and funding, which made this project possible.

Abbreviations

AR	Auto-Regression
MA	Moving Average
ARMA	Auto-Regressive Moving Average
ARIMA	Auto-Regressive Integrated Moving Average

SARMA	Seasonal Auto-Regressive Moving Average
SARIMA	Seasonal Auto-Regressive Integrated Moving Average
OSCB	Osborn, Chui, Smith, and Birchenhall
MAPE	Mean Absolute Percentage Error
RMSE	Root Mean Squared Error
SBC	Schwarz information criterion
AIC	Akaike Information Criterion
BIC	Bayesian Information Criterion
AC	Auto-Correlation
PAC	Partial Auto-Correlation
ACF	Auto-Correlation Function
PACF	Partial Auto-Correlation Function
Reg	Regressor
SVM	Support Vector Machine

References

Anvari S, Tuna S, Canci M, Turkay M (2016) Automated Box–Jenkins forecasting tool with an application for passenger demand in urban rail systems. J. Adv Transp 1(50):25–49

Basso B, Liu L (2019) Chapter four—seasonal crop yield forecast: methods, applications, and accuracies. In: Advances in agronomy, vol 154. Academic Press, New York pp 201–255. https://doi.org/10.1016/bs.agron.2018.11.002

Bates JM, Granger CWJ (1969) The combination of forecast. Oper Res Q 20(4):451–468

Bougas C (2013) Forecasting air passenger traffic flows in Canada: an evaluation of time series models and combination methods. Master's thesis, Laval University

Bowerman BL, Connell RT (1979) Forecasting and time series. Duxbury Press

Brown BG, Murphy AH (1996) Improving forecasting performance by combining forecasts: the example of road-surface temperature forecasts. Meteorol Appl 3(3):257–265

Burkom HS, Murphy SP, Shmueli G (2007) Automated time series forecasting for biosurveillance. Stat Med 26:4202–4218

Chakrabarti D, Faloutsos C (2002) F4: Large-scale automated forecasting using fractals. In: CIKM '02: Proceedings of the eleventh international conference on information and knowledge management

Chan YM (1993) Forecasting tourism: a sine wave time series regression approach. J Travel Res 32(2):58–60

Chopra S, Meindl P (2016) Supply chain management, strategy planning and operation. Pearson Education

Coshall J (1996) Time series analyses of UK outbound travel by air. J Travel Res 44(3):335–347

Davide Burba (2019) Towards data science tutorial. https://towardsdatascience.com/an-overview-of-time-series-forecasting-models-a2fa7a358fcb/. Accessed 6 Dec 2019

De Livera AM, Hyndman RJ, Snyder RD (2011) Forecasting time series with complex seasonal patterns using exponential smoothing. J Am Stat Assoc 106(496):513–1527

Emiray E, Rodríguez G (2003) Evaluating time series models in short and long-term forecasting of Canadian air passenger data. Master's thesis, Department of Economics, University of Ottawa

Fang H, Liang S (2014) Leaf area index models. Elsevier, Amsterdam. https://doi.org/10.1016/B978-0-12-409548-9.09076-X

Fong-Lin C (1998) Forecasting tourist arrivals: nonlinear sine wave or arima? J Travel Res 36(3):79–84

Fong-Lin C (2009) Forecasting tourism demand with arma-based methods. Tour Manag 30(5):740–751

Hamed MM (1999) Stochastic modeling of airlines' scheduled services revenue. J Air Transp World Wide 4(2):32–48
Hwang S, Park M, Lee HS, Kim H (2012) Automated time-series cost forecasting system for construction materials. J Construct Eng Manag 138(11):1259–1269
Hyndman R, Khandakar (2007) Automatic time series for forecasting: the forecast package for r. Tech. rep., Monash University, Department of Econometrics and Business Statistics
AFS Inc. (1998) Autobox software user manual, automatic forecasting systems. Tech. rep., Hatboro
Kazi-Tani LM, Bannari A (2022) Assessing multi-temporal changes of natural vegetation cover between 1987–2018 using serial NDVI: a case study of Tlemcen National Park (north-west of Algeria). Int J Environ Stud 79(1):37–53 https://doi.org/10.1080/00207233.2021.1875302
Krasic D, Gatti P (2009) Forecasting methodology of maritime passenger demand in a tourist destination. PROMET Traffic Transp 3(21):183–190
Kwiatkowski D, Phillips PC, Schmidt P, Shin Y (1992) Testing the null hypothesis of stationarity against the alternative of a unit root: how sure are we that economic time series have a unit root? J Econom 54(1):159–178
LP DAAC (2023) Earth data, LP DAAC. https://lpdaac.usgs.gov/
Lu Y, Abourizk SM (2009) Automated Box–Jenkins forecasting modelling. Autom Construct 18(5):547–558
Mwakalonge JL, Badoe DA (2014) Trip generation modeling using data collected in single and repeated cross-sectional surveys. J Adv Transp 48(4):318–331
Nguyen H, Hansen CK (2017) Short-term electricity load forecasting with time series analysis. In: 2017 IEEE international conference on prognostics and health management (ICPHM)
Oh CO, Morzuch BJ (2005a) Evaluating time-series models to forecast the demand for tourism in Singapore comparing within-sample and post sample results. J Travel Res 4(43):404–413
Oh CO, Morzuch BJ (2005b) Evaluating time-series models to forecast the demand for tourism in Singapore comparing within-sample and postsample results. J Travel Res 43(4):404–413
Omane-Adjepong M, Oduro FT, Oduro SD (2013) Determining the better approach for short-term forecasting of Ghana's inflation: seasonal-arima vs. Holt-Winters. J Am Stat Assoc 3(1):69–79
Perktold J et al (2009) Runs test statsmodels python library. https://www.statsmodels.org. Accessed 14 Dec 2019
Sharda B, Patil RB (1992) Connectionist approach to time series prediction: an empirical test. J Intell Manuf 3(1):317–323
Timmermann A (2013) Forecast combinations. Tech. rep., Oxford
Uysal M, Crompton JL (1985) An overview of approaches used to forecast tourism demand. J Travel Res 4(23):7–15
Yaffee RA, McGee M (2000) An introduction to time series analysis and forecasting: with applications of SAS and SPSS. Academic Press, New York

Part IV
Logistics in Sustainable Food Supply Chains

Chapter 10
Operational Practices Catalogue for Sustainable Citrus Supply Chains

Ahmed Kassem, Emad Alzubi, Denis Daus, Elizabeth Geraghty, Ani Melkonyan-Gottschalk, and Bernd Noche

1 Introduction

The food supply chain (FSC) is a complex network that involves various stakeholders and stages, such as production, processing, distribution, retailing, consumption, and waste management. Each stage of the supply chain (SC) requires human and/or natural resources, as well as equipment and/or machines. The food chains are interdependent processes, and the sequence of operations mirrors this complex interaction process. When one part of the SC is affected, it has a cascading impact on the entire agricultural supply chain (ASC), which closely resembles a domino effect. This effect becomes even more evident during political conflicts, pandemics, and climate change. Adopting the best operational practices industry-wide is essential to understanding and improving the FSC operations, ensuring global food security and sustainability.

Ahmed Kassem and Emad Alzubi contributed equally with all other contributors.

A. Kassem (✉) · E. Alzubi · B. Noche
Transports Systeme und -Logistik, Fakultät für Ingenieurwissenschaften, Universität Duisburg–Essen, Duisburg, Germany
e-mail: ahmed.kassem@uni-due.de

D. Daus
Zentrum für Logistik und Verkehr (ZLV), Universität Duisburg–Essen, Duisburg, Germany

E. Geraghty
University of Cincinnati, Cincinnati, OH, USA

A. Melkonyan-Gottschalk
Clausthal University of Technology, Clausthal-Zellerfeld, Germany

A. Melkonyan-Gottschalk, D. Daus (eds.), *Transformation Towards Circular Food Systems*, Greening of Industry Networks Studies 13,
https://doi.org/10.1007/978-3-031-63793-3_10

With the entire FSC sustainability, proper disposal of waste produced during the processing of oranges is a critical issue. Citrus waste endangers the environment and creates severe disposal problems and logistical difficulties for the citrus sector. In response to that, innovative solutions have been investigated, including biogas production (Ortiz et al. 2019; Rosas-Mendoza et al. 2020), orange peel waste incineration (Ortiz et al. 2019), or green extraction methods (Satari and Karimi 2018). However, the importance of sustainability initiatives in citrus networks goes beyond waste management. Researchers have investigated various aspects of citrus SCs to comprehensively explore best practices and practical implications, targeting sustainability and circularity concepts. Besides inventory management (Peña-Orozco et al. 2020), pull system approaches (Amer et al. 2018), innovative SC designs (Alzubi and Noche 2022a), seamless integration of SC stages (Aslan et al. 2020), and incorporation of closed-loop supply chains (CLSC) and reverse logistics (RL) (Liao et al. 2020; Roghanian and Cheraghalipour 2019; Goodarzian et al. 2023) have emerged as fundamental strategies to make citrus networks more sustainable, economically viable, and environmentally responsible, while fostering food security by reducing the food loss and waste (FLW).

The differences in citrus SC networks from country to country hinder the determination of standard operational practices. Thus, reflecting on the challenges and opportunities at all the SC stages, a systematic literature review has been conducted in Egypt, Italy, Brazil, China, Greece, India, Iran, Jordan, Colombia, and the United States. For instance, in the context of Brazilian citrus SC, eight papers were identified. Among these, four papers (Fermo et al. 2021; Barbosa et al. 2018; Barrales et al. 2018; Ferreira et al. 2020) specifically examined the processing stage. One paper (Gehrke et al. 2021) solely focused on the farm stage, and two papers (Caixeta-Filho 2006; Ferreira et al. 2016) discussed practices in both the farm and processing stages.

In China, five papers were discovered, with an equal distribution among different stages: one paper (Gehrke et al. 2021) centered on the farm stage, another (Ahrisa et al. 2021) on retailing, one (Chen et al. 2019) on processing, and one (Liu 2010) on citrus transportation.

In Egypt, the research primarily concentrated on the farm stage, with papers such as Abobatta (2018a, b, 2019), Abobatta (2021), and Alnaimy (2017). However, one study (Al-Sammarraie et al. 2022) specifically examined the processing stage. Other papers, like Galal and El-Kilany (2016), explored practices related to retailing and transportation, while Amer et al. (2018) discussed practices encompassing processing, retailing, and transportation.

From Greece, two papers were found, both focusing solely on the processing stage (Al-Sammarraie et al. 2022; Mamma and Christakopoulos 2008). Research conducted in India primarily centered around valorizing citrus loss and waste (CLW), with papers (Chavan et al. 2018; Suri et al. 2022; Magare et al. 2020; Panwar et al. 2021; Visakh et al. 2022; Joglekar et al. 2019) specifically addressing this topic. However, only one study analyzed practices on the farm stage, while two studies focused on the processing stage.

In Iran, the research was focused on the citrus SC network and optimizing models that encompassed all stages. Studies such as Cheraghalipour et al. (2018), Roghanian and Cheraghalipour (2019), and Liao et al. (2020) can be regarded as commendable efforts in adopting circular economy (CE) practices due to their consideration of the closed-loop citrus SC network. Additionally, Goodarzian et al. (2022) and Goodarzian et al. (2023) specifically examined citrus transportation.

Research conducted in Jordan aimed to provide solutions for minimizing CLW in various stages. For example, Alzubi et al. (2023b) analyzed operational practices during the harvesting season and their relationship to the quantity of citrus supplied to the market. Meanwhile, Alzubi and Noche (2022a) presented a model for optimizing citrus SC in Jordan to reduce transportation costs and CO_2 emissions. Furthermore, Alzubi et al. (2022) considered citrus packaging crates and transportation.

Research conducted in Colombia primarily focused on practices in the farm stage, as discussed by Peña Orozco et al. (2021) and Ortiz-Sanchez et al. (2021). However, Ortiz et al. (2019, 2020) and Ortiz-Sanchez et al. (2021), specifically examined the processing stage. Castro and Jaimes (2017) covered practices across all stages, including the consumption stage, while Rojas Palacios et al. (2022) analyzed the farm and retailing stages.

In Italy, researchers concentrated on specific stages of the value chain. Pergola et al. (2013), De Luca et al. (2015), and Scuderi et al. (2022) focused solely on the farm stage, while Lanza (2003), Raimondo et al. (2018), and Zema et al. (2019) examined the processing stage exclusively. On the other hand, Beccali et al. (2010) discussed practices in both the farm and processing stages. Similarly, Beccali et al. (2009) considered practices in the farm and processing stages, as well as logistics activities such as transportation, packaging, and storage.

In Mexico, only two papers were identified that discussed practices within the citrus SC. Rosas-Mendoza et al. (2020) focused on practices in the processing stage, while Lambert et al. (2014) examined practices in the farm, transportation, and citrus export stages. Additionally, two papers from Mexico (Larios-Cruz et al. 2018; Navarro-Pineda et al. 2022) discussed practices related to valorizing.

In Morocco, Lahlali et al. (2021) conducted research on the farm stage, while Meryem et al. (2023) discussed various options for valorizing CLW.

In Bangladesh, only one paper was published (Begum et al. 2022), which considered practices in the farm and retailing stages. Similarly, Afloukou et al. (2020) discussed practices in the farm and retailing stages in Benin.

In the United States, research primarily centered around agricultural practices, as evident in papers (Morgan et al. 2022; Boman 2012; Boman et al. 2020; Kadyampakeni and Duncan 2022; Li et al. 2020; Abbas and Fares 2009). Additionally, Thornsbury et al. (2020) examined the farm and retailing stages, while Bell and Horvath (2020) focused on the retailing stage and transportation. Furthermore, Chen and House (2021) specifically analyzed the consumption stage.

In Spain, research was distributed across different stages of citrus SC. Unfortunately, the specific papers and their focus within each stage and sustainability dimensions are not provided. For a comprehensive overview of the research, focus should be based on stages and sustainability dimensions within citrus SC in each country.

This review shows that the operational practices in citrus SCs are multifaceted, encompassing a variety of activities across different stages. Each stage plays a pivotal role in determining the quality, efficiency, and sustainability of the SC. Farming practices significantly impact the yield, quality, and health of citrus crops. These practices are critical as they directly influence the initial quality of the citrus fruits. In the processing stage, operational practices focus on methods to enhance efficiency, minimize waste, and ensure product quality. These practices are crucial in ensuring that the processed citrus products meet the required standards.

Logistics practices, which include packaging and transportation, are essential in maintaining the freshness of citrus products, while transporting them. These practices ensure that the fruits are transported efficiently from farms to processing units and eventually to downstream stages. Furthermore, distribution practices involve the coordination of logistics and transportation to ensure timely delivery to retailers, as well as consumers.

Retailing practices, such as storage and display, further contribute to preserving the freshness of citrus products. These practices can help extend the shelf life of citrus. The interplay of these operational practices is critical in reducing CLW throughout the SC, addressing factors such as spoilage, damage, and inefficient handling. Understanding and optimizing these practices can significantly contribute to the sustainability and success of the citrus industry, aligning with broader goals of enhancing sustainability.

1.1 Agricultural Practices

The extensive research conducted at the farm level addressed various practices that can be classified into agricultural practices related to planning, nutrition, pesticides, and irrigation, as well as operational practices that involve harvesting activities. Increased efficiency in these practices is essential to improve the quality and quantity of fruits (Caixeta-Filho 2006). For this, identification and then implementation of the best management practices (BMP) in the field is the key. These practices encompass various actions, such as educating and training farm workers on proper handling of irrigation, fertilizer, and pesticide practices. It also involves developing a nutrient plan tailored to the specific requirements of the trees, utilizing suitable equipment, and determining the appropriate amount of water needed for irrigation (Boman et al. 2020).

Soil-borne diseases can have a significant impact on citrus trees and the quality of their fruits (Paudel et al. 2018). To mitigate the risk of these diseases, it is recommended to skirt citrus trees up to 75 cm above the ground level. This practice helps to minimize the contact between the soil and the tree, reducing the chances of infection. Additionally, thinning the canopy is necessary to improve sunlight penetration and enhance the rate of transpiration. These measures contribute to maintaining the overall health of citrus trees, resulting in the production of high-quality fruits. In the case of high-density planting, specific guidelines should be followed. The maximum height of the hedge should not exceed 2.5 m, while the maximum width should not exceed 1 m. The minimum base width of the hedge should be 0.7 m (Chhetri and Kandel 2019). Furthermore, the planting area per tree in high-density planting systems should be 2.74 × 2.74 m. In contrast, for traditional planting methods, the recommended planting area typically falls within the range of 25–35 square meters (Alzubi and Noche 2022b).

Implementing good agricultural practices, such as canopy management, top netting, soil fertility, and water conservation, can contribute to mitigating the risks associated with climate change (Abobatta 2020). Pruning practices should be carried out at the appropriate time, such as after harvesting or before a new flowering cycle. This helps optimize sunlight exposure and air circulation, resulting in controlled humidity levels and reduced presence of pests and pathogens.

Removing low-lying and dried branches allows for adequate sunlight penetration. Rejuvenating old trees creates favorable conditions for new branch growth. This can be achieved by cutting off the head of the trees and maintaining an appropriate height, followed by grafting the main stem or utilizing the cleft method to encourage the development of new shoots.

In the context of climate change and natural disasters, Scuderi et al. (2022) proposed adopting sustainable strategies to understand and assess climate variations and associated conditions. This knowledge can help farmers better cope with these changes. For instance, replanting damaged areas following extreme natural disasters can be a favorable option in terms of sustainability. However, repairing damaged citrus plants should be avoided due to the high labor requirements and lack of economic viability. To further improve sustainability in farms, Alnaimy (2017) suggested alternative solutions such as returning CLW as animal feed. CLW is rich in nutrient values for animals like cows and sheep. Reusing CLW as animal feed or organic fertilizer enhances environmental sustainability by reducing CO_2 emissions (Ribal et al. 2019; Yang et al. 2020; Chen and House 2021; Mariana et al. 2021; Goodarzian et al. 2023; Nicolò et al. 2018). Additionally, CLW reuse can improve the economic dimension by reducing the costs of agricultural inputs (Usman et al. 2018; Strano et al. 2017) and transportation expenses (Goodarzian et al. 2023). Moreover, farmers can potentially increase their income by selling agricultural residues to waste treatment plants (Ghani et al. 2021; Strano et al. 2017; Paudel et al. 2018).

1.1.1 Fertilizer Practices

Adequate land preparation, including fertilization and weed protection, is the basis for ensuring proper nutrition for trees, which supports the increased tree health and thus increase in the quantity and quality of yield (Abobatta 2018b). Beccali et al. (2009) estimated the required amount of herbicide for weed control to be 1.08 kg/ha, while the pesticide with phosphorus (P) content needed was 3.26 kg/ha.

Additionally, replacing older or diseased citrus trees with younger ones can lead to an increase in yield (Abobatta 2018a, b). Weeding practices involve the removal of harmful species surrounding citrus trees, which can effectively reduce yield loss and enhance fruit quality. Furthermore, weeding activities not only improve the quantity and quality of citrus yield but also enhance the efficiency of other agricultural practices, while reducing risks such as fire, pests, pathogens, and frost effects (Abobatta 2018b). According to Chhetri and Kandel (2019), gourmands (vigorous shoots) around trees consume a significant amount of water and nutrients. Therefore, removing these gourmands is necessary causes in reduction of water consumption and the nutrient requirements per tree.

The utilization of organic fertilizer and pesticide offers several advantages. Studies such as Pergola et al. (2013) and Gehrke et al. (2021) have shown that these practices have a positive influence on citrus yield, biodiversity, and human health. Furthermore, they contribute to improving soil quality and nutrition (Cheraghalipour et al. 2018; Roghanian and Cheraghalipour 2019), reducing agricultural input costs (Usman et al. 2018), and decreasing CO_2 emissions on farms (Ribal et al. 2019; Nicolò et al. 2018). The choice of farming system depends on the specific area and population, as stated by De Luca et al. (2015). In most cases, conventional farming systems have demonstrated better social impacts. However, according to Afloukou et al. (2020), a combination of organic and chemical fertilizers can be beneficial, particularly in substituting missing minerals like magnesium and zinc. In terms of fertilizer application, Morgan et al. (2022) discussed important points. For example, using controlled-release fertilizers and ensuring an adequate and constant supply of nitrogen (N) are crucial for promoting tree growth and early fruit production, especially for trees in the initial 3 years after planting. Additionally, supplying calcium to the soil helps maintain a pH value within the optimal range of 5.5–6.5. If this value is already achieved, there is no need to add additional calcium. Furthermore, applying fertilizer in several small doses can enhance fertilizer efficiency by maintaining high nutrient levels.

Implementing proper fertilizer practices can help mitigate the risk of salinity in cases of reduced rainfall. Additionally, incorporating arbuscular mycorrhizal as a soil amendment supports to enhancement of the overall health and crop yield of citrus plants, especially in unfavorable climatic conditions (Abobatta 2021). To minimize the risk of nitrate leaching from fertilizers into groundwater, it is essential to specify the sources of N fertilizer, determine annual N rates, establish maximum N rates per application, consider fertilizer application timing, manage irrigation effectively, and maintain detailed records for citrus crops (Ortiz et al. 2020; Boman et al. 2020).

1.1.2 Pest Management Practices

Integrated pest management is the most efficient approach for crop protection, encompassing biological controls, cultural controls, pest monitoring, and other relevant practices. It is essential to select appropriate strategies that collectively contribute to effective pest management. Moreover, accurate application of agricultural chemicals requires the correct calibration and maintenance of crop protection product application equipment.

In pest management practices, it is recommended to use nozzles that produce larger droplets, as this ensures sufficient plant coverage and effective pest control. Spraying crop protection products is based on calibrated speed in relation to movement speed, guaranteeing uniform, and proper and efficient pest management. Maintaining accurate records and documentation of crop protection practices is crucial, as it helps meet legal responsibilities and provides reliable information and log-sheets for future applications (Boman 2012).

Employing a rotation strategy based on the mode of action helps to prevent resistance buildup, reduces toxicity to non-target species, and considers product solubility, persistence, leaching potential, irrigation schedules, soil type, and other site characteristics. Controlled release (encapsulated) formulations offer the additional benefit of reduced leachability. Furthermore, the timing for pest management practices is crucial and should be determined based on the population biology and abundance of pests, as well as the growth periods of the trees (Kadyampakeni and Duncan 2022).

The use of copper compounds can be effective in protecting citrus fruits from diseases and insects such as mealybugs, Mediterranean fruit flies (medflies), and mites (Li et al. 2020). Nevertheless, in cases where rainfall is expected, adjustments should be made to pesticide practices, and irrigation water must be carefully regulated, while considering the precipitation. Repeated irrigation or rainfall events can cause soluble nutrients and pesticides to leach below the root zone, leading to potential groundwater contamination and economic losses (Kadyampakeni and Duncan 2022). Therefore, proper water and pesticide management is essential to minimize these risks.

Applying the correct compounds as foliar applications can be effective in protecting against spreading insects, pests, and diseases. For instance, using copper fungicide during fruit set can help reduce the risk of back rot (Abobatta 2018b, 2019; Afloukou et al. 2020; Mahawer et al. 2023). Synergistic effects can be achieved by combining treatments, such as nutrients and antibiotics, insecticides and foliar nutrients, as well as biological control with chemicals. This approach enhances control over insect vectors like the Asian citrus psyllid and leads to increased yield improvement (Li et al. 2020).

1.1.3 Irrigation Practices

To ensure optimal growth and yield in citrus crops, it is important to avoid excessive irrigation, as citrus is not tolerant to flooded soil conditions (Abobatta 2021). Various studies have addressed irrigation practices in citrus cultivation. For example, Beccali et al. (2009) highlighted that irrigation requirements increase with decreasing rainfall, and the specific water needs depend on soil characteristics, typically ranging from 3×10^6 to 5×10^6 kg/ha/year. Furthermore, irrigation water should not contain high levels of chlorine, boron, and sodium, as they can negatively affect crop yield (Beccali et al. 2010).

According to Beccali et al. (2010), reusing purified water for irrigation and employing railway transportation for delivering final products can improve the eco-profile of citrus cultivation. It is essential to optimize irrigation practices to prevent water stress and delayed flowering, which can reduce the potential yield. Micro-irrigation systems, such as drips or bubblers, are effective in optimizing water usage and improving fertilizer utilization by minimizing yield losses. This, in turn, enhances soil health, ultimately improves fruit quality (Abobatta 2018a, b), and helps prevent over-irrigation (Lahlali et al. 2021).

1.1.4 Technologies in Agricultural Practices

The adoption of new technologies in farming can yield various benefits. For instance, the implementation of artificial inoculation has shown positive results in protecting citrus plants, such as grapefruit and lemon, by increasing their resistance against citrus canker (Mahawer et al. 2023). Automated liquid injection during irrigation can help reduce costs. For grapefruit, the recommended N rate ranges from 120 to 160 lb/acre, while for oranges, it ranges from 120 to 200 lb/acre. Research conducted by Braham et al. (2023) revealed that applying a triple agronomic recommended dose of N can have a positive impact on the growth parameters of clementine compared to controlled N applications. Studies such as Costa et al. (2021) and Fermo et al. (2021) have demonstrated the utility of image processing technology for improving disease diagnosis and management in citrus crops.

1.1.5 Managing Agricultural Practices

Sustainable citrus production may be achieved by implementing an integrated citrus production system, which incorporates regulated management practices. The integrated of expert system models, utilizing fuzzy logic, into agricultural practices can be enhance citrus yield. These systems provide valuable insights into unpredictable events such as rainfall, wind, or frost (Lambert et al. 2014). Expert systems play a significant role in managing various agricultural practices, including irrigation water management, fertilizer application, pest control, tree pruning, and planting density. Furthermore, these expert systems contribute to improving fruit quality by establishing relationships and correlations (Lambert et al. 2014). As per Abobatta (2018a), the implementation of an integrated management system can significantly enhance fruit quality, leading to the opening of new export markets.

Expert systems are deployed to control the inventory level of fresh fruit under uncertain scenarios such as drought, lack of wind for palletization, and more. They use if-then statements to guide farmers on the best next steps. Based on data collection of an array of behavioral patterns, this will lead to reduction in the probability of defective fruit before reaching the destination (Lambert et al. 2014).

The efficient SC efficiency significantly impacts the retail, consumption, and exporting of citrus fruits. Identifying the best practices for transportation monitoring, cooling monitoring, mitigating food waste and post-harvest loss, and quality control of products is crucial for delivering fresh and safe products. Post-harvest loss can be reduced by implementing proper training policies, ensuring fair pricing, improving storage facilities, ensuring proper transportation and communication, and enhancing regional cooperation (Begum et al. 2022).

1.2 Operational Practices in Farm

Other practices in citrus farms include harvesting activities, which contribute significantly to the amount of CLW. Harvesting, a pivotal phase, involves the meticulous picking of citrus fruits, a task influenced by factors like fruit maturity, precision in harvesting entire fruits, and the size of the harvested produce. Workers' familiarity with identifying optimal fruit ripeness plays a pivotal role in minimizing wastage. Instances include workers' unfamiliarity with fruit maturity, imprecise harvesting resulting in incomplete fruit gathering, picking smaller-sized fruits, and contamination due to mud (Al Hiary 2012). Aside from increasing labor income to motivate workers and improve profitability, it is essential to hire a sufficient number of workers for effective citrus cultivation. This reduces citrus losses and waste on the farm, resulting in increased citrus supply to the market and improved self-sufficiency index (SSI) (Alzubi et al. 2023b). Moreover, the performance of farm workers significantly impacts the quality and quantity of yield (Naseer et al. 2019a), which highlights the need for relevant training and education programs for farm workers (Aslan et al. 2020).

2 Operational Practices at the Processing Stage

Citrus processors can be divided into two categories: Primary processors are those who produce citrus products from fresh oranges or other citrus fruits. They usually have large-scale facilities with advanced equipment and technology to produce efficiently and consistently. Secondary processors are those who use the by-products of primary citrus processing, such as peel, pulp, pectin, essences, and animal feed. They can add value to these by-products by refining them into different products or finding new uses for them.

The process generally starts with feeding the primary processor with the citrus fruits for the extraction of juice, oil, or citrus products in several forms (Sarbatly et al. 2023; Tetra Pak 2017; Bates et al. 2001; Kimball 1991). This is followed by washing and sorting, which involves cleaning and inspecting the fruit (Mérillon et al. 2023; Akyýldýz et al. 2023). The next step is extraction, which involves squeezing or extracting juice using mechanical pressure (Sarbatly et al. 2023). The processing types and methods are mainly based on the form and use of the citrus products of processed citrus fruits (Mushtaq 2018).

Citrus processors primarily focus on measuring and improving the efficiency and effectiveness of key performance indicators (KPIs) that are essential for their operations. These KPIs include:

1. Yield Efficiency: The ratio of the amount of juice extracted to the total weight of the fruit processed, including the amount of juice obtained from a given amount of fruit (Tetra Pak 2017; Akyýldýz et al. 2023).
2. Oil Recovery: The amount of oil obtained from a given amount of peel or fruit (Tetra Pak 2017; Akyýldýz et al. 2023).
3. Product Quality: The consistency and quality of the final product, including physical, chemical, microbiological, and sensory characteristics that meet customer expectations and specifications (Strano et al. 2022).
4. Waste Management: The amount of waste generated from a given amount of fruit, including the efficiency of waste disposal and recycling processes (Echavarría et al. 2011; Mérillon et al. 2023).
5. Cost reduction: the amount of money spent on a given amount of product (Strano et al. 2022).

2.1 Packhouse Facility

The packhouse facility is where citrus fruits are received from the farm to be prepared and packed. The operations of citrus packhouses include culling, washing, grading, sorting, waxing, refrigeration, packing, inspection, and transporting the fruit to the market or distributor. The packhouse process can be divided into three stages (CRW 2023): the initial processes, which include receiving, culling, and washing the fruits; the preparation processes, which involve grading, sorting, waxing, and refrigeration; and the final processes, which consist of packing and inspection before transporting the fruit (CRW 2023).

The integration of machine vision and digital signal processors, as highlighted in the studies by Aleixos et al. (2002), revolutionizes citrus processing by enabling real-time multispectral inspection, grading, and sorting of citrus fruits. This cost-effective and versatile technology optimizes the entire fruit quality control and sorting process, rapidly detecting fruit size, shape, color, and external defects, thus ensuring consistent quality control (Al-Sammarraie et al. 2022; Cubero et al. 2016; Aleixos et al. 2000). It allows for nondestructive examination of both internal and external features of the fruits (Tian et al. 2020; Aleixos et al. 2002). An artificial intelligence-based method for predicting the sweetness of orange fruits based on their RGB values (Al-Sammarraie et al. 2022) can enhance the classification of oranges according to their ripeness or flavor. These technologies can significantly improve the quality control processes in the packhouse processing industry, leading to the production of high-quality products with standard classification and potentially higher customer satisfaction.

2.2 Juice Extraction

Citrus juice extraction is a process that depends on the type of the final product. A concentration method or pasteurization follows the extraction. The concentration method involves reducing the water content of the juice using thermal or membrane processes (Sarbatly et al. 2023), while the pasteurization method involves heating citrus juices to a certain temperature to kill harmful microorganisms and extend the shelf life of the products (Sarbatly et al. 2023; Tetra Pak 2017). This entire process ensures getting the best quality of citrus juice products.

Juice processors produce different types of citrus juice based on their processing methods, including fresh juice, concentrated juice, not-from-concentrate (NFC) juice, from-concentrate (FC) juice, and reconstituted juice. Juice is the most popular and widely traded type of processed citrus fruit (Akyýldýz et al. 2023; Tetra Pak 2017). Fresh juice is consumed immediately after extraction, without any further processing or preservation. It has the highest quality and nutritional value, but also the shortest shelf life. Concentrated juice is evaporated to reduce its water content and increase its sugar content. It can be stored at ambient or frozen temperatures for several months or years. NFC juice is clarified and filtered to remove pulp and other solids and then pasteurized and stored in bulk tanks under aseptic conditions. FC juice is produced by reconstituting the concentrated juice with water before packaging and selling. It is cheaper and easier to transport and store than NFC juice, but it may have lower quality and flavor due to the processing methods. Reconstituted juice is created by adding water to concentrated juice to bring it back to its original strength. It is similar to FC juice but is often labeled as "reconstituted" to indicate the addition of water.

2.3 Oil Extraction

Citrus oil extraction is a process that involves recovering the essential oil from the peel of citrus fruits using mechanical or steam distillation methods (Tetra Pak 2017; Bates et al. 2001). Various techniques can be employed for this process, such as hydro-distillation, cold pressing, solvent extraction, or supercritical fluid extraction (Rajput et al. 2023; Fakayode and Abobi 2018; Palazzolo et al. 2013).

Citrus oil producers extract the oil from citrus peels using several extraction methods, which include cold pressing, steam or hydro-distillation, water, steam and organic solvent extraction, and supercritical CO_2 extraction (Ganem et al. 2020; Palazzolo et al. 2013; Arce and Soto 2008). These are some of the common methods of citrus oil extraction. Each method has its advantages and disadvantages in terms of yield, quality, cost, energy consumption, environmental impact, and safety. Therefore, choosing the best method depends on various factors, such as the availability of raw materials, market demand, production capacity, and quality standards. The pressing method is one of the most common methods used for extracting all citrus oils. Specific types of essential oil are extracted using the cold press method (Ganem et al. 2020).

During juice production, the cold-pressed method is used to simultaneously extract juice from the interior of the fruit and oil from the peel surface (Arce and Soto 2008). Additionally, cold-pressed citrus seed oils are commonly used for edible purposes and for cosmetic, pharmaceutical, or oleochemical industries (Yilmaz and Güneşer 2017). A study by Yilmaz and Güneşer (2017) compared the cold press and solvent (hexane) extraction methods for lemon seed oils. The cold-pressed oil produced a higher quality oil in terms of total phenolics, α-tocopherol, and antioxidant capacity and lower concentrations of free fatty acids, peroxides, and p-Anisidine (or para-anisidine), while yielding less oil than hexane extraction. This highlights the importance of considering both yield and quality when choosing an extraction method for lemon seed oils.

2.4 Pectin Extraction

Citrus pectin extraction involves several methods, each with its advantages and disadvantages in terms of yield, quality, cost, energy consumption, environmental impact, and safety. Therefore, the choice of the best method depends on various factors, such as the availability of raw materials, market demand, production capacity, and quality standards. Pectin, a food hydrocolloid, is utilized in the food and pharmaceutical industries as a gelling, emulsifying, thickening, and stabilizing agent (Khedkar 2024). Citrus fruit peels are a key source of pectin, and most pectin products are made from apples or citrus peels, both of which are rich sources of this fiber.

Microwave-assisted extraction is a green technology that uses microwaves to heat the citrus peels and extract pectin (Arora et al. 2023). The ultrasound-assisted extraction process uses acoustic cavitation effects that result in the damage of plant cell walls, enhancing contact between the solvent and the target content of the cell (Panwar et al. 2023). This leads to an increase in mass transfer, thereby resulting in high pectin yields (Hundie 2020; Misra et al. 2018). It is recognized as an efficient and quick method for enhancing the efficiency of pectin yield (Vathsala et al. 2024). Shaking water bath extraction involves extracting pectin from citrus peel powder in a shaking water bath at different temperatures, pH levels, and times (Kamal et al. 2021). Conventional extraction involves using hot acidified water with either mineral acids such as sulfuric, hydrochloric, nitric, or organic acids like citric or tartaric (Gurev et al. 2023; Gołębiewska et al. 2022; Belkheiri et al. 2021; Wicker et al. 2014). Bioactive compound extraction involves collecting citrus residues, which are then washed, dried, and ground before bioactive compound extraction

(Anoopkumar et al. 2023). During the extraction of bio-compounds, essential oils, polyphenols, and flavonoids are removed to improve pectin's quality (David Durán-Aranguren et al. 2022).

3 Logistics Practices

Logistics activities hold significant importance for fruit safety and quality as well. Castro and Jaimes (2017) discussed logistics practices to improve the responsiveness of citrus SC systems, which can decrease CLW and enhance inventory performance through increased efficiency and reduced inventory levels. The implementation of lean and flexible SC structures can improve food security by ensuring better access for the different stakeholders in the downstream stages. Agile and available structures can enhance product availability and service levels by reducing delivery times and minimizing CLW.

On one hand, managing marketing and distribution channels through processors or contractors can positively impact farmers' profitability. On the other hand, traditional marketing through intermediaries may have negative effects on profitability (Naseer et al. 2019b). Having packaging and storage facilities in close proximity to production regions can improve marketing performance and reduce the risk of perishability (Naseer et al. 2019a). However, the presence of numerous intermediaries within the citrus SC system and low offered prices to farmers can negatively impact citrus marketing (Usman et al. 2018). In decentralized citrus SC systems, backup agreement contracts can enhance stakeholders' income (Rojas Palacios et al. 2022; Munir et al. 2022).

Numerous scholars have examined packaging materials such as crates, bags, and other options. For instance, in a study by Munir et al. (2022), various packaging materials were tested, including colored polyethylene bags (PM1), transparent polyethylene bags (PM2), intermediate polyethylene bag (PM3), hard polyethylene bag (PM4), gunny bags (PM5), card boxes (PM6), plastic boxes (PM7), and plastic net (PM8). The main finding of the study highlighted the superior performance of polyethylene bags, particularly PM1, as they maintained the physiochemical characteristics of the fruit for a longer duration compared to other packaging materials. Similarly, in another research (Dabargainya et al. 2022), experiments were conducted to explore different packaging scenarios. These scenarios included no packaging materials, cardboard packaging, newspaper wrapping, plastic packaging without any perforation, plastic packaging with three perforations, plastic packaging with six perforations, and rice straw packaging. The study concluded that plastic packaging exhibited better performance in terms of color, physiological weight loss, juice content, total soluble solids, pH, and total acidity, particularly when six perforations were present. For instance, the weight loss was reduced by 8.56%, and the extracted juice content was approximately 41.07%.

Improper selection of packaging materials during transportation can result in an increase in CLW. Alzubi et al. (2022) conducted an experiment comparing two types of packaging crates made from different materials: polypropylene and polystyrene. The study concluded that the use of polypropylene crates can reduce CLW during transportation by 69.2%. Similar findings were also reported by Liao et al. (2020) and Afloukou et al. (2020). Additionally, Afloukou et al. (2020) suggested storing cultivated citrus in polyethylene bags. Based on the analysis conducted by Alzubi et al. (2022), the recommendation was to opt for plastic crates due to their reusability. In case of any damage, these crates can be recycled (Liao et al. 2020). The utilization of reusable packaging crates would contribute to a decrease in environmental impacts by maximizing return trips and reducing natural resource consumption (Ribal et al. 2019).

From another point of view, Wu et al. (2019a) conducted a comparison among three packaging boxes in a cold storage context, namely, a fully closed standard box, super vent box also known as ventilated carton box, and open-top box, focusing on their environmental impacts. The study found that the use of super vent boxes resulted in the least environmental impacts, as it required less energy for cooling in the precooling unit and during transportation in cold containers.

Optimizing storage conditions with appropriate temperature control is crucial for extending the marketing period of citrus. For oranges, the recommended storage temperature range is 0–6.7 °C, while for mandarins, it is 6–7 °C. It is important to note that lower temperatures contribute to a longer lifespan (Lambert et al. 2014). However, according to Lambert et al. (2014) and Liu (2010), an optimum temperature of 15 °C is recommended for storing citrus for up to 3 months. Yet, in the absence of packaging, it is necessary to maintain a high humidity level, such as 85–95% (Lambert et al. 2014; Roghanian and Cheraghalipour 2019). Implementing technologies like the electric nose detects odor changes in stored citrus products (Jiang et al. 2017). Treating oranges with 24-epibrassinolide (Br) before storage extends their lifespan and reduces losses (Habibi et al. 2021). Cold storage and transportation minimize citrus deterioration (Lambert et al. 2014). Cave or refrigerated storage maintains citrus quality but has high construction and maintenance costs.

In transport, the optimum cooling degree for truck transportation is 6–7 °C (Ribal et al. 2019). Increasing precooling hours for exported citrus reduces temperature variability and ensures proper cold treatment (Khumalo et al. 2023a). Using solar energy for the precooling stage in cold storage can reduce environmental impacts (Wu et al. 2019a). Similarly, Alzubi et al. (2023a) discussed the use of solar energy to charge vehicles for transportation between citrus farms, which reduces environmental impacts and operating costs for packing stores and collection points.

3.1 *Reverse Logistics and Closed-Loop Supply Chains*

Improving responsiveness within the citrus SC minimizes losses throughout all stages and transportation activities (Castro and Jaimes 2017). Likely, optimizing

CLSCs for citrus fruit crates can yield favorable environmental and economic outcomes.

Several authors (Alzubi et al. 2023b; Liao et al. 2020; Roghanian and Cheraghalipour 2019; Goodarzian et al. 2023) discussed the importance of CLSCs and RL in addressing environmental issues and waste management in SCs. The studies highlighted the need for recycling plastic citrus crates in CLSC networks. Returning empty crates (Liao et al. 2020) and using CLW as animal feed (Alzubi and Noche 2022b) for the farm can also contribute to citrus SC optimization.

Optimizing transportation routes results in reduced diesel consumption, associated costs, and CO_2 emissions (Beccali et al. 2009). Effective planning among stakeholders also lowers transportation and logistics expenses. Adequate truck capacity is necessary to enhance responsiveness (Goodarzian et al. 2022). Minimizing transportation costs and CO_2 emissions is directly linked to the minimum order quantity (Galal and El-Kilany 2016). Similarly, based on a simulation study of a two-echelon orange SC, Amer et al. (2018) analyzed the effect of changing order quantities on sustainability measures. Implementing a pull system, in which supply depends on demand, and setting order quantities improves overall sustainability of the citrus SC and substantially impacts various SC performance measures. However, a suitable value needs to be chosen carefully, since small variations in the order quantity could double food waste.

Choosing the appropriate transportation mode can significantly decrease CO_2 emissions. For instance, rail transport from Florida to New York City emits five times less CO_2 than truck transport (Bell and Horvath 2020). However, stacking excessive citrus based on transportation and packaging capacity can stress the bottom layers and cause damage (Alzubi et al. 2022). Therefore, implementing closed-loop systems to recycle CLW decreases CO_2 emissions (Goodarzian et al. 2023; Roghanian and Cheraghalipour 2019; Alzubi and Noche 2022b).

Several studies have applied different algorithms and simulation models to develop sustainable citrus SC networks. Roghanian and Cheraghalipour (2019) focused on optimizing closed-loop citrus SC using meta-heuristic algorithms. Designing a suitable SC network, including both forward and RL, helps reduce environmental impact and improve efficiency and responsiveness as well as customer satisfaction. To transport products throughout the network, various types of vehicles using different fuel sources can be employed. Additionally, employing a fleet of vehicles for citrus transportation, implementing RL operations to transport organic waste to recycling facilities, and closing the loop by returning compost as fertilizer can be integrated (Roghanian and Cheraghalipour 2019; Alzubi et al. 2023a).

The use of multiple transportation vehicles can reduce CO_2 emissions, minimize total costs, and maximize the responsiveness of both forward and reverse demands. Hence, implementing RL in the citrus SC not only adds a value to returned products but also helps to reduce pollution and environmental damage while optimizing the allocation and inventory of products.

3.2 Cold Chain

The cold chain, a critical component, involves maintaining a controlled temperature environment during transportation and storage, ensuring the freshness and quality of citrus products (Vrat et al. 2018). This is particularly relevant to citrus, which is sensitive to temperature fluctuations. Citrus fruits are highly perishable and require proper handling and storage to maintain their quality and shelf life. One of the key aspects of the citrus cold chain is the use of temperature monitoring devices to ensure that the fruits are kept at the optimal temperature for transportation and consumption. However, there are different methods of temperature monitoring, each with its own advantages and disadvantages. There are mainly three methods used for the citrus cold chain: forced-airflow precooling (Wu et al. 2019b), ambient loading (Berry et al. 2021; Defraeye et al. 2015), and ambient cooling (Berry et al. 2021; Defraeye et al. 2015).

The forced-airflow precooling involves forcing cold air horizontally through the already packaged fruits at high airflow rates. It is the most common technique used in the cold chain, especially in South Africa where a significant amount of citrus is exported (Wu et al. 2019a, b). However, this cooling process often experiences delays if there are long distances from the field to the precooling center or if the center is already at capacity (Wu et al. 2019b).

The ambient loading palletized fruit bypass a precooling facility and are loaded warm into the reefer container, after which they are cooled within the container (Berry et al. 2020). This approach is a highly valued solution toward reducing stress on facilities, costs, and waiting times (Defraeye et al. 2015; Berry et al. 2020).

The ambient cooling, also known as static cooling, is the process where fruits are cooled in a large cold room before shipment (Wu et al. 2019b). This method can result in larger losses compared to the two other methods due to heat transfer from other products or equipment in the room (Berry et al. 2013, 2021). In this method, pallets of citrus fruits are placed in a cold room for a certain period to achieve the desired pulp temperature of 4 °C (Berry et al. 2021).

After post-processing and palletization, citrus fruits are transported directly into precooling storage centers to rapidly remove heat. This process, often referred to as the cold chain, is crucial for maintaining a standard temperature, which is imperative to exporting and consumption. A decrease in product temperature by 10 °C from ambient conditions typically doubles the shelf life (Wu et al. 2019b). Ambient loading and ambient cooling are gaining traction. Ambient loading, where fruits are loaded into a refrigerated container at an ambient, warm, temperature, produces similar results in fruit quality to precooling. Ambient cooling, also called static cooling, refers to the process, where fruits are cooled in a large cold room before shipment results in larger losses compared to the two other methods (Wu et al. 2019a).

All three cooling methods use monitors for consistency. Typical wired temperature monitoring loggers are low cost but lack consistency or readily available data across transit. These wired monitors also need to be calibrated before use. The fruit at the welcoming port can be rejected if one of three USDA probes is above the protocol temperature (Khumalo et al. 2023b). Newer innovations in cellular monitors show promise for consistent temperature tracking but are restricted in effectiveness by the proximity of cellular towers and need to be manually reset if malfunctioning (Khumalo et al. 2023b).

3.3 Waste Management Practices

Sustainability operational practices in the context of citrus SCs involve a dynamic collaboration of various stakeholders and activities (Sharma et al. 2023; Fritz et al. 2022). Waste management practices are integral, addressing issues such as by-product utilization, recycling, and composting to minimize environmental impact (Raimondo et al. 2018; Gaudillat et al. 2018). Valorizing citrus processing waste via green extraction techniques and biochemical processes holds significant potential (Satari and Karimi 2018) to manage orange peel waste effectively. Biomass can serve as a sustainable resource for biorefineries, producing various products such as pectin. Additionally, citrus waste can be utilized as a biosorbent (for instance, for the removal of contaminants from polluted water), for the production of nanomaterials, and as organic fertilizer. Green extraction techniques can be applied to recover pectin, bioactive compounds, and essential oils. Furthermore, citrus waste is useful to obtain derivatives that serve as a source for energy generation.

Ortiz et al. (2019) evaluated two strategies to transition toward cleaner, small-scale orange juice industries: biogas production and incineration of peel waste and coal. By comparing these scenarios, the authors found that burning a mixture of coal (20%) and dried orange peels (80%) is the optimal solid fuel combination, as it reduces environmental impacts and disposal costs. However, while biogas production is the most environmentally friendly option, its implementation is not feasible for small- and medium-sized companies due to high investment costs.

Nevertheless, biogas can be economically generated through the anaerobic digestion of citrus effluents discarded during citrus processing (Rosas-Mendoza et al. 2020). Anaerobic digestion refers to the biological conversion of organic matter into different products, such as biogas. In a case study approach, the authors discuss the process of electrical energy production from biogas in a citrus processing plant in Veracruz, Mexico. The liquid residues from citrus processing cause environmental, social, and economic problems but possess physiochemical characteristics that are favorable for anaerobic digestion. By integrating sourcing, citrus production, processing, and distribution along the SC, electricity generated by biogas can be used in the citrus processing plants. This way, bioenergy SCs can be established as an important pillar for renewable energy. Due to these practices, 3962 kWh of net electricity per day could be produced from citrus effluents, which would meet 5.9% of the total daily demand. On a national scale, wastewater from the Mexican citrus industry treated by anaerobic digestion could cover 3% of national energy production.

Despite anaerobic digestion being more environmentally friendly compared to incineration and using coal and landfill as sources of energy, it has higher costs (Ortiz et al. 2020). A suitable waste management practice needs to be defined based on local circumstances (Alzubi et al. 2023a), such as technological and economic feasibility.

3.4 International Trade and Market Entrance

Based on interviews and questionnaires with citrus exporting firms in the eastern Mediterranean region of Türkiye, Aslan et al. (2020) analyzed the effects of integrated planning within the citrus sector. Small-scale farmers often face challenges in collecting and processing fruits with the required product compliance processes requested by international markets. To address the challenge of meeting export standards, horizontal integration at regional and national levels can enhance the performance of stakeholders and the competitiveness of a region. This integration is characterized by consolidated citrus harvesting, purchasing, and processing operations, which are closely supervised by associations or cooperatives to ensure adherence to required practices. For example, citrus harvesting within the framework of a regional association helps producers implement and maintain proper growing and pesticide practices ensuring sustainability of production and product quality. A horizontally integrated purchasing strategy helps reduce the time to procure fruits with certain standards demanded by international markets. These approaches strengthen suppliers' positions to meet changing market demands as well as eliminate the risk of unavailability of products with desired qualities.

Furthermore, backup agreements can cultivate commercial relationships between farmers and retailers, enhancing profits, particularly for smaller-scale farmers. Favorable demand conditions frequently encourage increased production and reduced losses. However, smaller-scale farmers may struggle with these demand conditions that favor increased production, making it challenging for them to absorb losses and remain financially viable (Rojas Palacios et al. 2022). Therefore, these agreements can provide a safety net for smaller producers, ensuring their continued operation in the face of fluctuating market conditions.

4 Conclusions

The citrus SC presents a diverse array of stakeholders and activities, covering farming, processing, retailing, and transportation. This complex network faces challenges associated with CLW at various stages, leading to sustainability decrease across the entire chain. Research conducted globally showcases diverse strategies aimed at addressing these challenges. Different countries emphasize distinct approaches—Egypt and the United States focus on agricultural practices, and Italy leans toward LCA to evaluate those practices in the farming stage, while Brazil and China balance efforts across farming and processing stages.

Significant steps in managing CLW have been made in countries like India and Mexico, where waste is transformed into valuable products like pectin, bioactive compounds, oils, and energy sources. Technological innovations, including green extraction techniques and biochemical processes, offer promising ways to reduce environmental impact and enhance efficiency. Yet, implementing these advancements faces obstacles such as high costs, technological feasibility, and local unfavorable infrastructure and political conditions.

Sustainable practices encompass a wide spectrum, including proper nutrition for trees, efficient land preparation, organic fertilizer utilization, and efficient irrigation management systems. These practices contribute significantly to tree health, yield quality, and the citrus industry's overall sustainability. Optimizing agricultural and operational practices pre-harvesting and during harvesting is pivotal for improving fruit quality and quantity. Combining organic and chemical fertilizers maximizes nutritional value, while precise management practices help to utilize resources effectively.

Efficient logistics operations are the key to increase sustainability and resilience across the entire citrus SC. Selecting suitable packaging materials, adopting cold logistics technology, and integrating renewable energy sources contribute effectively to reducing CLW and associated costs. Optimizing transportation routes, particularly in CLSCs, enhances efficiency and decreases fuel consumption. SC practices are key in sustainability efforts, focusing on biomass waste management and valorization. Identifying feasible valorization procedures and implementing circular practices like CLSCs significantly reduce environmental impact and operational costs within SCs. Horizontal integration of SCs at regional and national levels boosts competitiveness and ensures compliance with international sustainability standards.

In conclusion, while several advancements have been made at the various stages of the SCs, increased sustainability across the entire citrus SC needs continued collaboration, socio-technical innovation, and knowledge-sharing of best practices. Therefore, continuous research and development are needed to improve these methods and find innovative solutions for the citrus SC. This can be achieved by leveraging diverse strategies and technologies, while promoting a CE to unlock the full potential in sustainability transformation within the citrus industries.

References

Abbas F, Fares A (2009) Best management practices in citrus production. Available online at https://api.semanticscholar.org/CorpusID:198976110

Abobatta W (2018a) Challenges for citrus production in Egypt. Acta Sci Agric 2:40–41

Abobatta W (2018b) Development growth and productivity of orange orchards (Citrus Sinensis L) in Egypt (delta region). Adv Agric Technol Plant Sci 1:1–11

Abobatta W (2019) Management of alternative bearing in citrus varieties-review. Adv Agric Technol Plant Sci 2:180028

Abobatta WF (2020) Citriculture and climate change. Mod Concepts Dev Agron 6(3). https://doi.org/10.31031/MCDA.2020.06.000639

Abobatta WF (2021) Managing citrus orchards under climate change. MOJ Ecol Environ Sci 6(2):43–44. https://doi.org/10.15406/mojes.2021.06.00212

Afloukou F, Zinsou V, Onelge N (2020) Citrus in Benin Republic: past, present, and future challenges. Citrus Res Technol 41:e1060. https://doi.org/10.4322/crt.20820

Ahrisa D, Arhim M, Arsyad M (2021) Collaboration practice in supply chain of unshu citrus in Muchacha-En, Ltd, Japan. IOP Conf Ser Earth Environ Sci 681(1):12073. https://doi.org/10.1088/1755-1315/681/1/012073

Akyýldýz, A, Burcu DK, Erdal A (2023) Orange juice processing and quality. In: Mérillon J-M, Riviere C, Lefèvre G (eds) Natural products in beverages. Springer International Publishing, Cham (Reference Series in Phytochemistry), pp 1–29.

Al Hiary M (2012) Estimation of post harvest losses of some fruit crops in Jordan. J King Saud Univ Sci 24:151–162

Aleixos N, Blasco J, Molto E, Navarron F (2000) Assessment of citrus fruit quality using a real-time machine vision system. In: Proceedings 15th international conference on pattern recognition. ICPR-2000, Barcelona, Spain, 3–7 September 2000. IEEE Computer Society, pp 482–485

Aleixos N, Blasco J, Navarrón F, Moltó E (2002) Multispectral inspection of citrus in real-time using machine vision and digital signal processors. Comput Electron Agric 33(2):121–137. https://doi.org/10.1016/S0168-1699(02)00002-9

Alnaimy A (2017) Using of citrus by-products in farm animals feeding. Open Access J Sci 1(3). https://doi.org/10.15406/OAJS.2017.01.00014

Al-Sammarraie MAJ, Gierz Ł, Przybył K, Koszela K, Szychta M, Brzykcy J, Baranowska HM (2022) Predicting fruit's sweetness using artificial intelligence—case study: orange. Appl Sci 12(16):8233. https://doi.org/10.3390/app12168233

Alzubi E, Noche B (2022a) A multi-objective model to find the sustainable location for citrus hub. Sustainability 14(21):14463. https://doi.org/10.3390/su142114463

Alzubi E, Noche B (eds) (2022b) Improving sustainability of orange supply chain: a system dynamics model to eliminating pre-harvesting loss, increase workers, to improve farmer's profit. In: The international conference on industrial engineering and operations management, Istanbul, Turkey, 7–10 March 2022. Available online at https://www.researchgate.net/publication/361983366_Improving_Sustainability_of_Orange_Supply_Chain_A_System_Dynamics_Model_to_Eliminating_Pre-Harvesting_Loss_Increase_Workers_to_Improve_Farmer's_Profit. Checked on 5/12/2023

Alzubi E, Kassem A, Noche B (2022) A comparative life cycle assessment: polystyrene or polypropylene packaging crates to reduce citrus loss and waste in transportation? Sustainability 14(19):12644. https://doi.org/10.3390/su141912644

Alzubi E, Kassem A, Melkonyan A, Noche B (2023a) Enhancing economic-social sustainability through a closed-loop citrus supply chain: a life cycle cost analysis. Resour Conserv Recyl Adv:200199. https://doi.org/10.1016/j.rcradv.2023.200199

Alzubi E, Shbikat N, Noche B (2023b) A system dynamics model to improving sustainable performance of the citrus farmers in Jordan Valley. Cleaner Prod Lett 4:100034. https://doi.org/10.1016/j.clpl.2023.100034

Amer HH, Galal NM, El-Kilany KS (2018) A simulation study of sustainable agri-food supply chain

Anoopkumar AN, Aneesh EM, Sirohi R, Tarafdar A, Kuriakose LL, Surendhar A et al (2023) Bioactives from citrus food waste: types, extraction technologies and application. J Food Sci Technol. https://doi.org/10.1007/s13197-023-05753-9

Arce A, Soto A (2008) Citrus essential oils: extraction and deterpenation. Available online at https://api.semanticscholar.org/CorpusID:33776612

Arora S, Kataria P, Ahmad W, Mishra R, Upadhyay S, Dobhal A et al (2023) Microwave assisted green extraction of pectin from citrus maxima albedo and flavedo, process optimization, characterisation and comparison with commercial pectin. Food Anal Methods. https://doi.org/10.1007/s12161-023-02547-9

Aslan H, Zereyak E, Tüzün Rad S (2020) Modelling of citrus production and export process: eastern Mediterrenean region of Turkey. Karadeniz Uluslararası Bilimsel Dergi 1(48):169–187. https://doi.org/10.17498/kdeniz.801203

Barbosa PDPM, Ruviaro AR, Macedo GA (2018) Comparison of different Brazilian citrus by-products as source of natural antioxidants. Food Sci Biotechnol 27(5):1301–1309. https://doi.org/10.1007/s10068-018-0383-4

Barrales FM, Silveira P, Barbosa PDPM, Ruviaro AR, Paulino BN, Pastore GM et al (2018) Recovery of phenolic compounds from citrus by-products using pressurized liquids – an application to orange peel. Food Bioprod Process 112:9–21. https://doi.org/10.1016/J.FBP.2018.08.006

Bates RP, Morris JR, Crandall PG (2001) Principles and practices of small- and medium-scale fruit juice processing. Food and Agriculture Organization of the United Nations (FAO agricultural services bulletin, 146), Rome

Beccali M, Cellura M, Iudicello M, Mistretta M (2009) Resource consumption and environmental impacts of the agrofood sector: life cycle assessment of Italian citrus-based products. Environ Manag 43(4):707–724. https://doi.org/10.1007/s00267-008-9251-y

Beccali M, Cellura M, Iudicello M, Mistretta M (2010) Life cycle assessment of Italian citrus-based products. Sensitivity analysis and improvement scenarios. J Environ Manag 91(7):1415–1428. https://doi.org/10.1016/j.jenvman.2010.02.028

Begum M, Marium B, Farid MS, Hasan M (2022) Post-harvest loss assessment and marketing practices of fruits: an empirical study of Maulvibazar District in Bangladesh. J Econ Manage Trade:15–27. https://doi.org/10.9734/JEMT/2022/v28i130385

Belkheiri A, Forouhar A, Ursu AV, Dubessay P, Pierre G, Delattre C et al (2021) Extraction, characterization, and applications of pectins from plant by-products. Appl Sci 11(14):6596. https://doi.org/10.3390/app11146596

Bell EM, Horvath A (2020) Modeling the carbon footprint of fresh produce: effects of transportation, localness, and seasonality on US orange markets. Environ Res Lett 15(3):34040. https://doi.org/10.1088/1748-9326/ab6c2f

Berry T, Delele MA, Opara UL, Cronjé PJR (2013) Spatial variation of fruit cooling rate and relative humidity inside 'super-vent' packaging for citrus during static cooling. Acta Hortic 1007:207–213. https://doi.org/10.17660/ActaHortic.2013.1007.20

Berry TM, Defraeye T, Sibiya GM, North J, Cronje PJR (2020) Ambient loading – increasing citrus cold chain efficiency. Acta Hortic 1275:323–328. https://doi.org/10.17660/ActaHortic.2020.1275.44

Berry TM, Defraeye T, Wu W, Sibiya MG, North J, Cronje PJR (2021) Cooling of ambient-loaded citrus in refrigerated containers: what impacts do packaging and loading temperature have? Biosyst Eng 201:11–22. https://doi.org/10.1016/j.biosystemseng.2020.11.002

Boman B (2012) Citrus best management practices. In: Srivastava AK (ed) Advances in citrus nutrition. Springer Netherlands, Dordrecht, pp 391–413

Boman BJ, Obreza TA, Morgan KT (2020) Nutrition of Florida citrus trees, 3rd edition: Chapter 10. Environmental issues and best management practices. EDIS 2020(2). https://doi.org/10.32473/edis-ss677-2020

Braham M, Boulahia-Kheder S, Kahia M, Nouira S (2023) Aphids and citrus responses to nitrogen fertilization. J Saudi Soc Agric Sci. https://doi.org/10.1016/j.jssas.2023.03.003

Caixeta-Filho JV (2006) Orange harvesting scheduling management: a case study. J Oper Res Soc 57(6):637–642. https://doi.org/10.1057/palgrave.jors.2602041

Castro JAO, Jaimes WA (2017) Dynamic impact of the structure of the supply chain of perishable foods on logistics performance and food security. J Ind Eng Manage 10(4):687. https://doi.org/10.3926/jiem.2147

Chavan P, Singh AK, Kaur G (2018) Recent progress in the utilization of industrial waste and by-products of citrus fruits: a review. J Food Process Eng 41(8):e12895. https://doi.org/10.1111/jfpe.12895

Chen LA, House L (2021) An overview of the grapefruit market in the U.S. EDIS 2021(3). https://doi.org/10.32473/edis-fe1095-2021

Chen Y, Barzee TJ, Zhang R, Pan Z (2019) Citrus. In: Integrated processing technologies for food and agricultural by-products. Elsevier, pp 217–242

Cheraghalipour A, Paydar MM, Hajiaghaei-Keshteli M (2018) A bi-objective optimization for citrus closed-loop supply chain using Pareto-based algorithms. Appl Soft Comput 69:33–59. https://doi.org/10.1016/j.asoc.2018.04.022

Chhetri L, Kandel B (2019) Intensive fruit cultivation technology of citrus fruits: high density planting: a brief review. J Agric Stud 7:63–74. https://doi.org/10.5296/jas.v7i2.14865

Costa L, Archer L, Ampatzidis Y, Casteluci L, Caurin GAP, Albrecht U (2021) Determining leaf stomatal properties in citrus trees utilizing machine vision and artificial intelligence. Precis Agric 22(4):1107–1119. https://doi.org/10.1007/s11119-020-09771-x

CRW (2023) Citrus Packhouse Best Practice Handbook. Workflow 4 - Packhouse Management. Citrus Resource Warehouse. Available online at https://www.citrusresourcewarehouse.org.za/home/document-home/best-practice-handbook/citrus-packhouse-best-practice-handbook/workflow-4-packhouse-management, checked on 9/30/2023

Cubero S, Lee WS, Aleixos N, Albert F, Blasco J (2016) Automated systems based on machine vision for inspecting citrus fruits from the field to postharvest—a review. Food Bioprocess Technol 9(10):1623–1639. https://doi.org/10.1007/s11947-016-1767-1

Dabargainya B, Acharya B, Acharya P (2022) Effect of different packaging materials on post-harvest life of mandarin (Citrus reticulata blanco). Rev Food Agric 3(2):87–91. https://doi.org/10.26480/rfna.02.2022.87.91

De Luca AI, Iofrida N, Strano A, Falcone G, Gulisano G (2015) Social life cycle assessment and participatory approaches: a methodological proposal applied to citrus farming in Southern Italy. Integr Environ Assess Manag 11(3):383–396. https://doi.org/10.1002/ieam.1611

Defraeye T, Verboven P, Opara UL, Nicolai B, Cronjé P (2015) Feasibility of ambient loading of citrus fruit into refrigerated containers for cooling during marine transport. Biosyst Eng 134:20–30. https://doi.org/10.1016/j.biosystemseng.2015.03.012

Durán-Aranguren DD, Ramírez CJA, Díaz LCV, Valderrama MA, Sierra R (2022) Production of pectin from citrus residues: process alternatives and insights on its integration under the biorefinery concept. In: Masuelli MA (ed) Pectins – the new-old polysaccharides. IntechOpen

Echavarría AP, Torras C, Pagán J, Ibarz A (2011) Fruit juice processing and membrane technology application. Food Eng Rev 3(3–4):136–158. https://doi.org/10.1007/s12393-011-9042-8

Fakayode OA, Abobi KE (2018) Optimization of oil and pectin extraction from orange (Citrus sinensis) peels: a response surface approach. J Anal Sci Technol 9(1). https://doi.org/10.1186/s40543-018-0151-3

Fermo IR, Cavali TS, Bonfim-Rocha L, Srutkoske CL, Flores FC, Andrade CMG (2021) Development of a low-cost digital image processing system for oranges selection using hopfield networks. Food Bioprod Process 125:181–192. https://doi.org/10.1016/j.fbp.2020.11.012

Ferreira JO, Batalha MO, Domingos JC (2016) Integrated planning model for citrus agribusiness system using systems dynamics. Comput Electron Agric 126:1–11. https://doi.org/10.1016/j.compag.2016.04.029

Ferreira G, Sobral L, Barreto DW, Calado V (2020) Thermal and compositional analysis of orange essential oil obtained from citrus industry waste. J Adv Therm Sci Res 7:48–55. https://doi.org/10.15377/2409-5826.2020.07.6

Fritz MMC, Silva ME, Touboulic A (2022) Practicing sustainability in operations and supply chain management. Supply Chain Forum Int J 23(4):323–328. https://doi.org/10.1080/16258312.2022.2138160

Galal NM, El-Kilany KS (2016) Sustainable agri-food supply chain with uncertain demand and lead time. Int J Simul Model 15(3):485–496. https://doi.org/10.2507/IJSIMM15(3)8.350

Ganem F, Mattedi S, Rodríguez O, Rodil E, Soto A (2020) Deterpenation of citrus essential oil with 1-ethyl-3-methylimidazolium acetate: a comparison of unit operations. Sep Purif Technol 250:117208. https://doi.org/10.1016/j.seppur.2020.117208

Gaudillat P, Antonopoulos IS, Canfora P, Dri M (2018) Best environmental management practice for the waste management sector. In: Learning from frontrunners, EUR, Scientific and technical research series, 29136. Publications Office, Luxembourg

Gehrke V, Maron GK, da Silva Rodrigues L, Alano JH, de Pereira CMP, Orlandi MO, Carreño NLV (2021) Facile preparation of a novel biomass-derived H3PO4 and $Mn(NO_3)_2$ activated carbon from citrus bergamia peels for high-performance supercapacitors. Mater Today Commun 26:101779. https://doi.org/10.1016/j.mtcomm.2020.101779

Ghani A, Mohtashami S, Jamalian S (2021) Peel essential oil content and constituent variations and antioxidant activity of grapefruit (Citrus × paradisi var. red blush) during color change stages. Food Meas 15(6):4917–4928. https://doi.org/10.1007/s11694-021-01051-0

Gołębiewska E, Kalinowska M, Yildiz G (2022) Sustainable use of apple pomace (AP) in different industrial sectors. Materials (Basel, Switzerland) 15(5). https://doi.org/10.3390/ma15051788

Goodarzian F, Kumar V, Ghasemi P (2022) Investigating a citrus fruit supply chain network considering CO2 emissions using meta-heuristic algorithms. Ann Oper Res. https://doi.org/10.1007/s10479-022-05005-7

Goodarzian F, Ghasemi P, Santibanez Gonzalez E, Tirkolaee EB (2023) A sustainable-circular citrus closed-loop supply chain configuration: pareto-based algorithms. J Environ Manag 328:116892. https://doi.org/10.1016/j.jenvman.2022.116892

Gurev A, Cesko T, Dragancea V, Ghendov-Mosanu A, Pintea A, Sturza R (2023) Ultrasound- and microwave-assisted extraction of pectin from apple pomace and its effect on the quality of fruit bars. Foods (Basel, Switzerland) 12(14). https://doi.org/10.3390/foods12142773

Habibi F, Serrano M, Zacarías L, Valero D, Guillén F (2021) Postharvest application of 24-epibrassinolide reduces chilling injury symptoms and enhances bioactive compounds content and antioxidant activity of blood orange fruit. Front Plant Sci 12:629733. https://doi.org/10.3389/fpls.2021.629733

Hundie KB (2020) Ultrasound-assisted optimization of pectin extraction from orange peel using response surface methodology (RSM) and artificial neural network (ANN). Int J Appl Sci Eng Technol 8(1). https://doi.org/10.30954/2322-0465.2.2020.1

Jiang X, Jia P, Qiao S, Duan S (2017) Odor change of citrus juice during storage based on electronic nose technology. In: Liu D, Xie S, Li Y, Zhao D, El-Alfy E-SM (eds) Neural information processing, Lecture notes in computer science, vol 10638. Springer International Publishing, Cham, pp 317–326

Joglekar SN, Pathak PD, Mandavgane SA, Kulkarni BD (2019) Process of fruit peel waste biorefinery: a case study of citrus waste biorefinery, its environmental impacts and recommendations. Environ Sci Pollut Res Int 26(34):34713–34722. https://doi.org/10.1007/s11356-019-04196-0

Kadyampakeni DM, Duncan LW (2022) 2022–2023 Florida citrus production guide: best management practices for soil-applied agricultural chemicals. EDIS. https://doi.org/10.32473/edis-cg027-2022

Kamal MM, Kumar J, Mamun MAH, Ahmed MNU, Shishir MRI, Mondal SC (2021) Extraction and characterization of pectin from Citrus sinensis peel. J Biosyst Eng 46(1):16–25. https://doi.org/10.1007/s42853-021-00084-z

Khedkar R (2024) A novel sustainable approach for extraction of pectin from citrus and dairy waste. In: Thakur M (ed) Sustainable food systems, World sustainability series, vol II. Springer Nature Switzerland, Cham, pp 47–68

Khumalo G, Goedhals-Gerber L, Cronje P, Berry T (2023a) Inefficiency in land-side cold-chain logistics: solutions to improve the handling of citrus during preparation for cold-treatment protocols. Soc Sci Humanit 8(1):100500. https://doi.org/10.1016/j.ssaho.2023.100500

Khumalo G, Goedhals-Gerber LL, Cronje P, Berry T (2023b) Product visibility in the South African citrus cold chain: examining the efficacy of temperature loggers. Heliyon 9(1):e12732. https://doi.org/10.1016/j.heliyon.2022.e12732

Kimball D (1991) Citrus processing varieties. In: Kimball D (ed) Citrus processing. Springer Netherlands, Dordrecht, pp 180–210

Lahlali R, Jaouad M, Moinina A, Mokrini F, Belabess Z (2021) Farmers' knowledge, perceptions, and farm-level management practices of citrus pests and diseases in Morocco. J Plant Dis Prot 128(5):1213–1226. https://doi.org/10.1007/s41348-021-00479-2

Lambert GF, Lasserre AAA, Ackerman MM, Sánchez CGM, Rivera BOI, Azzaro-Pantel C (2014) An expert system for predicting orchard yield and fruit quality and its impact on the Persian lime supply chain. Eng Appl Artif Intell 33:21–30. https://doi.org/10.1016/j.engappai.2014.03.013

Lanza CM (2003) Citrus fruits | Processed and derived products of oranges. In: Encyclopedia of food sciences and nutrition. Elsevier, pp 1346–1354

Larios-Cruz R, Rodríguez-Jasso RM, Ruiz HA, Prado-Barragán A, Wong-Paz JE, Rodríguez-Herrera R et al (2018) Utilization of citrus waste biomass for antioxidant production by solid-state fermentation. In: Singhania RR, Agarwal RA, Praveen Kumar R, Sukumaran RK (eds) Waste to wealth, Energy, environment, and sustainability. Springer Singapore, Singapore, pp 83–96

Li S, Wu F, Duan Y, Singerman A, Guan Z (2020) Citrus greening: management strategies and their economic impact. HortScience 55(5):604–612. https://doi.org/10.21273/HORTSCI14696-19

Liao Y, Kaviyani-Charati M, Hajiaghaei-Keshteli M, Diabat A (2020) Designing a closed-loop supply chain network for citrus fruits crates considering environmental and economic issues. J Manuf Syst 55:199–220. https://doi.org/10.1016/j.jmsy.2020.02.001

Liu F-W (2010) Development and application of citrus storage technologies with concurrent consideration of fruit quality preservation, energy use, and costs

Magare ME, Sahu N, Kanade GS, Chanotiya CS, Thul ST (2020) An Integrated process of value addition to citrus waste and performance of fenton process for its conversion to biogas. Waste Biomass Valoriz 11(1):165–172. https://doi.org/10.1007/s12649-018-0385-8

Mahawer AK, Sharma RM, Dubey AK, Awasthi OP, Singh D, Dahuja A et al (2023) Effect of weather parameters and citrus genotypes on the occurrence of citrus canker incited by Xanthomonas citri pv. citri. Indian Phytopathol. https://doi.org/10.1007/s42360-023-00606-z

Mamma D, Christakopoulos P (2008) Citrus peels: an excellent raw material for the bioconversion into value-added products

Mariana O-S, Alzate C, Ariel C (2021) Comparative environmental life cycle assessment of orange peel waste in present productive chains. J Clean Prod 322:128814. https://doi.org/10.1016/j.jclepro.2021.128814

Mérillon J-M, Riviere C, Lefèvre G (eds) (2023) Natural products in beverages. Springer International Publishing, Cham (Reference Series in Phytochemistry)

Meryem S, Mohamed D, Nour-eddine C, Faouzi E (2023) Chemical composition, antibacterial and antioxidant properties of three Moroccan citrus peel essential oils. Sci Afr 20:e01592. https://doi.org/10.1016/j.sciaf.2023.e01592

Misra NN, Martynenko A, Chemat F, Paniwnyk L, Barba FJ, Jambrak AR (2018) Thermodynamics, transport phenomena, and electrochemistry of external field-assisted nonthermal food technologies. Crit Rev Food Sci Nutr 58(11):1832–1863. https://doi.org/10.1080/10408398.2017.1287660

Morgan KT, Kadyampakeni DM, Zekri M, Schumann AW, Vashisth T, Obreza TA, Ferrarezi R (2022) 2022–2023 Florida citrus production guide: nutrition management of citrus trees. EDIS. https://doi.org/10.32473/edis-cg091-2022

Munir A, Khan S, Jamal A, Mihoub A, Saeed M, Ahmad S et al (2022) Effect of different packaging storage techniques on physiological and chemical changes in Citrus sinensis, vol L 44, pp 45–53

Mushtaq M (2018) Chapter 8: Extraction of fruit juice: an overview. In: Rajauria G, Tiwari BK (eds) Fruit juices. Academic Press, San Diego, pp 131–159

Naseer M, Ashfaq M, Hassan S, Abbas A, Razzaq A, Mehdi M et al (2019a) Critical issues at the upstream level in sustainable supply chain management of agri-food industries: evidence from Pakistan's citrus industry. Sustainability 11(5):1326. https://doi.org/10.3390/su11051326

Naseer M, Mehdi M, Ashfaq M, Hassan S, Abid M (2019b) Effect of marketing channel choice on the profitability of citrus farmers: evidence form Punjab-Pakistan. Pak J Agric Sci 56:1003–1011. https://doi.org/10.21162/PAKJAS/19.8671

Navarro-Pineda FS, Equihua-Sánchez M, Grande-Acosta GK, Sacramento Rivero JC, Islas-Samperio JM, Bonales Revuelta J et al (2022) Assessment of the environmental and economic performance of heat generation from orange peels and sugarcane straw. Bioenergy Res 15(4):1747–1763. https://doi.org/10.1007/s12155-021-10326-9

Nicolò BF, De Salvo MC, Ramirez-Sanz C, Estruch V, Sanjuan N, Falcone G, Strano A (2018) Life cycle assessment applied to different citrus farming systems in Spain and Italy. Agroecol Sustain Food Syst 42(10):1092–1105. https://doi.org/10.1080/21683565.2018.1490942

Ortiz DL, Ospina S, Orrego CE (2019) Orange peel management can make the shift to cleaner small-scale orange juice industries. A Colombian case study

Ortiz DL, Batuecas E, Orrego CE, Rodríguez LJ, Camelin E, Fino D (2020) Sustainable management of peel waste in the small-scale orange juice industries: a Colombian case study. J Clean Prod 265:121587. https://doi.org/10.1016/j.jclepro.2020.121587

Ortiz-Sanchez M, Solarte-Toro JC, Orrego-Alzate CE, Acosta-Medina CD, Cardona-Alzate CA (2021) Integral use of orange peel waste through the biorefinery concept: an experimental, technical, energy, and economic assessment. Biomass Convers Biorefinery 11(2):645–659. https://doi.org/10.1007/s13399-020-00627-y

Palazzolo E, Laudicina VA, Germanà MA (2013) Current and potential use of citrus essential oils. Curr Org Chem 17(24):3042–3049. https://doi.org/10.2174/13852728113179990122

Panwar D, Saini A, Panesar PS, Chopra HK (2021) Unraveling the scientific perspectives of citrus by-products utilization: progress towards circular economy. Trends Food Sci Technol 111:549–562. https://doi.org/10.1016/J.TIFS.2021.03.018

Panwar D, Panesar PS, Chopra HK (2023) Ultrasound-assisted extraction of pectin from Citrus limetta peels: optimization, characterization, and its comparison with commercial pectin. Food Biosci 51:102231. https://doi.org/10.1016/j.fbio.2022.102231

Paudel I, Bar-Tal A, Levy GJ, Rotbart N, Ephrath JE, Cohen S (2018) Treated wastewater irrigation: soil variables and grapefruit tree performance. Agric Water Manag 204:126–137. https://doi.org/10.1016/J.AGWAT.2018.04.006

Peña Orozco DL, Gonzalez-Feliu J, Rivera L, Mejía Ramirez CA (2021) Integration maturity analysis for a small citrus producers' supply chain in a developing country. Bus Process Manag J 27(3):836–867. https://doi.org/10.1108/BPMJ-05-2020-0237

Peña-Orozco DL, Cano-Larrahondo AF, Acosta-Imbachi CG, Rivera-Cadavid L, Bravo-Bastidas JJ (2020) Perishable product sensitivity analysis in the design of an inventory control system in a three-echelon fruit supply chain. In: García-Alcaraz JL, Sánchez-Ramírez C, Avelar-Sosa L, Alor-Hernández G (eds) Techniques, tools and methodologies applied to global supply chain ecosystems, Intelligent systems reference library, vol 166. Springer International Publishing, Cham, pp 309–332

Pergola M, D'Amico M, Celano G, Palese AM, Scuderi A, Di Vita G et al (2013) Sustainability evaluation of Sicily's lemon and orange production: an energy, economic and environmental analysis. J Environ Manag 128:674–682. https://doi.org/10.1016/j.jenvman.2013.06.007

Raimondo M, Caracciolo F, Cembalo L, Chinnici G, Pecorino B, D'Amico M (2018) Making virtue out of necessity: managing the citrus waste supply chain for bioeconomy applications. Sustainability 10(12):4821. https://doi.org/10.3390/su10124821

Rajput S, Kaur S, Panesar PS, Thakur A (2023) Supercritical fluid extraction of essential oils from Citrus reticulata peels: optimization and characterization studies. Biomass Convers Biorefinery 13(16):14605–14614. https://doi.org/10.1007/s13399-022-02807-4

Ribal J, Estruch V, Clemente G, Fenollosa ML, Sanjuán N (2019) Assessing variability in carbon footprint throughout the food supply chain: a case study of Valencian oranges. Int J Life Cycle Assess 24(8):1515–1532. https://doi.org/10.1007/s11367-018-01580-9

Roghanian E, Cheraghalipour A (2019) Addressing a set of meta-heuristics to solve a multi-objective model for closed-loop citrus supply chain considering CO2 emissions. J Clean Prod 239:118081. https://doi.org/10.1016/j.jclepro.2019.118081

Rojas Palacios MN, Peña Orozco DL, Gonzalez-Feliu J (2022) Backup agreement as a coordination mechanism in a decentralized fruit chain in a developing country. Games 13(3):36. https://doi.org/10.3390/g13030036

Rosas-Mendoza ES, Palacios-Ríos JH, Méndez-Contreras JM, Vallejo-Cantú NA, Alvarado-Lassman A (2020) Designing a supply chain for the generation of bioenergy from the anaerobic digestion of citrus effluents. In: García-Alcaraz JL, Sánchez-Ramírez C, Avelar-Sosa L, Alor-Hernández G (eds) Techniques, tools and methodologies applied to global supply chain ecosystems, Intelligent systems reference library, vol 166. Springer International Publishing, Cham, pp 209–235

Sarbatly R, Sariau J, Krishnaiah D (2023) Recent developments of membrane technology in the clarification and concentration of fruit juices. Food Eng Rev 15(3):420–437. https://doi.org/10.1007/s12393-023-09346-2

Satari B, Karimi K (2018) Citrus processing wastes: environmental impacts, recent advances, and future perspectives in total valorization. Resour Conserv Recycl 129:153–167. https://doi.org/10.1016/j.resconrec.2017.10.032

Scuderi A, Matarazzo A, La Via G, Cammarata M, Timpanaro G (2022) Multi-criteria analysis of investment choices following flood damage: the case study of sustainable citrus farming in Sicily. Agriculture 12(8):1122. https://doi.org/10.3390/agriculture12081122

Sharma M, Antony R, Tsagarakis K (2023) Green, resilient, agile, and sustainable fresh food supply chain enablers: evidence from India. Ann Oper Res:1–27. https://doi.org/10.1007/s10479-023-05176-x

Strano A, Falcone G, Nicolò BF, Stillitano T, de Luca AI, Nesci FS, Gulisano G (2017) Eco-profiles and economic performances of a high-value fruit crop in southern Italy: a case study of bergamot (Citrus bergamia Risso). Agroecol Sustain Food Syst 41(9–10):1124–1145. https://doi.org/10.1080/21683565.2017.1357064

Strano MC, Altieri G, Allegra M, Di Renzo GC, Paterna G, Matera A, Genovese F (2022) Postharvest technologies of fresh citrus fruit: advances and recent developments for the loss reduction during handling and storage. Horticulturae 8(7): 612. https://doi.org/10.3390/horticulturae8070612

Suri S, Singh A, Nema PK (2022) Current applications of citrus fruit processing waste: a scientific outlook. Appl Food Res 2(1):100050. https://doi.org/10.1016/j.afres.2022.100050

Tetra Pak (2017) The orange book. In: A unique guide to orange juice production, 3rd edn. Tetra Pak International S.A, Lund. Available online at https://orangebook.tetrapak.com/

Thornsbury S, Buzby J, Hitaj C, Kantor L, Kuchler F, Ellison B et al (2020) Economic drivers of food loss at the farm and pre-retail sectors: a look at the produce supply chain in the United States

Tian H, Wang T, Liu Y, Qiao X, Li Y (2020) Computer vision technology in agricultural automation—a review. Inf Process Agric 7(1):1–19. https://doi.org/10.1016/j.inpa.2019.09.006

Usman M, Ashraf I, Chaudhary KM, Talib U (2018) Factors impeding citrus supply chain in Central Punjab, Pakistan. Int J Agric Ext 6(1):1–5. https://doi.org/10.33687/ijae.006.01.2301

Vathsala V, Singh SP, Bishnoi M, Varghese E, Saurabh V, Khandelwal A, Kaur C (2024) Ultrasound-assisted extraction (UAE) and characterization of citrus peel pectin: comparison between pummelo (Citrus grandis L. Osbeck) and sweet lime (Citrus limetta Risso). Sustain Chem Pharm 37:101357. https://doi.org/10.1016/j.scp.2023.101357

Visakh NU, Pathrose B, Narayanankutty A, Alfarhan A, Ramesh V (2022) Utilization of pomelo (Citrus maxima) peel waste into bioactive essential oils: chemical composition and insecticidal properties. Insects 13(5). https://doi.org/10.3390/insects13050480

Wicker L, Kim Y, Kim M-J, Thirkield B, Lin Z, Jung J (2014) Pectin as a bioactive polysaccharide – extracting tailored function from less. Food Hydrocoll 42:251–259. https://doi.org/10.1016/j.foodhyd.2014.01.002

Wu W, Beretta C, Cronje P, Hellweg S, Defraeye T (2019a) Environmental trade-offs in fresh-fruit cold chains by combining virtual cold chains with life cycle assessment. Appl Energy 254:113586. https://doi.org/10.1016/j.apenergy.2019.113586

Wu W, Cronjé P, Verboven P, Defraeye T (2019b) Unveiling how ventilated packaging design and cold chain scenarios affect the cooling kinetics and fruit quality for each single citrus fruit in an entire pallet. Food Packag Shelf Life 21:100369. https://doi.org/10.1016/j.fpsl.2019.100369

Yang M, Long Q, Li W, Wang Z, He X, Wang J et al (2020) Mapping the environmental cost of a typical citrus-producing county in China: hotspot and optimization. Sustainability 12(5):1827. https://doi.org/10.3390/su12051827

Yilmaz E, Güneşer BA (2017) Cold pressed versus solvent extracted lemon (Citrus limon L.) seed oils: yield and properties. J Food Sci Technol 54(7):1891–1900. https://doi.org/10.1007/s13197-017-2622-8

Zema DA, Calabro PS, Folino A, Tamburino V, Zappia G, Zimbone SM (2019) Wastewater management in citrus processing industries: an overview of advantages and limits. Water 11(12):2481. https://doi.org/10.3390/w11122481

Chapter 11
Optimizing a Closed-Loop Agricultural Supply Chain: A Case Study from Jordan

Emad Alzubi, Ahmed Kassem, and Bernd Noche

1 Introduction

The food supply chain (FSC) is a complex network of activities involved in producing, processing, distributing, and consuming food. It includes various actors, from farmers and food manufacturers to retailers and consumers. The amount of CO_2 emissions generated from the FSC varies depending on several factors, such as the mode of transportation, processing methods, and packaging materials. According to the Intergovernmental Panel on Climate Change (IPCC) report, food systems are responsible for around 21–37% of global greenhouse gas (GHG) emissions. This includes emissions from land use, deforestation, livestock production, food processing, transportation, and packaging. In general, transportation is one of the largest contributors to CO_2 emissions in the FSC, with air travel generating the highest amount of emissions per unit of distance travelled (IPCC 2022).

One major challenge facing the FSC is the massive sum of food loss and waste (FLW). According to the Food and Agriculture Organization (FAO), approximately one-third of all global food is lost or wasted yearly (FAO 2022). This represents a significant economic and environmental cost and a missed opportunity to provide food for those in need. Reverse logistics (RL) can address this issue by creating systems for efficient and sustainable management of surplus food, including redistribution to food banks, composting, and recycling. Minimizing FLW can create a more sustainable and equitable food system. Therefore, RL is crucial in the context of FLW, as effective RL can help reduce waste and minimize the environmental impact of discarded products (Krikke et al. 2013).

E. Alzubi (✉) · A. Kassem · B. Noche
Transportssysteme und -Logistik, Fakultät für Ingenieurwissenschaften, Universität Duisburg–Essen, Duisburg, Germany
e-mail: emad.alzubi@uni-due.de

A. Melkonyan-Gottschalk, D. Daus (eds.), *Transformation Towards Circular Food Systems*, Greening of Industry Networks Studies 13,
https://doi.org/10.1007/978-3-031-63793-3_11

1.1 Agri Supply Chain (Fruits and Vegetables)

To enhance food security, it is imperative to improve the sustainability of FSC (Friedman and Ormiston 2022). The sustainability of FSC is affected by FLW in various ways, prompting researchers to seek solutions to eliminate its causes (Vilariño et al. 2017; Corrado and Sala 2018; Frei et al. 2020; Mousa 2020; Dora et al. 2021). FLW has been identified as a driver of hunger, poverty, and price inflation (Pingali et al. 2017). Additionally, FLW has negative environmental impacts on FSCs (Brancoli et al. 2020). While some scholars, such as Eriksson et al. (2014) view FLW as a competitiveness issue due to limited natural resources, others discuss the relationship between competitiveness and the economic performance of SCs (Alzubi and Akkerman 2022). Reducing FLW aligns with the three-bottom line of sustainability and would enhance the sustainability performance of FSC.

The intricate and inefficient nature of FSC has placed increased pressure on all stakeholders involved, ranging from growers to retailers, to meet the demand for food. This pressure has led to the emergence of various sustainability issues, which are intertwined with the broader sustainability concerns of FSC (Sala et al. 2017). The configurations of agricultural supply chains (ASCs) are vital for ensuring the sustainability of FSC. Decentralized SCs offer numerous advantages, including the implementation of distributed decision models (Qu et al. 2010) to meet customer needs faster (Martin et al. 2022). While Tirado-Kulieva et al. (2022) identified centralization as a drawback for FSC due to concerns related to stability and food security, Alzubi and Noche (2022a) argued in favor of centralized SCs for effectively managing stakeholders and addressing the causes of FLW.

Thomé et al. (2021) categorized various structures under the term short FSC, with a focus on producers and consumers. Despite research indicating that local marketplaces can enhance food resilience (Kurtz et al. 2020), short FSCs are often associated with crowded logistics, resulting in high operational costs (Hübner et al. 2016; Melkonyan et al. 2020) and a significant carbon footprint (Alzubi and Noche 2022a). Moreover, FLW is a crucial key performance indicator (KPI) that has a significant impact on the sustainability of FSC.

The circular economy (CE) is a modern approach to production that emphasizes sustainability and innovation, aiming to recycle waste and reduce the use of raw materials (Fragoso and Figueira 2021). It offers solutions for achieving a sustainable food system by addressing the issue of wasted food and related products such as packaging plastics, crates, and agricultural residues (AR). This approach involves reuse, recycling, and transformation into energy, fertilizers, or other industrial applications to minimize waste and promote resource efficiency, as well as RL (Martin-Rios et al. 2018; de Oliveira et al. 2021; Alzubi et al. 2023).

Hence, it is necessary to revamp ASCs to incorporate the principles of the CE and closed-loop SCs to reduce FLW and enhance the sustainability of FSC. To achieve this, a mathematical model was built and integrated with a system dynamic model in this study aiming to reduce FLW and facilitate the collection of AR at different stages of the SC. Furthermore, it would also enhance economic performance in citrus SCs, while decreasing CO_2 emission.

1.2 Citrus Supply Chain and Closed-Loop Supply Chains

Citrus is a crucial subtropical and tropical fruit globally. However, a significant amount of citrus fruits may be discarded from citrus SC and become inedible due to decay, resulting in substantial losses for stakeholders. This occurs annually in farms, packing houses, processors, and fruit markets. The costs associated with this problem are not limited to transportation costs but also include the expenses of collecting and disposing of the rotten fruit by human resources.

In 2021, global citrus production amounted to approximately 162 million tons (knoema 2023), with the Mediterranean region accounting for roughly 80–85% of fresh citrus trade and 20% of citrus production (Zacarias et al. 2020). Oranges, mandarins, and lemons are all native to the Mediterranean region (Alzubi et al. 2022). Various causes of citrus losses on farms were discussed by Defraeye et al. (2013). Alzubi et al. (2023) elaborated on the causes of losses during postharvest stages, including citrus transportation. Researchers have studied the effects of vibration and stress during citrus transportation due to inappropriate packaging. For example, Zheng et al. (2022) developed a new packaging design with a divider to eliminate the vibration effect on citrus fruits such as oranges during transportation. Defraeye et al. (2013) also examined the performance of packaged citrus during cold transportation, testing two different packaging materials stacked on a pallet. Berry et al. (2021) investigated how using appropriate packaging material for citrus can reduce CLW during pre-shipping.

Around 45–55% of fresh citrus is discarded from the SC, which puts great pressure on all stakeholders to find solutions to eliminate the associated causes, recycle this amount of CLW, or direct it other industrial applications (Alzubi and Noche 2022b). However, CLW generated in all stages in citrus SC could be directed to generate value. For instance, CLW could be returned as fertilizer to close the loop at the farm (Cheraghalipour et al. 2018), as feed to a biogas plant to close the loop at the processing stage (Ortiz et al. 2020), and as input materials for cosmetics and chemical industries (Del Osorio et al. 2021) or could be directed to other industrial applications as presented in Fig. 11.1, which illustrates the circular CSC that could create value from CLW, wastewater generated from washing citrus, packaging materials, and other mechanical components within CSC. The innovation of this study stems from considering CLW to close the loop within CSC in Mediterranean countries. The circular FSC was analyzed to propose a network model based on a CE to recycle/reuse the CLW.

1.3 Reverse Logistics and Circular Economy

RL is becoming increasingly important in the ASC to manage waste, reduce costs, and promote sustainability. It can include processes such as collecting, transporting, and processing surplus or damaged products and managing used or expired agricultural inputs such as fertilizers, pesticides, and seedlings. Farmers

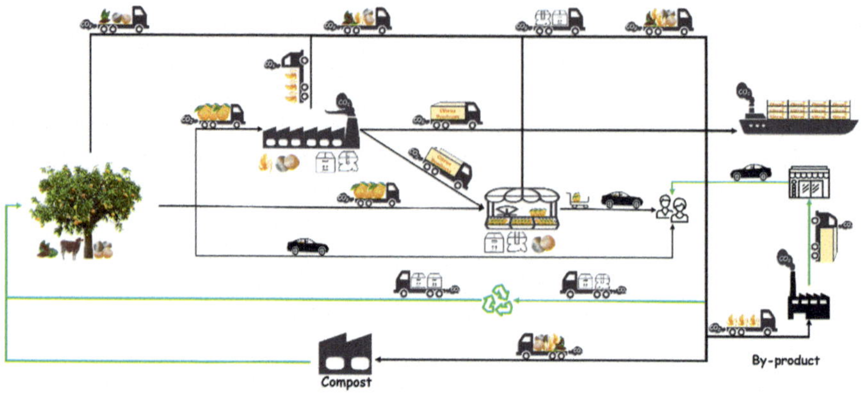

Fig. 11.1 Circular citrus SC

and producers can reduce food waste by developing efficient RL, improving resource utilization, and generating new revenue streams by recovering and reusing materials (Alzubi et al. 2023). Moreover, RL can play a critical role in reducing the environmental impact of agriculture by promoting circularity, which entails keeping materials in use for as long as possible, reducing the need for virgin resources, and minimizing waste (Alzubi and Akkerman 2022). Overall, RL has the potential to create a more sustainable and resilient ASC that benefits producers, consumers, and the environment, since it is an essential component of any efforts to reduce FLW, making it an important consideration for the agricultural industry (Dabees et al. 2023).

RL and CLSC piqued the interest of several researchers, particularly in ASCs and citrus SCs. Cheraghalipour et al. (2018) and Roghanian and Cheraghalipour (2019) incorporated RL to establish a closed loop within citrus SC by redirecting CLW to recycling centers to create organic fertilizer. In another study, Alzubi et al. (2023) integrated RL within a system dynamics model to reuse packaging crates in the farm in a citrus SC. However, it showed great potential to increase farmers' profits by 20%. The high value contained in CLW has led to increased attention on this topic. RL is associated with the CLSC (Govindan et al. 2015) and aims at increased utilization of the return trips instead of empty returning trips.

1.4 Optimizing ASCs

The optimization of ASCs is a recurring topic in literature. Ahumada and Villalobos (2009) conducted one of the early studies on designing models for ASCs, reviewing literature prior to 2009. In general, the economic dimension of ASCs has been the focus of extensive research, with a focus on profit maximization or cost minimization. For example, Tsao (2013) presented a nonlinear programming model for ASCs that maximizes profits. Similarly, the problem of transporting and storing bulk wheat across multiple periods within a two-stage SC having the objective of overall cost minimization was examined. However, other optimization models related to ASCs, such as those by Cheraghalipour et al. (2018), Sahebjamnia et al. (2020), Alzubi and Noche (2022a), Gharye Mirzaei et al. (2022), and Goodarzian et al. (2023), consistently incorporate costs or profits as a critical element.

Given the complex and multifaceted nature of ASCs, it is essential to utilize multi-objective optimization techniques to balance and reconcile conflicting objectives, considered in several studies. For instance, Sarker and Ray (2009) used a metaheuristic algorithm in combination with ε-constrained method to solve their multi-objective model for crop planning. Others like Teimoury et al. (2013) utilized system dynamics approach to optimize their multi-objective model within fruits and vegetables imports. Alzubi et al. (2023) employed system dynamics to improve citrus supply, reduce CLW, and maximize farmers profits. Cheraghalipour et al. (2019) applied classic and hybrid algorithms, such as genetic algorithm (GA) and particle swarm optimization (PSO) to propose a bi-level mixed-integer linear programming (MILP) model for rice SC in Iran.

Other studies also took into account the reduction of CO_2 emissions, which is a significant environmental impact. Manzoor and Aryanpur (2017) optimized fuel consumption to reduce the carbon footprint in the power sector and found that altering the power generation mix can lead to long-term cost reduction. Atabaki and Aryanpur (2018) employed a hybrid weighted method and analytical hierarchy process (AHP) to optimize their multi-objective model. The model aimed to improve the economic and social dimensions, such as job creation, while minimizing CO_2 emissions. Similarly, Alzubi and Noche (2022b) introduced an innovative design for a citrus SC, which aimed to mitigate CLW and decrease CO_2 emissions. The study findings indicate that implementing their model would result in a 65% reduction in CO_2 emissions. Several other researchers, such as Zhen et al. (2019) endeavored to reduce CO_2 emissions in their models.

However, mathematical optimization models such as MILP are utilized to address large-scale logistical challenges. These models can achieve high efficiency and enhance transportation efficiency and resource utilization (Sahebjamnia et al. 2020). MILP has been used in literature to solve complex optimization problems to minimize FLW within the framework of CE (Bieber et al. 2018). Lütke Entrup et al. (2005) also employed MILP to enhance profit by minimizing costs within the yoghurt SC. Furthermore, Wangsa et al. (2023) provided an optimization mode to minimize CO_2 emissions in an FSC using MILP. In this paper, a multi-objective system dynamics model to optimize the citrus SC was developed. The proposed model aims to minimize CLW, reduce CO_2 emissions, and maximize profits. This study provides insights into both FSC and CE by addressing these critical issues in the citrus industry.

Against this background, the given chapter aims to provide a simulation model to optimize citrus SC considering sustainability aspects and to compare a linear and circular citrus SC, which is based on the data collected from a case study. Therefore, the aim of this research can be accomplished by addressing the following research questions (RQ):

RQ1: What is the most contributing stage in FSC to FLW?
RQ2: How do the structures of citrus SC affect the variation in CO_2 emissions?
RQ3: How can CLW be used to create value and increase stakeholders' profits?

The structure of the paper is as follows: Sect. 2 describes the problem being addressed, while Sect. 3 describes the process of model formulation, testing, and validation processes. Section 4 introduced the scenarios and discussed the results. Finally, Sect. 5 concludes the paper and provides suggestions for future research.

2 Problem Definition

The paper introduces a novel approach to citrus SC in Mediterranean nations, which involves embracing circular thinking to extract value from waste and create a more sustainable food system. Typically, the variations in citrus SCs between different countries pose a challenge to achieving an optimized standard SC. Likewise, there are differences present within citrus SCs in the Mediterranean region. Therefore, the suggested citrus SC framework can select the included stages in the SC and work to optimize citrus SC according to the proposed objective functions.

The quantity of food waste generated by the SC leads to various environmental issues and results in significant expenses, which are not limited to waste transportation, disposal, and incineration. Therefore, it is necessary to reconsider FLW to extract value from it. For the purpose of transition to a circular citrus SC, it is necessary to redirect the CLW back to a specific stage within the SC, where it can be classified as a closed-loop citrus SC. Consequently, as a result, a new stage, namely, the waste treatment facility (WF), has been incorporated into the system.

As depicted in Fig. 11.1, in the circular citrus SC, citrus products are transported from farms to consumers, while CLW are transported in reverse to the WF, as input material. The WF processes these materials and transports its outputs, based on their specific types, to a certain stage in the same SC. For instance, if the WF represents a biogas plant, the output would consist of energy and compost. The energy generated from biogas can be used in any stage in the SC, while compost, in most cases, can be used as fertilizer in citrus farm. This closed-loop system reduces waste and enhances the sustainability and productivity of the entire citrus SC.

3 Formulation

The objective functions and constraints presented in Appendix A were transformed and integrated into a system dynamics model using Vensim® presented in Fig. 11.2, which was tested under the extreme conditions in Sect. 3.1 and validated based on the data collected from the case study in Sects. 3.2, 3.3, and 3.4.

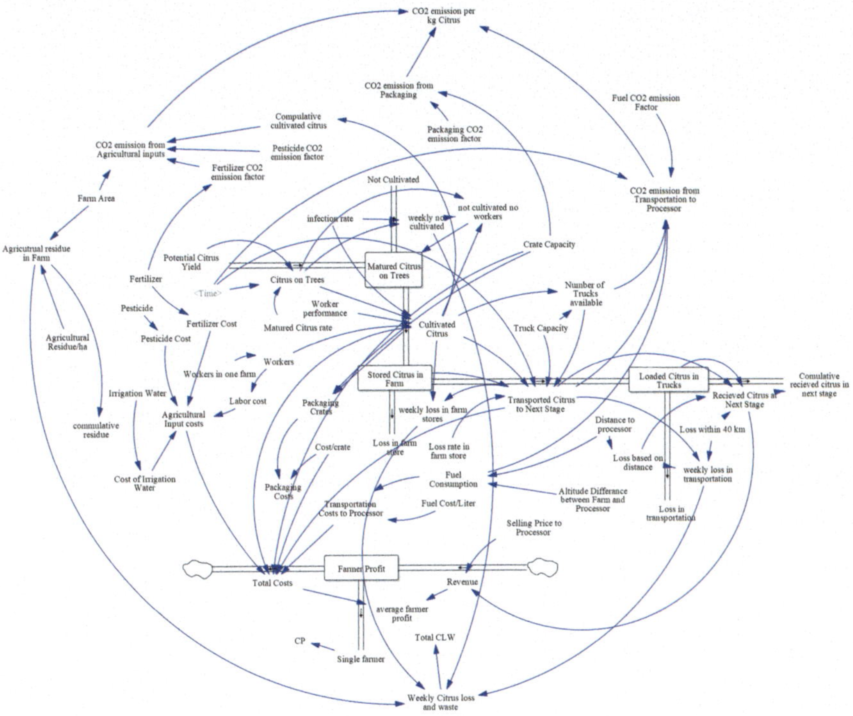

Fig. 11.2 Linear citrus supply chain model

Following Sedlacko et al. (2014), the collected information while observing the harvesting process, transportation activities, and discussion with farmers and farmworkers, causal loop diagram (CLD) was created as an essential step to understand the expected behavior of model. However, discussion with farmers and farmworkers helped to identify the variables in the system, which have been classified into stock, flow, and auxiliary variables. The stock variables represent the accumulated variables while flow variables represent the inflow and outflow variables from the accumulated ones. This aid to draw the causal connections and how these relations could be affected by the connected variables (Binder et al. 2004). The CLD was structured following the participatory of system mapping (PSM) and then transformed into a stock and flow diagram (SFD) as presented in Fig. 11.2.

In addition to discussing stakeholders, we also reviewed relevant literature and case studies on citrus SCs, such as Dusch Silva (2018), to identify influencing factors and variables. The primary influencing factors include farm size, the number of citrus trees, potential yield, the number of workers hired during the harvesting process, storage capacity, the quantity and capacities of trucks, the stages present in the citrus SC, distances between these stages, input costs, market demand, and selling prices. Considering these factors within different sustainability dimensions would facilitate the design and development of a model capable of assessing the existing linear citrus SC and evaluating various scenarios, including a closed-loop citrus SC.

3.1 Model Testing

The model had been tested under extreme conditions to verify the expected behavior of the causal-loop relationships (Das and Dutta 2013). For instance, in the scenario of no workers being hired to cultivate citrus in a specific time period, it is expected that no citrus will be cultivated and the whole matured should be considered as a loss in the farm. Consequently, the objective functions are expected to give extreme outputs such as high values of CO_2 emissions from the use of agricultural inputs, while zero CO_2 emissions from transportation. Furthermore, given that no citrus fruits will be sold to subsequent stages, stakeholders such as farmers are likely to experience a negative profit. This is due to the expenses incurred during the preharvest period, which includes preparation for the season and the purchase of agricultural inputs. Similarly, entering a long distance for transporting citrus will lead to a high loss in transportation (around 80%), high CO_2 emissions from transportation (11.06 kg CO_2/kg citrus), and negative profit (−11,043 Euro) due to the high transportation costs. However, the results of the testing process reflected the expected behavior from the model.

3.2 Case Study: Citrus SC in Jordan

Around 90% of citrus farms in Jordan are located in the northern part of Jordan Valley (JV), where the average total production is estimated by the Department of Statistics (DoS) at 124,600 tons (DoS 2022). The cultivated citrus is transported to the central market (CM) of fruits and vegetables, with distances varying from 40 to 140 km (Alzubi et al. 2022), with high elevation differences that lead to an increase the fuel consumption (Alzubi and Noche 2022a). In addition, agricultural inputs such as fertilizers, pesticides, water, and electricity are usually purchased from the local market. Data related to the farming stage were collected from various sources. For instance, the costs of the agricultural inputs, labor, and packaging crates were collected from farmers in JV, while fuel costs from tank stations. In addition, the CO_2 emission factors per kg of fertilizer, kg of pesticide, and L of fuel were collected from the literature (Table 11.1).

3.3 General Assumptions

- Simulation runs for 1 year (52 weeks).
- Week 1 in the model is the beginning of the season, which is the second week in September, and the season lasts for 35 weeks (mid of May).
- Farms operate 7 days a week.
- No exceptional cases like rainfall delays and the COVID-19 pandemic.

Table 11.1 Summary of the input parameters

Parameter	Source	Data	Unit
Planted area	Farmer	590	Ha
Potential yield	Farmer	150,000	Tons/year
Labor cost	Farmer	0.07	Euro/kg citrus
Fertilizer cost	Farmer	0.02	Euro/kg citrus
Pesticide cost	Farmer	0.02	Euro/kg citrus
Water cost	Farmer	0.02	Euro/kg citrus
Packaging	Farmer	0.23	Euro/crate
Diesel costs	Local market	0.91	Euro/L
CO_2 emission factor:			
Fertilizer	Hoxha and Christensen (2019)	7.39	Kg CO_2e/ha
Pesticide	Soheilifard et al. (2020)	1.98	Kg CO_2/ha
Diesel	Alzubi and Noche (2022a)	2.64	Kg CO_2/L
Crate	Abejón et al. (2020)	1.63	Kg CO_2/box
Crate capacity	Farmer	15	Kg citrus
Truck capacity	Farmer	2000	Kg citrus
Trip(s)	Farmer	1	Trip/day
Worker performance	Farmer	300–400	Kg citrus/day
Available workers	Farmer	2–10	Worker/day
JV altitude	SolarGIS	−200	Meter
Selling price	CM	0.45–1.25	Euro/kg citrus

- Return trips in linear case are empty trips, in which trucks return to the original starting point.

3.4 *Model Validation*

Following Alzubi et al. (2019) and Altarazi et al. (2012), the model was validated based on the input parameters listed in Table 11.1. The results from the simulation were compared to the current situation to reflect reality. The comparison was done based on the data collected from farmers or retrieved data from DoS. According to the results summarized in Table 11.2, the percentage difference between the results from the model and the retrieved data from DoS or collected from farmers was modest.

4 Results and Discussions

Accordingly, several scenarios were created to enhance and trade-off between the three objective functions of the model.

Table 11.2 Comparison between SD results and Department of Statistics data and farmer observations

Parameter	Data (source)	SD model result	Difference
Supplied citrus to CM	124,600 tons (DoS 2022)	127,741 tons	2.5%
CLW	45–55% (Alzubi and Noche 2022a)	42%	6.6%
Profit	39,000 Euro (farmer)	36,866 Euro	5.5%
Transportation CO_2 emission	0.95 kg CO_2/kg citrus (Alzubi and Noche 2022a)	0.92 kg CO_2/kg citrus	3.2%
Citrus CPC	13.10 kg/person/year (DoS 2022)	12.78 kg/person/year	2.4%

4.1 *Scenario 1 (S1): Considering a Processing Stage*

If a processing stage exists, such as a juice extractor, and 50% of the total yield is transformed to produce juice, it is expected that peel waste constitutes around 50% of the total citrus weight as mentioned by Ortiz et al. (2020). Therefore, additional assumptions were made to support S1 as follows:

- Citrus purchasing costs at the processing stage are 0.35 Euro/kg citrus.
- The labor cost per L of juice produced is 0.36 Euro/L.
- Overhead cost of 1 L of juice produced is 0.11 Euro/L.
- The average selling price of 1 L of juice is 1.25 Euro/L.
- CO_2 emissions from transporting juice to the next stage are assumed to have the same values as in Fig. 11.2. Therefore, the average CO_2 emissions/L of juice transported is 0.82 kg CO_2.
- CO_2 emissions from electricity and refrigerator were collected from a company, which preferred not to mention their information, and considered to be 0.07 and 0.02 kg CO_2/L juice, respectively.

The sub-model in Fig. 11.3 illustrates the considered processing stage with the relevant variables and parameters. The result indicated that the CO_2 emissions of 1 kg of citrus from the farm to the processor have been reduced by almost 5%, while a new measurement for CO_2 emissions of producing and transporting 1 L of juice in the processing stage is to be on average 0.99 kg CO_2/L juice. However, when including the CO_2 emission from citrus cultivation and transportation to the processing stage, the total CO_2 emissions are around 2.11 kg CO_2/L juice in the market. In addition, the average profit of one farmer has been reduced by around 6%, because of the fixed selling price to processors at 0.35 Euro/kg citrus, which is lower than the selling price in CM. The annual profit of the processor is 73,261.4 Euros.

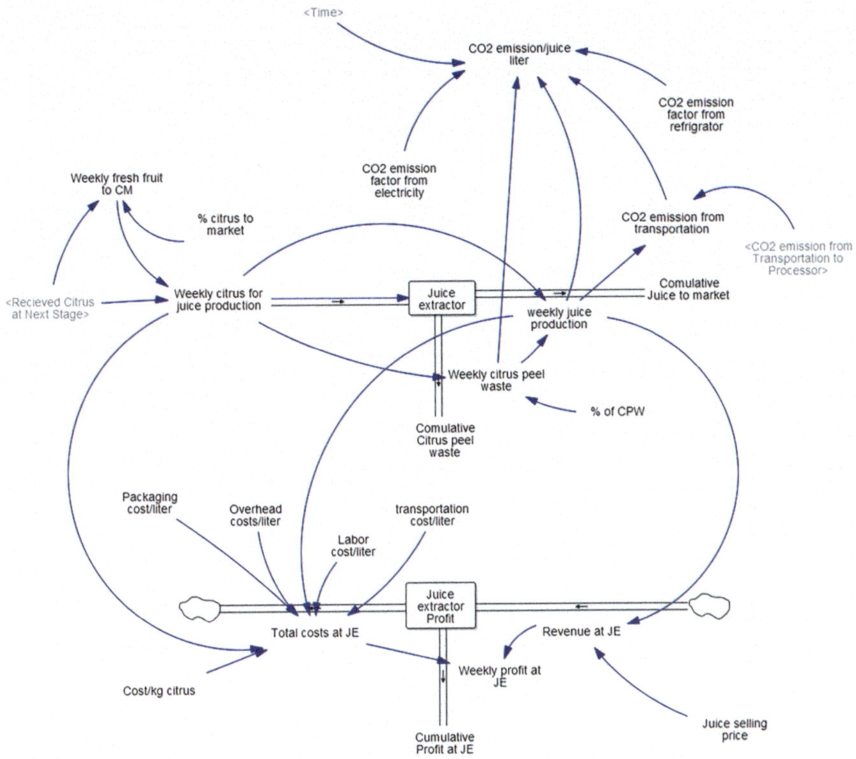

Fig. 11.3 The considered processing stage in S1

4.2 *Scenario 2 (S2): Closed-Loop Citrus SC for the Current State*

The model was developed using scenarios from two different cases, which are linear citrus SC (Fig. 11.2) and CLCSC by integrating the model presented in Fig. 11.4. In the closed-loop case, the composting option has been considered, following Roghanian and Cheraghalipour (2019). The behavior of the model was analyzed to find out how closed-loop SC would change the objective functions.

Recycling CLW, as a result of the current scenario presented in Fig. 11.2 and S1, closes the loop as illustrated in Fig. 11.4. This enhances the results of the objective functions by increasing the farmers' profit by approximately 2449.9 Euros, representing a 7% increase. It is expected that any advantages the farmer gains will reflect on the SC stakeholders. For instance, since the costs of agricultural inputs have decreased, the selling price to subsequent stages should also decrease. Additionally, CO_2 emissions per kg of citrus are expected to decrease by around 29% on average. The increase in profit is due to the decrease in the cost of agricultural inputs such as fertilizer.

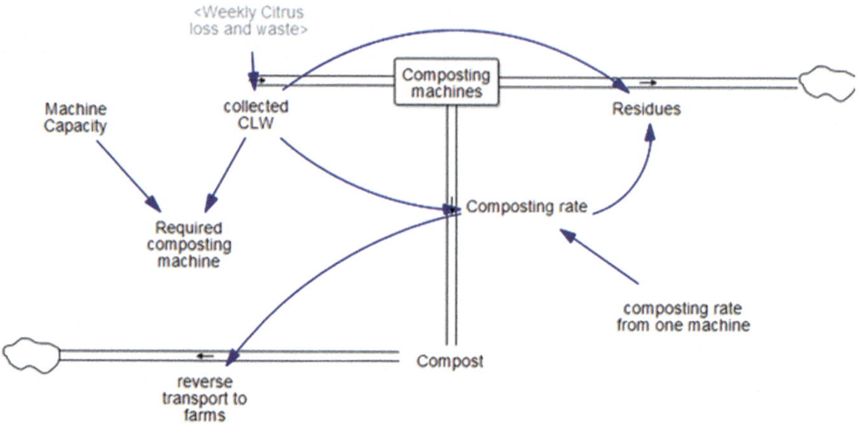

Fig. 11.4 Closing the loop within the citrus supply chain (considering compost option) for S2

The decrease in CO_2 emissions is due to enhanced utilization of return trips from downstream stages to the farm, which were previously empty, and the replacement of chemical fertilizer with compost.

4.3 Scenario 3 (S3): Upcycling CLW at Processing Stage

Other valorization options were considered in the literature on CLW. For instance, extracting essential oil from peels would lead to an increase in profit. Since peel waste is the output at the processing stage, S3 provides analysis on extracting the essential oil at the processing stage, where a modification on Fig. 11.3 was made to illustrate the essential oil extracting machinery in Fig. 11.5.

The average essential oil that would be extracted from peels is estimated at 0.3 L/ton (Bates et al. 2001), which was integrated into the model to estimate the amount of essential oil yield. An assumption was made, that is, the selling price of 1-L essential oil varies from 300 to 400 Euro.

The results from this scenario showed an improvement in the processor profit as presented in Fig. 11.6 compared to S1. Extracting the essential oil increased the processor profit by 21%. Furthermore, closed-loop citrus SC improved farmer profit, as can be seen in Fig. 11.7. With regard to the CO_2 emissions for both functional units (1 kg citrus from the farming stage and 1 L juice from the processing stage), it can be seen that the closed-loop citrus SC (S2) performed better for 1 kg citrus because the CO_2 emissions were reduced by 29%, as illustrated in Fig. 11.8. Nevertheless, CO_2 emissions for 1 L of juice from the processing stage were reduced by 26% in the closed-loop citrus SC (S2) when compared to the linear citrus SC in S1, which can be seen in Fig. 11.9.

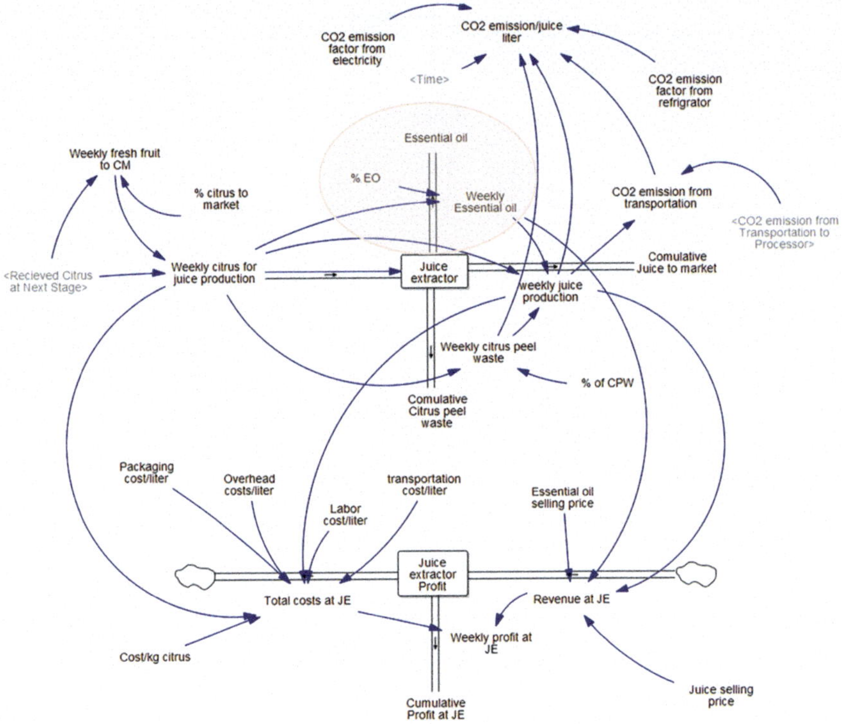

Fig. 11.5 S3 upcycling CLW at the processing stage

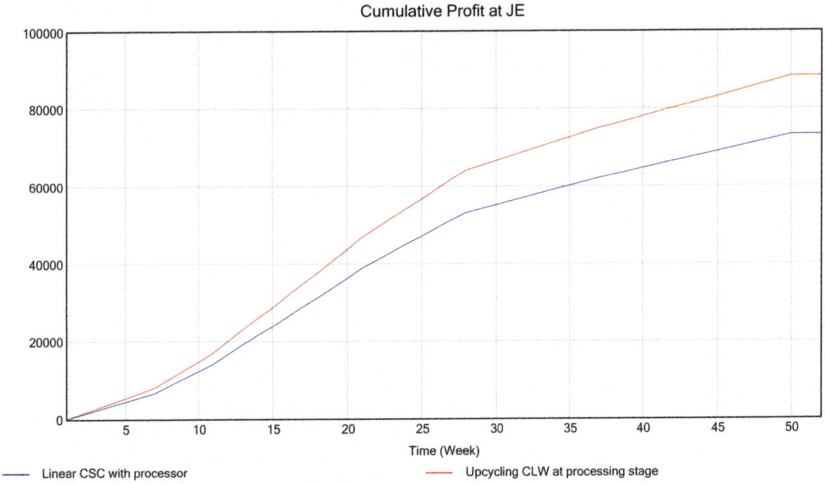

Fig. 11.6 Cumulative processor profit for S1 and S2

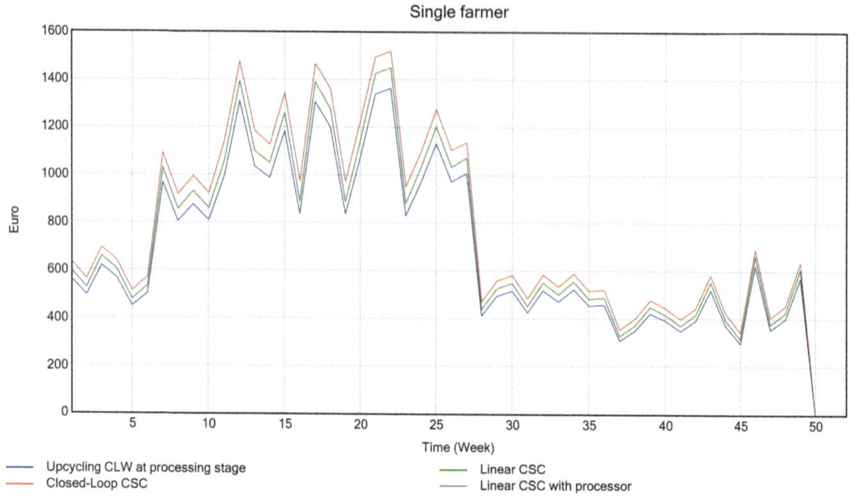

Fig. 11.7 Average weekly profit of a single farmer in citrus SC

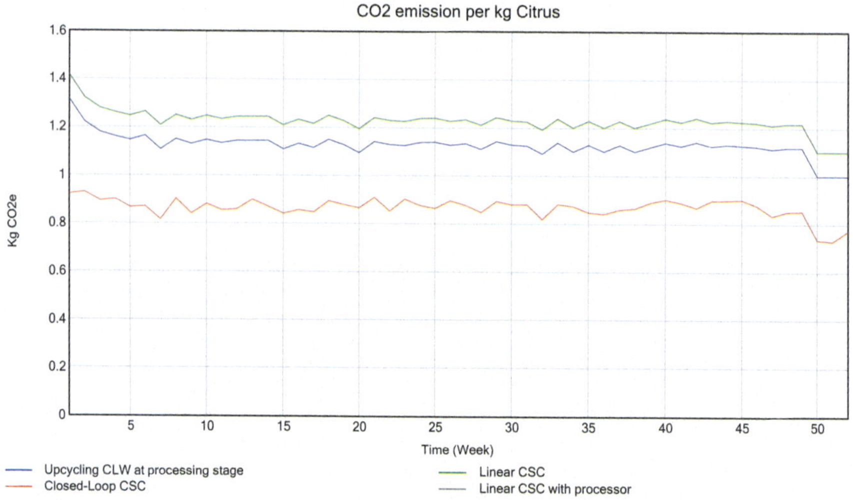

Fig. 11.8 CO_2 emissions/kg citrus from the farming stage in all scenarios

In terms of profit and CO_2 emissions, the general results indicate that the closed-loop citrus SC performs significantly better than the linear one. This is especially true when it is combined with other valorization options, such as essential oil extraction. These combinations can lead to improvements in the overall sustainability of the citrus SC. The results from this simulation can be compared with those from Cheraghalipour et al. (2018), who conducted optimization models

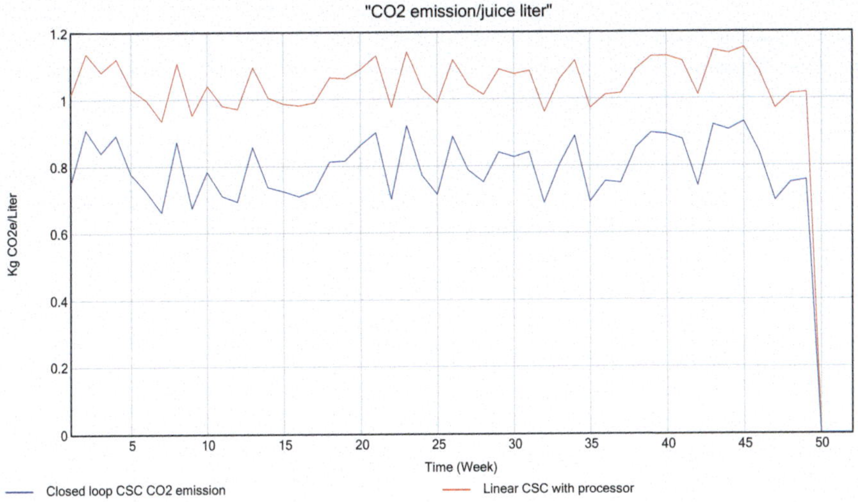

Fig. 11.9 CO_2 emission for 1 L of juice from the processing stage in S1 and S2

for citrus SC in Iran and found that compost would lead to reduction of the CO_2 emissions and reduce the agricultural inputs costs.

Other valorization options such as the essential oil would also create more value from CLW, which is an emerging SC that deals with another by-product from citrus SC and has a high impact on the overall citrus SC. For instance, due to the high purchasing price per 1 L of essential oil at around 300–400 Euro, the profit of the stakeholders would be increased. However, the essential oil extraction process mainly depends on two main factors, which are the temperature and the extraction time (Rajabi-Kafshgar et al. 2023), and in this paper, only the case of cold press extraction is considered.

5 Conclusion

The paper studied the economic and environmental dimensions within citrus SC, for which citrus SC in Jordan was selected as a case study to validate the model and evaluate these dimensions within three different scenarios. S1 dealt with considering a processing stage to reduce CLW and CO_2 emissions, while improving stakeholder profitability, S2 dealt with collecting CLW to recycle it and return as compost to close the loop at the farming stage, and S3 considered the upcycling option by creating value from waste within the closed loop citrus SC. In general, S2 improved farmers profit by almost 7% (2449.4 Euro) and reduced the CO_2 emissions of 1 kg of citrus and 1 L of juice by 29% and 26%, respectively.

Limitations of this study might be considered the essential oil process with the cold press and data from the processing stage was estimated by seeking experts' opinion. Therefore, future research might include the effect of temperature and extraction time effect on the extraction process of essential oil. Moreover, future research might include CO_2 emissions from 1 L of essential oil and consider the juice as a byproduct of essential oil. Finally, the model considered citrus SC in Jordan; therefore, future research might use other citrus SC from other countries to study the differences of the different citrus SCs from country to country.

Appendix A

This section presents the mathematical equations and constraints included in the system dynamics simulation, including indices, parameters, and variables, as follows:

Indices:	
i	A node in CSC, referring to facilities that include farms, processors, distributors, exporters, export destinations, and waste treatment facilities
i,j	Pair of nodes referring to an arc from node *i* to node *j*.
t	Index of transportation trucks
k	Index of transportation mode
Sets definitions:	
F	Set of citrus farms
P	Set of processors
D	Set of distributors
V	Set of exporters
O	Set of export destinations
W	Set of waste treatment facilities
T	Set of transportation trucks
FP_a	Set of arcs between citrus farms and processors, where the citrus farm is the starting point
FW_a	Set of arcs between citrus farms and waste treatment facilities, where the citrus farm is the starting point
PV_a	Set of arcs between processors and exporters, where a processor is the starting point
PD_a	Set of arcs between processors and distributors, where a processor is the starting point
PW_a	Set of arcs between processors and waste treatment facilities, where a processor is the starting point
DW_a	Set of arcs between distributors and waste treatment facilities, where a distributor is the starting point
VO_a	Set of arcs between exporters and export destinations, where the exporter is the starting point

$T_{i,j}$	Set of trucks that can transport citrus products, processed citrus, and waste for arc $i,j \in FP_a \cup FW_a \cup PV_a \cup PD_a \cup PW_a$
$K_{i,j}$	Set of transportation modes that can transport citrus and its by-products using arc $i,j \in VO_a$
Parameters:	
α_f	Matured citrus in farm $i \in F$ (in %)
Y_f	Citrus yield in farm $i \in F$ (in ton)
capacity$_f$	Capacity of farm $f \in F$ (ton)
capacity$_i$	Node $i \in P \cup D \cup V$ inventory capacity (ton)
$Q_{i,j}$	Citrus transported from node $i \in F \cup P \cup V$ to node $j \in P \cup D \cup V \cup O$ using arc $i,j \in FP_a \cup PV_a \cup PD_a \cup VO_a$ (ton)
β_i	Citrus cultivation rate in farm $i \in F$ (kg/resource time)
Y_i	Citrus stored in node $i \in P \cup D \cup V$ (in ton)
α_i	Processed citrus in node $i \in P$ (in %)
L_i	Citrus loss rate in node $i \in F \cup P \cup V \cup D \cup O$ (in %)
$L_t^{i,j}$	Citrus loss rate while transporting citrus from node $i \in F \cup P \cup V$ to node $j \in P \cup D \cup V$ using arc $i, j \in FP_a \cup PV_a \cup PD_a$ using truck $t \in T$ (in %)
ρ	Selling price of citrus (Euro/ton)
h_i	Citrus harvesting cost in farm $i \in F$ (Euro/ton)
$\varnothing_i$	Citrus processing cost in processor $i \in P$ (Euro/ton)
r_i	Costs of collecting citrus loss and waste in node $i \in F \cup P \cup D$ (Euro/ton)
$C_t^{i,j}$	Cost of transporting citrus or citrus loss and waste from node $i \in F \cup P \cup D$ to node $j \in P \cup D \cup V \cup W$ using arc $i,j \in FP_a \cup FW_a \cup PD_a \cup PV_a \cup PW_a \cup DW_a$, *using* truck $t \in T$ (Euro/ton)
X_i	Cost of storing one ton of citrus or citrus loss and waste at node $i \in F \cup P \cup D \cup V$ for a period of time λ (Euro/ton)
δ_i	Selling price of citrus loss and waste from node $i \in F \cup P \cup D$ (Euro/ton)
$d_{i,j}$	Traveling distance using arc $i,j \in FP_a \cup FW_a \cup PD_a \cup PV_a \cup PW_a \cup DW_a$, using truck $t \in T$ (km)
ω_i	Fuel consumptions rate in node $i \in F \cup P \cup D \cup W \cup V$ (L)
ε_i	CO_2 emission rate in node $i \in F \cup P \cup D \cup W \cup V$ ($kgCO_2$/L)
$\omega_t^{i,j}$	Fuel consumptions rate from transporting citrus (by-products) or citrus loss and waste using arc $i,j \in FP_a \cup FW_a \cup PD_a \cup PV_a \cup PW_a \cup DW_a$, using truck $t \in T$ (L)
$\varepsilon_t^{i,j}$	CO_2 emission rate from transporting citrus (by-products) or citrus loss and waste using arc $i,j \in FP_a \cup FW_a \cup PD_a \cup PV_a \cup PW_a \cup DW_a$, using truck $t \in T$ ($kgCO_2$/L)
$N_t^{i,j}$	A number of trips for truck $t \in T$ from node $i \in F \cup P \cup D \cup W \cup V$ to node $j \in P \cup D \cup W \cup V$, while using arc $i,j \in FP_a \cup FW_a \cup PD_a \cup PV_a \cup PW_a \cup DW_a$.
μ_t	Capacity of truck $t \in T$ (ton).
$U_t^{i,j}$	Utilization of truck $t \in T$, in the case of not fully loaded, while using arc $i,j \in FP_a \cup FW_a \cup PD_a \cup PV_a \cup PW_a \cup DW_a$ (in %)
$\omega_k^{i,j}$	Fuel consumption rate from transporting citrus or citrus by-products using arc $i,j \in VO_a$, using transportation mode $k \in K$ (L)

$\varepsilon_k^{i,j}$	CO_2 emission rate from transporting citrus or citrus by-products using arc $i,j \in VO_a$, using mode $k \in K$ ($kgCO_2/L$)
M	Is a big positive number
Decision variables:	
$\varphi_{i,j}$	Is a binary variable that equals 1; truck $t \in T$ is fully loaded while using arc $\in FP_a \cup FW_a \cup PD_a \cup PV_a \cup PW_a \cup DW_a$; otherwise 0.
$\tau_{i,j}$	Is a binary variable that equals 1; truck $t \in T$ is empty while using arc $\in FP_a \cup FW_a \cup PD_a \cup PV_a \cup PW_a \cup DW_a$; otherwise 0.
$\theta_{i,j}$	Is a binary variable that equals 1; truck $t \in T$ is not fully loaded while using arc $\in FP_a \cup FW_a \cup PD_a \cup PV_a \cup PW_a \cup DW_a$; otherwise 0.
∂_i	Is a binary variable equal to 1 if node $i \in F \cup P \cup D \cup V$ stores citrus (by-products) or waste; otherwise, 0.
∞_i	Is a binary variable that equals 1 if node $i \in P \cup D \cup V$ exists in CSC; otherwise, 0.
$S_{i,j}$	Is a binary variable that equals 1 if node $i \in F \cup P \cup D \cup V$ sends citrus (by-products) or waste to node $j \in P \cup D \cup V$; otherwise, 0.

Objectives

$$Z_1 = \sum_{i \in F} \alpha_i Y_i \ (1 - L_i \beta_i) - \sum_{t \in T} \sum_{i \in j} \sum_{i \neq j} L_t^{i,j} \infty_i Q_{i,j}$$

$$Z_{2.1} = \left[\sum_{i \in F} \sum_{j \in P} Q_{i,j} (1 - L_i) (\rho_i - h_i - r_i - \partial_i X_i) + \sum_{t \in T} \sum_{i \in F} \sum_{j \in P} \delta_i Q_{i,j} \left(L_i + L_t^{i,j} (1 - L_i) \right) \right] - \left[\sum_{t \in T} \sum_{i \in F} \sum_{j \in P} C_t^{i,j} (1 - L_i) Q_{i,j} + \sum_{t \in T} \sum_{i \in F} \sum_{j \in W} C_t^{i,j} \left(L_i + L_t^{i,j} \right) \right]$$

$$Z_{2.2} = \left[\sum_{i \in P \cup D} \sum_{i \neq j} \infty_i Q_{i,j} (1 - L_i) (\rho_i - \varnothing_i - r_i - \partial_i X_i) + \sum_{t \in T} \sum_{i \in P} \sum_{j \in W} \delta_i Q_{i,j} \left(L_i + L_t^{i,j} (1 - L_i) \right) \right] - \left[\sum_{t \in T} \sum_{i \in j} \sum_{i \neq j} C_t^{i,j} (1 - L_i) Q_{i,j} + \sum_{t \in T} \sum_{i \in P} \sum_{j \in W} C_t^{i,j} \left(L_i + L_t^{i,j} \right) \right]$$

$$Z_3 : \sum_{i\in j} Q_i\omega_i\varepsilon_i + \left[\sum_{t\in T}\sum_{i\in j}\sum_{i\neq j} \infty_i\varphi_{i,j}d_{i,j}\omega_t^{i,j}\varepsilon_t^{i,j} + \sum_{t\in T}\sum_{i\in j}\sum_{i\neq j} \infty_i\theta_{i,j}U_t^{i,j}d_{i,j}\omega_t^{i,j}\varepsilon_t^{i,j} + \sum_{t\in T}\sum_{j\in i}\sum_{j\neq i} \tau_{j,i}d_{j,i}\omega_t^{j,i}\varepsilon_t^{j,i} + \sum_{t\in T}\sum_{j\in i}\sum_{j\neq i} \infty_i\theta_{j,i}U_t^{i,j}d_{j,i}\omega_t^{j,i}\varepsilon_t^{j,i}\right] + \left[\sum_{k\in K}\sum_{i\in V}\sum_{j\in O} \infty_i d_{i,j}\omega_k^{i,j}\varepsilon_k^{i,j}\right]$$

The objective function (Z_1) minimizes CLW, which might be generated in a certain stage in CSC and comprise the stakeholder profit in ($Z_{2.1}$) and ($Z_{2.2}$). Objective functions ($Z_{2.1}$) and ($Z_{2.2}$) maximize the profit according to the general equations of (revene − total costs). On the other hand, the objective function (Z_3) minimizes CO_2 emissions when transporting citrus from one stage to another in CSC. The objective function (Z_3) is considered per 1 kg of citrus transported, which can be done by letting the dominator be $Q_{i,j}$.

Constraints

$$\sum_{i\in F}\sum_{j\in P} Q_{i,j} = \sum_{i\in F} \beta_i\alpha_i Y_i, \qquad \forall i \in F, \forall J \in P \tag{11.1}$$

$$\sum_i\sum_j S_{i,j} \leq 1, \qquad \forall i \in F \cup P \cup D \cup V, j \in P \cup D \cup V \tag{11.2}$$

$$\sum_{i\in F} Q_{i,j}\left(1 + L_t^{i,j}\right) + \sum_{f\in F} L_f\alpha_f\beta_f \leq Y_i, \qquad \forall j \in P, \forall t \in T \tag{11.3}$$

$$\sum_{i\in j}\sum_{i\neq j} Q_{i,j}\left(1 + L_t^{i,j}\right) + \sum_i Y_i\left(L_i + \alpha_i\right) \leq \sum_j Q_{f,j}, \qquad \forall i \text{ and } j \in P \cup D \cup V, \ \forall f \in F \tag{11.4}$$

$$Y_{i,\lambda} + \sum_{i\in F}\sum_j \alpha_i\beta_i - Q_{i,j} = Y_{i,\lambda+1}, \qquad \forall j \in P \tag{11.5}$$

$$Y_{i,\lambda} + \sum_i\sum_{i\neq j} \partial_i\infty_i\left(\alpha_i Y_i - Q_{i,j}\right) = Y_{i,\lambda+1}, \qquad \forall i \in P \cup D \cup V, \forall j \in D \cup V \tag{11.6}$$

$$\sum_{i\in V}\sum_{j\in O} Q_{i,j} \le \sum_{i\in F}\sum_{j\in P} Q_{i,j} - \sum_{i\in P}\sum_{j\in D} Q_{i,j}, \ \ \forall i \in F\cup P\cup V,\ \ \forall j \in P\cup D\cup O \tag{11.7}$$

$$N_t^{i,j} \le \text{int}\left[\sum_{i\in j}\sum_{i\ne j}\frac{Y_i}{\mu_t}\right] + 1, \qquad \forall t \in T,\ \ \forall i \in F\cup P,\ \ \forall j \in P\cup D\cup V \tag{11.8}$$

$$\sum_{i\in j}\sum_{i\ne j} Q_{i,j} = \sum_{t\in T}\mu_t\left(N_t^{i,j} + U_t^{i,j}\right), \quad \forall t \in T,\ \ \forall i \in F\cup P,\ \ \forall j \in P\cup D\cup V\cup W \tag{11.9}$$

$$\forall i, j \in \text{FP}_\text{a}\cup\text{FW}_\text{a}\cup\text{PD}_\text{a}\cup\text{PV}_\text{a}\cup\text{PW}_\text{a}\cup\text{DW}_\text{a}$$

$$\sum_t \theta_{i,j} \le 1, \qquad \forall i, j \in \text{FP}_\text{a}\cup\text{FW}_\text{a}\cup\text{PD}_\text{a}\cup\text{PV}_\text{a}\cup\text{PW}_\text{a}\cup\text{DW}_\text{a} \tag{11.10}$$

$$Y_i - \sum_{t\in T}\mu_t N_t^{i,j} < \theta_{i,j}\mu_t, \qquad \forall i, j \in \text{FP}_\text{a}\cup\text{FW}_\text{a}\cup\text{PD}_\text{a}\cup\text{PV}_\text{a}\cup\text{PW}_\text{a}\cup\text{DW}_\text{a} \tag{11.11}$$

$$U_t^{i,j} = \frac{Y_i - \sum_{t\in T}\mu_t N_t^{i,j}}{\mu_t}, \qquad \forall i, j \in \text{FP}_\text{a}\cup\text{FW}_\text{a}\cup\text{PD}_\text{a}\cup\text{PV}_\text{a}\cup\text{PW}_\text{a}\cup\text{DW}_\text{a} \tag{11.12}$$

$$\sum_{n\in N}\sum_{i\in j}\sum_{i\ne j}\infty_i \text{S}_{i,j} Q_{i,j} \le \mu_t, \quad \forall t \in T, \forall i \in F\cup P,\ \ \forall j \in P\cup D\cup V\cup W \tag{11.13}$$

$$\forall i, j \in \text{FP}_\text{a}\cup\text{FW}_\text{a}\cup\text{PD}_\text{a}\cup\text{PV}_\text{a}\cup\text{PW}_\text{a}\cup\text{DW}_\text{a}$$

$$\omega_{\text{not\_full}}^{i,j} = \left(\theta_{i,j}\omega_{\text{empty}}^{i,j}\right) + U_t^{i,j}\left(\varphi_{i,j}\omega_{\text{full}}^{i,j} - \theta_{i,j}\omega_{\text{empty}}^{i,j}\right), \quad \forall i, j \in \text{FP}_\text{a}\cup\text{FW}_\text{a}\cup\text{PD}_\text{a}\cup\text{PV}_\text{a}\cup\text{PW}_\text{a}\cup\text{DW}_\text{a} \tag{11.14}$$

$$\sum_{f,\lambda} Y_{f,\lambda} \le \text{capacity}_f, \qquad \forall f \in F \tag{11.15}$$

$$\sum_{i\in j}\sum_{i\ne j}\infty_i S_{i,j} Q_{i,j} \le \text{capacity}_j, \qquad \forall i \in F\cup P, \forall j \in P\cup D\cup V \tag{11.16}$$

$$Y_f \ge 0, \quad \forall f \in F \tag{11.17}$$

$$\alpha_f \geq 0, \quad \forall f \in F \tag{11.18}$$

$$\beta_f \geq 0, \quad \forall f \in F \tag{11.19}$$

$$\text{capacity}_f \geq 0, \quad \forall f \in F \tag{11.20}$$

$$\text{capacity}_i \geq 0, \quad \forall i \in P \cup D \cup V \tag{11.21}$$

$$\mu_t > 0, \quad \forall t \in T \tag{11.22}$$

$$Y_i \geq 0, \quad \forall i \in P \cup D \cup V \tag{11.23}$$

$$L_i \geq 0, \quad \forall i \in F \cup P \cup D \cup V \tag{11.24}$$

$$Q_{i,j} \geq 0, \quad \forall i,j \in \text{FP}_\text{a} \cup \text{FW}_\text{a} \cup \text{PD}_\text{a} \cup \text{PV}_\text{a} \cup \text{PW}_\text{a} \cup \text{DW}_\text{a} \cup \text{VO}_\text{a} \tag{11.25}$$

$$L_t^{i,j} \geq 0, \quad \forall i,j \in \text{FP}_\text{a} \cup \text{PD}_\text{a} \cup \text{PV}_\text{a} \tag{11.26}$$

Constraint (11.1) ensures that the citrus flow from one stage to another must equal citrus cultivated in the farm stage. Constraint (11.2) ensures that each node $i \in F \cup P \cup D \cup V$ is transporting their citrus (by-products) or waste to at most one node $j \in P \cup D \cup V$. Constraint (11.3) restricts capacity on cultivated and transported citrus in farm $i \in F$ to processor $j \in P$, while constraint (11.4) restricts capacity on processed and transported citrus (by-products) in node $i \in P \cup D \cup V$ to processor $j \in P \cup D \cup V$. Constraint (11.5) balances the citrus yield in farm $i \in F$ over the time. Constraint (11.6) balances the citrus (by-products) stored in node $i \in P \cup D \cup V$ over time. Constraint (11.7) restricts exports by meeting the domestic demand first. Therefore, exported citrus (by-products) must not exceed the gap between flow quantities from farms to processors (supply) and from processors and the local distributors (demand). Constraint (11.8) is related to constraint (11.9), which ensures the flow from node $i \in F \cup P \cup D \cup V$ to node $j \in P \cup D \cup V$, using truck $t \in T$, considering truckload. However, only one not fully loaded truck $t \in T$ is allowed to be in arc $i,j \in \text{FP}_\text{a} \cup \text{FW}_\text{a} \cup \text{PD}_\text{a} \cup \text{PV}_\text{a} \cup \text{PW}_\text{a} \cup \text{DW}_\text{a}$, following Soysal (2015). Constraints (11.10, 11.11, and 11.12) ensure that only one not fully loaded truck $t \in T$ is allowed to be in arc $i,j \in \text{FP}_\text{a} \cup \text{FW}_\text{a} \cup \text{PD}_\text{a} \cup \text{PV}_\text{a} \cup \text{PW}_\text{a} \cup \text{DW}_\text{a}$. Constraint (11.13) ensures that the flow of material in arc

$i, j \in FP_a \cup FW_a \cup PD_a \cup PV_a \cup PW_a \cup DW_a$ will not exceed the capacity of truck $t \in T$. Since the fuel consumption rate depends on the load, distance, and elevation, constraint (11.14) calculates the fuel consumption rate in the case of a not fully loaded truck $t \in T$ while using arc $i,j \in FP_a \cup FW_a \cup PD_a \cup PV_a \cup PW_a \cup DW_a$. Constraint (11.15) ensure that the citrus yield will come to an endpoint; a restriction should be made based on the farm $f \in F$ capacity. Constraint (11.16) restricts inventory capacity at node $j \in P \cup D \cup V$, to ensure that citrus supply will not exceed it. Constraints (11.17–11.26) are non-negative constraints.

References

Abejón R, Bala A, Vázquez-Rowe I, Aldaco R, Fullana-i-Palmer P (2020) When plastic packaging should be preferred: life cycle analysis of packages for fruit and vegetable distribution in the Spanish peninsular market. Resour Conserv Recycl 155:104666. https://doi.org/10.1016/j.resconrec.2019.104666

Ahumada O, Villalobos JR (2009) Application of planning models in the agri-food supply chain: a review. Eur J Oper Res 196(1):1–20. https://doi.org/10.1016/j.ejor.2008.02.014

Altarazi S, Ammouriova M, Alzubi E (2012) Concurrent manual-order picking warehouses design via simulation

Alzubi E, Akkerman R (2022) Sustainable supply chain management practices in developing countries: an empirical study of Jordanian manufacturing companies. Cleaner Prod Lett 2:100005. https://doi.org/10.1016/j.clpl.2022.100005

Alzubi E, Noche B (2022a) A multi-objective model to find the sustainable location for citrus hub. Sustainability 14(21):14463. https://doi.org/10.3390/su142114463

Alzubi E, Noche B (2022b) Improving sustainability of orange supply chain: a system dynamics model to eliminating pre-harvesting loss, increase workers, to improve farmer's profit

Alzubi E, Atieh AM, Abu Shgair K, Damiani J, Sunna S, Madi A (2019) Hybrid integrations of value stream mapping, theory of constraints and simulation: application to wooden furniture industry. Processes 7(11):816. https://doi.org/10.3390/pr7110816

Alzubi E, Kassem A, Noche B (2022) A comparative life cycle assessment: polystyrene or polypropylene packaging crates to reduce citrus loss and waste in transportation? Sustainability 14(19):12644. https://doi.org/10.3390/su141912644

Alzubi E, Shbikat N, Noche B (2023) A system dynamics model to improving sustainable performance of the citrus farmers in Jordan Valley. Cleaner Prod Lett 4:100034. https://doi.org/10.1016/j.clpl.2023.100034

Atabaki MS, Aryanpur V (2018) Multi-objective optimization for sustainable development of the power sector: an economic, environmental, and social analysis of Iran. Energy 161:493–507. https://doi.org/10.1016/j.energy.2018.07.149

Bates RP, Morris JR, Crandall PG (2001) Principles and practices of the small- and medium-scale fruit juice processing. FAO, Rome

Berry TM, Defraeye T, Wu W, Sibiya MG, North J, Cronje P (2021) Cooling of ambient-loaded citrus in refrigerated containers: what impacts do packaging and loading temperature have? Biosyst Eng 201:11–22. https://doi.org/10.1016/j.biosystemseng.2020.11.002

Bieber N, Ker JH, Wang X, Triantafyllidis C, van Dam KH, Koppelaar RH, Shah N (2018) Sustainable planning of the energy-water-food nexus using decision making tools. Energy Policy 113:584–607. https://doi.org/10.1016/j.enpol.2017.11.037

Binder T, Vox A, Belyazid S, Haraldsson H, Zusammenfassung MS (2004) Developing system dynamics models from causal loop diagrams

Brancoli P, Bolton K, Eriksson M (2020) Environmental impacts of waste management and valorisation pathways for surplus bread in Sweden. Waste Manage (New York, NY) 117:136–145. https://doi.org/10.1016/j.wasman.2020.07.043

Cheraghalipour A, Paydar MM, Hajiaghaei-Keshteli M (2018) A bi-objective optimization for citrus closed-loop supply chain using Pareto-based algorithms. Appl Soft Comput 69:33–59. https://doi.org/10.1016/j.asoc.2018.04.022

Cheraghalipour A, Paydar MM, Hajiaghaei-Keshteli M (2019) Designing and solving a bi-level model for rice supply chain using the evolutionary algorithms. Comput Electron Agric 162:651–668. https://doi.org/10.1016/j.compag.2019.04.041

Corrado S, Sala S (2018) Food waste accounting along global and European food supply chains: state of the art and outlook. Waste Manage (New York, NY) 79:120–131. https://doi.org/10.1016/j.wasman.2018.07.032

Dabees A, Barakat M, Elbarky SS, Lisec A (2023) A framework for adopting a sustainable reverse logistics service quality for reverse logistics service providers: a systematic literature review. Sustainability 15(3):1755. https://doi.org/10.3390/su15031755

Das D, Dutta P (2013) A system dynamics framework for integrated reverse supply chain with three way recovery and product exchange policy. Comput Ind Eng 66(4):720–733. https://doi.org/10.1016/j.cie.2013.09.016

de Oliveira MM, Lago A, Dal'Magro GP (2021) Food loss and waste in the context of the circular economy: a systematic review. J Clean Prod 294:126284. https://doi.org/10.1016/j.jclepro.2021.126284

Defraeye T, Lambrecht R, Tsige AA, Delele MA, Opara UL, Cronjé P, Verboven P, Nicolai B (2013) Forced-convective cooling of citrus fruit: package design. J Food Eng 118(1):8–18. https://doi.org/10.1016/j.jfoodeng.2013.03.026

Del Osorio LR, Flórez-López E, Grande-Tovar CD (2021) The potential of selected agri-food loss and waste to contribute to a circular economy: applications in the food, cosmetic and pharmaceutical industries. Molecules (Basel, Switzerland) 26(2). https://doi.org/10.3390/molecules26020515

Dora M, Biswas S, Choudhary S, Nayak R, Irani Z (2021) A system-wide interdisciplinary conceptual framework for food loss and waste mitigation strategies in the supply chain. Ind Mark Manag 93:492–508. https://doi.org/10.1016/j.indmarman.2020.10.013

DoS (2022) Jordan in figure 2021. https://dosweb.dos.gov.jo/products/jordan-in-figure2021/. Accessed 20 Sept 2023

Dusch Silva S Squeezed behind the scenes of the juice industry (2018)

Eriksson M, Strid I, Hansson P-A (2014) Waste of organic and conventional meat and dairy products—a case study from Swedish retail. Resour Conserv Recycl 83:44–52. https://doi.org/10.1016/j.resconrec.2013.11.011

FAO (2022) Food balances (2010-). Available at https://www.fao.org/faostat/en/#data/FBS. Accessed 06 Nov 2022

Fragoso R, Figueira JR (2021) Sustainable supply chain network design: an application to the wine industry in SOuthern Portugal. J Oper Res Soc 72(6):1236–1251. https://doi.org/10.1080/01605682.2020.1718015

Frei R, Jack L, Krzyzaniak S-A (2020) Sustainable reverse supply chains and circular economy in multichannel retail returns. Bus Strateg Environ 29(5):1925–1940. https://doi.org/10.1002/bse.2479

Friedman N, Ormiston J (2022) Blockchain as a sustainability-oriented innovation?: opportunities for and resistance to Blockchain technology as a driver of sustainability in global food supply chains. Technol Forecast Soc Chang 175:121403. https://doi.org/10.1016/j.techfore.2021.121403

Gharye Mirzaei M, Goodarzian F, Maddah S, Abraham A, Abdelkareim Gabralla L (2022) Investigating a dual-channel network in a sustainable closed-loop supply chain considering energy sources and consumption tax. Sensors (Basel, Switzerland) 22(9). https://doi.org/10.3390/s22093547

Goodarzian F, Ghasemi P, Santibanez Gonzalez E, Tirkolaee EB (2023) A sustainable-circular citrus closed-loop supply chain configuration: Pareto-based algorithms. J Environ Manag 328:116892. https://doi.org/10.1016/j.jenvman.2022.116892

Govindan K, Soleimani H, Kannan D (2015) Reverse logistics and closed-loop supply chain: a comprehensive review to explore the future. Eur J Oper Res 240(3):603–626. https://doi.org/10.1016/j.ejor.2014.07.012

Hoxha A, Christensen B (2019) The carbon footprint of fertiliser production: regional reference values. In: Proceedings/International Fertiliser Society, vol 805. International Fertiliser Society, Colchester

Hübner A, Holzapfel A, Kuhn H (2016) Distribution systems in omni-channel retailing. Bus Res 9(2):255–296. https://doi.org/10.1007/s40685-016-0034-7

IPCC (2022) Climate Change and Land: an IPCC special report on climate change, desertification, land degradation, sustainable land management, food security, and greenhouse gas fluxes in terrestrial ecosystems. In: Climate Change and Land. Cambridge University Press, pp. 1–36. https://doi.org/10.1017/9781009157988.001

Knoema (2023) World citrus fruit production. https://knoema.com/atlas/World/topics/Agriculture/Crops-Production-Quantity-tonnes/Citrus-fruit-production. Accessed 20 Sept 2023

Krikke H, Hofenk D, Wang Y (2013) Revealing an invisible giant: a comprehensive survey into return practices within original (closed-loop) supply chains. Resour Conserv Recycl 73:239–250. https://doi.org/10.1016/j.resconrec.2013.02.009

Kurtz JE, Woodbury PB, Ahmed ZU, Peters CJ (2020) Mapping U.S. food system localization potential: the impact of diet on foodsheds. Environ Sci Technol 54(19):12434–12446. https://doi.org/10.1021/acs.est.9b07582

Lütke Entrup M, Günther H-O, van Beek P, Grunow M, Seiler T (2005) Mixed-Integer Linear Programming approaches to shelf-life-integrated planning and scheduling in yoghurt production. Int J Prod Res 43(23):5071–5100. https://doi.org/10.1080/00207540500161068

Manzoor D, Aryanpur V (2017) Power sector development in Iran: a retrospective optimization approach. Energy 140:330–339. https://doi.org/10.1016/j.energy.2017.08.096

Martin NL, Kononova N, Cerdas F, Herrmann C (2022) LCA-based framework to support planning of centralized vs. decentralized production of solid pharmaceuticals. Procedia CIRP 105:128–133. https://doi.org/10.1016/j.procir.2022.02.022

Martin-Rios C, Demen-Meier C, Gössling S, Cornuz C (2018) Food waste management innovations in the foodservice industry. Waste Manage (New York, NY) 79:196–206. https://doi.org/10.1016/j.wasman.2018.07.033

Melkonyan A, Gruchmann T, Lohmar F, Kamath V, Spinler S (2020) Sustainability assessment of last-mile logistics and distribution strategies: the case of local food networks. Int J Prod Econ 228:107746. https://doi.org/10.1016/j.ijpe.2020.107746

Mousa TY (2020) Food security, food waste, and food donation and redistribution in Jordan. Austin J Nutr Food Sci 8(3)

Ortiz DL, Batuecas E, Orrego CE, Rodríguez LJ, Camelin E, Fino D (2020) Sustainable management of peel waste in the small-scale orange juice industries: a Colombian case study. J Clean Prod 265:121587. https://doi.org/10.1016/j.jclepro.2020.121587

Pingali P, Mittra B, Rahman A (2017) The bumpy road from food to nutrition security – slow evolution of India's food policy. Glob Food Secur 15:77–84. https://doi.org/10.1016/j.gfs.2017.05.002

Qu T, Huang GQ, Zhang Y, Dai QY (2010) A generic analytical target cascading optimization system for decentralized supply chain configuration over supply chain grid. Int J Prod Econ 127(2):262–277. https://doi.org/10.1016/j.ijpe.2009.08.008

Rajabi-Kafshgar A, Gholian-Jouybari F, Seyedi I, Hajiaghaei-Keshteli M (2023) Utilizing hybrid metaheuristic approach to design an agricultural closed-loop supply chain network. Expert Syst Appl 217:119504. https://doi.org/10.1016/j.eswa.2023.119504

Roghanian E, Cheraghalipour A (2019) Addressing a set of meta-heuristics to solve a multi-objective model for closed-loop citrus supply chain considering CO2 emissions. J Clean Prod 239:118081. https://doi.org/10.1016/j.jclepro.2019.118081

Sahebjamnia N, Goodarzian F, Hajiaghaei-Keshteli M (2020) Optimization of multi-period three-echelon citrus supply chain problem. J Optim Ind Eng 13(1):39–53. https://doi.org/10.22094/joie.2017.728.1463

Sala S, Anton A, McLaren SJ, Notarnicola B, Saouter E, Sonesson U (2017) In quest of reducing the environmental impacts of food production and consumption. J Clean Prod 140:387–398. https://doi.org/10.1016/j.jclepro.2016.09.054

Sarker R, Ray T (2009) An improved evolutionary algorithm for solving multi-objective crop planning models. Comput Electron Agric 68(2):191–199. https://doi.org/10.1016/j.compag.2009.06.002

Sedlacko M, Martinuzzi A, Røpke I, Videira N, Antunes P (2014) Participatory systems mapping for sustainable consumption: discussion of a method promoting systemic insights. Ecol Econ 106:33–43. https://doi.org/10.1016/j.ecolecon.2014.07.002

Soheilifard F, Marzban A, Ghaseminejad Raini M, Taki M, van Zelm R (2020) Chemical footprint of pesticides used in citrus orchards based on canopy deposition and off-target losses. Sci Total Environ 732:139118. https://doi.org/10.1016/j.scitotenv.2020.139118

Soysal M (2015) Decision support modeling for sustainable food logistics management Mehmet Soysal. Wageningen School of Social Sciences. Online: https://core.ac.uk/download/pdf/29206052.pdf. Accessed on 21 May 2023

Teimoury E, Nedaei H, Ansari S, Sabbaghi M (2013) A multi-objective analysis for import quota policy making in a perishable fruit and vegetable supply chain: a system dynamics approach. Comput Electron Agric 93:37–45. https://doi.org/10.1016/j.compag.2013.01.010

Thomé KM, Cappellesso G, Ramos ELA, Duarte SCDL (2021) Food supply chains and short food supply chains: coexistence conceptual framework. J Clean Prod 278:123207. https://doi.org/10.1016/j.jclepro.2020.123207

Tirado-Kulieva VA, Miranda-Zamora WR, Hernández-Martínez E, Choque-Rivera TJ, Luque-Vilca OM (2022) The vulnerability of a centralized food system: an opportunity to improve food security in times of COVID-19-Peru perspective. Front Sustain Food Syst 6. https://doi.org/10.3389/fsufs.2022.901417

Tsao Y-C (2013) Designing a fresh food supply chain network: an application of nonlinear programming. J Appl Math 2013:1–8. https://doi.org/10.1155/2013/506531

Vilariño MV, Franco C, Quarrington C (2017) Food loss and waste reduction as an integral part of a circular economy. Front Environ Sci 5. https://doi.org/10.3389/fenvs.2017.00021

Wangsa ID, Vanany I, Siswanto N (2023) An optimization model for fresh-food electronic commerce supply chain with carbon emissions and food waste. J Ind Prod Eng 40(1):1–21. https://doi.org/10.1080/21681015.2022.2099473

Zacarias L, Cronje PJ, Palou L (2020) Postharvest technology of citrus fruits. In: The genus citrus. Elsevier, pp 421–446

Zhen L, Huang L, Wang W (2019) Green and sustainable closed-loop supply chain network design under uncertainty. J Clean Prod 227:1195–1209. https://doi.org/10.1016/j.jclepro.2019.04.098

Zheng D, Chen J, Lin M, Wang D, Lin Q, Cao J, Yang X, Duan Y, Ye X, Sun C, Di W, Wang J, Chen K (2022) Packaging design to protect Hongmeiren orange fruit from mechanical damage during simulated and road transportation. Horticulturae 8(3):258. https://doi.org/10.3390/horticulturae8030258

Part V
Innovation Potential of Circular Supply Chains: Towards New Business Models and Their Scalability

Chapter 12
Scenarios for Sustainable Citrus Supply Chains in the Mediterranean

Denis Daus and Sk. Riad Bin Ashraf

1 Introduction

In times of the global pandemic and extreme weather events due to climate change, our world becomes increasingly unstable and complex. Conducting business as usual in such an environment threatens the survival of companies and entire industries (wbcsd 2021). In addition, the discussion about circular economy in relation to the sustainability of supply chains is becoming increasingly relevant. This trend can be explained, for example, by increasing international regulations regarding sustainable business operations and the associated adaptation of CE concepts (World Manufacturing Foundation 2018).

In globalized economies, the interconnected supply chain networks are susceptible to disruptions that have significant consequences on a global scale. The vulnerability of global supply chains was particularly highlighted due to the COVID-19 pandemic and international commercial or military conflicts between nations. The complexity of supply networks coupled with such global complications revealed the weak links between industries (Fernandes 2020; Sarkis et al. 2020). The widespread impacts are highly noticeable in the food sector, and the risk of food insecurity increased severely (Guerrieri et al. 2022). Especially the people in developing countries are existentially threatened by famine. To overcome environmental challenges, unpredictable markets, and reduced public investment in agriculture, thus secure global food, it's crucial for those involved in the food system to adopt innovation, which is essential for ensuring global food (Bragdon and Smith 2015). Consequently, the central question is how to promote innovative and sustainable food supply chains (FSCs) under circular economy principles to address societal

D. Daus (✉) · S. R. B. Ashraf
Centre for Logistics and Traffic (ZLV), University of Duisburg-Essen, Duisburg, Germany
e-mail: denis.daus@uni-due.de; riad.ashraf@uni-due.de

A. Melkonyan-Gottschalk, D. Daus (eds.), *Transformation Towards Circular Food Systems*, Greening of Industry Networks Studies 13,
https://doi.org/10.1007/978-3-031-63793-3_12

challenges such as food security, regional economic growth and competitiveness, employment and quality of life, or environmental resilience.

Utilizing scenarios is a valuable method for steering innovation adoption and transformation in FSCs. Existing research highlights the uncertainties and vulnerabilities in globalized FSCs, discussing the need for alternative strategies amid technological, societal, and policy influences. In this view, a noticeable gap emerges, particularly in the absence of specific scenarios addressing sustainability in Mediterranean citrus supply chains (CSCs). The given chapter addresses this gap, aiming to mitigate uncertainties and understand sustainable practices tailored to the unique challenges of citrus production and distribution.

Citrus Supply Chains in the Mediterranean

Due to its importance in terms of economic, environmental, and social impacts, the Mediterranean region is playing a central role in the promotion of sustainable CSCs. While government regulations have contributed to a growth trajectory in the citrus fruits industry within this region, certain countries are encountering competitiveness challenges that hinder their ability to meet the ever-increasing demand and adapt their agricultural policies accordingly. Notably, North African and Near East countries have emerged as dynamic and competitive players in citrus production, while other countries, primarily EU members, lag significantly behind in terms of competitiveness (Schimmenti et al. 2013). While there are scientific studies that developed scenarios for FSCs and have addressed various sustainability aspects of CSCs in the Mediterranean, there is a lack of comprehensive future development paths and scenarios for Mediterranean CSCs, which could support businesses in developing sustainable business models and in decision-making under uncertainties.

Horizontal and vertical fragmentation as well as a lack of integration among the actors limit the adaptation of CSCs to both actual and emerging challenges. To achieve sustainable citrus production and consumption, a comprehensive understanding of the entire CSC is essential. This entails not only knowledge about agricultural structures and practices but also trade and transportation systems, factors that drive consumer demand, and the different methods of citrus treatment, processing, and disposal. This understanding is required for a successful supply chain transformation with a significant impact, e.g., in terms of increased agricultural output, higher environmental protection, risk mitigation, and food logistics resilience along CSCs. In particular, food logistics faces challenges due to underdeveloped infrastructure, problematic handling processes, and spoilage of fresh fruits due to long distances or the absence of cold chain transports.

Nutrition-rich foods like citrus fruits are disproportionately prone to food loss and waste. For example, citrus peel waste alone makes up around 50% of the total fruit mass. Since citrus waste possesses a high economic potential, the extraction of value-added (by-)products should be considered (Sharma et al. 2017), which could decrease food demand. The valorization of citrus waste and by-products not only avoids costly disposal and reduces the environmental burden, but also discloses new business opportunities for stakeholders along CSCs.

1.1 Aim of This Chapter

The aim of this chapter is to align macroeconomic parameters in global FSCs with the business priorities of actors in CSCs in the Mediterranean by means of strategic scenarios that integrate Mediterranean citrus production and distribution systems within a circular economy while considering regional conditions and operational requirements for sustainable FSCs. This should reduce uncertainties in strategic decision-making for stakeholders along the CSC and provide insights for developing sustainable and innovative business models. Strategic CSC scenarios represent a wide range of integrated sustainable solutions for the citrus sector. Integrated sustainable solutions support organizations and processes along the CSC to achieve an environmentally and socially favorable state embedded in a stable economic system. In this context, the entire CSC is considered in the scenario creation process while providing the full range of technical and social innovations and stakeholder impacts.

Thereby, this chapter aims to channel citrus production conditions toward future developments and establish a basis for innovating business models in the citrus sector from a sustainability and circularity perspective.

The remainder of the study is structured as follows: Sect. 2 provides an overview of related literature to set the groundwork for scenario generation by summarizing developed scenarios for FSCs, synthesizing the conditions in FSCs and CSCs, and linking those to global sustainability concepts. The literature section provides an overview of research areas that affect the construction of plausible CSC scenarios under circular economy principles, starting with a broad view on a global scale and narrowing down to the supply chain level. Different aspects of CSCs in the Mediterranean region were explored, which provide valuable insights and serve as a foundation for the creation of projections and subsequent scenarios. In Sect. 3, the research design for scenario generation is presented. Section 4 presents the results and final scenarios. We discuss the results in Sect. 5. Finally, conclusions are drawn, and an outlook for further research is provided in Sect. 6.

2 Literature Review

Future scenarios for sustainable CSCs consist of a wide range of factors that need to be considered. Research related to FSCs covers a broad spectrum of research themes as well: waste management, supply chain sustainability and impact assessment, decision support, operations management and optimization, sustainable business models, food quality and safety, innovation and technological solutions, supply chain strategy, and social sustainability (Kumar et al. 2022).

2.1 Scenarios for Food Supply Chains

Considering the land, water, and energy scarcity as well as the growing population, food safety is a global concern and will be increasing in relevance in the future. In this view, Ambler-Edwards et al. (2008) discuss four global FSC scenarios, which aim to address the challenges they pose and their effects on the European Union and the United Kingdom. The study highlights the urgency of developing alternative food supply systems in a participatory approach between governments, FSC stakeholders, and social groups, given the possible inadequacies of existing "business as usual" approaches. It also underscores the significance of acquiring raw materials and efficiently using resources in the design of these alternative systems.

Meyer (2007) presents an analysis of scenarios within the European food and agricultural sector, focusing on how technology, farming methods, government policies, and changes in society can shape the food sector. The author identified three major areas of uncertainty that influence the European food sector—technology, society, and politics—suggesting close collaboration between governments and supply chain associates. To make the food and farming sectors sustainable, more research is needed on the interplay of technology, social changes, and politics.

Addressing the resilience of global food systems, it is essential to consider the diverse potential shocks and systemic risks inherent in globalized supply chains. Scenario narratives developed by Hamilton et al. (2020) point out known and new sources of vulnerability in food systems, potential impacts of shocks, and factors that amplify these impacts, concluding that all three dimensions of the triple bottom line (social, economic, and environmental) are influenced by shocks. Vulnerability of food systems is furthermore fostered by increased connectivity across the food system, dependency on single crops, and technologies. The authors suggest that the creation of projections should take place in cooperation between research, industry, and policy, while avoiding scenario development based on a *business as usual* approach (Hamilton et al. 2020).

Huan-Niemi et al. (2010) evaluate the impact of various policy scenarios on global food production. The authors compared a business-as-usual scenario with an economic recession, enhanced climate change mitigation policies, the removal of EU agricultural subsidies, and global trade liberalization. Key findings reveal diverse impacts on food production: short-term effects of the recession, adverse outcomes of stringent climate policies on certain food sectors, minor effects from EU subsidy removal, and substantial shifts in meat production under global trade liberalization.

The evolution of food systems has resulted in adverse effects on human health and environmental issues. Benton (2019) states the reason for that being a significant gap in the alignment of production patterns, nutritional requirements, and sustainability goals since food systems are designed around consumption and productivity growth. This is manifested in an internationally expanded food system based on a "just in time" approach, which is inherently fragile, rendering it susceptible to a range of forces and events. The author developed four scenarios considering the ranges of global-local connectivity and degree of dietary shirts. Intensification of agriculture and concentration on few crops in specific areas drive climate change and contribute to a global homogenization of diets, resulting in rationalization of food waste and the overconsumption of calorie-dense products. The author suggests to move attention from agricultural productivity to the total productivity of the food system, focusing on feeding people with healthy food in a sustainable way (Benton 2019).

Similarly, Hubert et al. (2010) look at the problem of how to feed nine billion people by 2050, highlighting that global economic growth and ecological intensification constitute the main challenges in this endeavor. The study also looks at different possible futures for food security, agriculture supply, and demand, underlining the role of food technology in creating healthy meals and new ways to grow food. Technologies also help to reduce poverty by connecting the geographic areas of demand and supply. The authors stress the need for more studies of agriculture and food since new technologies and governmental hurdles make it hard to get into markets, especially in poor countries.

Moreover, Sundbo (2016) outlines two scenarios for the future of the food sector, focusing on global and regional contexts: high technology-based food production and aquatic-based food production. Besides highlighting the significance of a sustainable environment and animal welfare, the research suggests emphasizing the use of local foods and authenticity, enhancing service quality in e-business, and promoting innovative approaches in local businesses.

Caputo et al. (2014) explored food scenarios, particularly focusing on the local institutional food system in Italy, with the goal of providing sustainable local strategies that would optimize the FSC. Life Cycle Assessment was used to examine how energy and materials move with the goal of lowering energy use all along the supply chain. The study quantified food waste generation stages, suggesting the use of seasonal and energy-efficient products, an increase in local food production, and sensible measures to utilize food waste.

Melkonyan et al. (2019) developed scenarios for future food systems from the viewpoint of sustainable supply chain management, discussing transformation pathways for resource use in FSCs. The study highlights that the implementation of local and sustainable FSCs requires awareness for environmental and social responsibility, which would lead to innovative supply concepts and business models. The authors furthermore note the significance of strategic planning and scenario analysis in effectively managing and adjusting FSCs to address future problems, such as environmental and socioeconomic transformations.

2.2 *Sustainability in Global Supply Chains*

With increasing wealth and projected population growth, the planet faces enormous sustainability challenges and additional pressure on resources, which leads to social, economic, and especially environmental problems. In this view, the linkage between population growth and sustainable development should be from the perspective of climate change and other global environmental challenges (United Nations 2022). The development of sustainable practices is important to counteract the negative consequences of resource depletion and to promote the long-term sustainable use of natural resources (Banaitė and Tamošiūnienė 2016).

Sustainability-oriented innovations in supply chains have recently focused on green innovations (Abu Seman et al. 2019; Zailani et al. 2015). The United Nations emphasized the importance of the supply chain management (SCM) function within businesses to help create more sustainable products and processes (UNGC 2015). The negative environmental impacts of global supply chains due to increased international trade call for strategic planning of sustainable SCM practices (Carter and Liane Easton 2011). To define a scenario-based transformation process for sustainable and innovative production-consumption systems in complex value-creating networks, the SCM function is a suitable perspective. In that regard, this study combines the concept of sustainable development with SCM literature from a strategic point of view.

The notion of sustainable development became increasingly important in socioeconomic debates and is continuously expanding, as the linear model of economic development is facing many challenges. Moving away from a linear approach to more sustainable design and management of supply chains requires a focus on slowing the pace of growth and consumption, consideration of external environmental costs, and technological solutions for recycling and circular production processes (Skawińska and Zalewski 2018). Strategic planning of sustainable supply chain activities at operational level is closely linked to reverse logistics (Beh et al. 2016), preparedness for sustainability and logistics management (Wan Ahmad et al. 2016), as well as sustainable supply network design and life cycle assessment (Eskandarpour et al. 2015).

2.3 *Sustainability in Food Supply Chains*

The 17 goals for sustainable development adopted by the United Nations describe targets to be achieved until the year 2030. A major focus of the 12th goal, "Responsible Production and Consumption," is on the challenges in agriculture and FSCs. It stresses the need for research and development in the food sector to reduce food waste, which could enhance food security. The notion of food security has rapidly become one of the priorities in policy, societal, and science agendas around the world. Debates on food security often center around aspects of food production, which has long been the focus of research. Increasing production is an important strategy to help alleviate food insecurity. However, despite the fact that more than enough food is produced to adequately feed the global population, about 800 million people faced food insecurity in 2020 (FAO et al. 2020).

At the same time, it is assumed that one third of global food products is wasted along the FSC. This means that all resources that were used for processing and distributing these food products are wasted as well (Göbel et al. 2015). The resulting adverse impacts pose a significant burden, e.g., on natural resources, ecosystems, or human health. This is even more evident in developing countries due to a lack of production capabilities or ineffective processes along FSCs (Krishnan et al. 2020). This situation has far-reaching effects on the environment and leads to numerous problems (Banaitė 2016), such as the increasing CO_2 concentration in the atmosphere with the associated greenhouse effect, pollution of soils, water scarcity, the increasing amount of municipal and industrial waste, or a shortage of natural resources with the effect of rising prices or fluctuating markets (Bouchery et al. 2017, p. 15). However, most of the studies dealing with waste management focus on developed countries. Hence, there is the need for developing countries to evaluate waste treatment scenarios (Kumar et al. 2022). There is a significant demand for food systems to adapt to changes in climatic conditions, population growth, and challenges pertaining to ecosystems, such as water scarcity, deforestation, or loss of biodiversity (UNEP 2017).

In comparison with other supply chains, FSCs are complex to manage due to perishability, long production lead times, seasonality in production and consumption, as well as variability of product quality and yield (Amorim et al. 2014). In addition, changing customer behavior and policy regulations require considering the environmental and social impacts in FSCs (Zhu et al. 2018).

A better resource utilization can be achieved by introducing circular economy concepts, such as closed-loop processes, reuse, recycle, or remanufacturing (Vlajic et al. 2018). In this view, circular economy should be integrated in the production and distribution planning in FSCs while considering barriers, challenges, and critical factors (Kumar et al. 2022). In contrast to other products or industries (e.g., rare earths), food cycles are better to be closed on a local level, instead of aiming for a global solution (Ibn-Mohammed et al. 2021).

Krishnan et al. (2021) analyzed farmer producer organizations and highlighted that introducing new product variants and using waste to create value, selecting crops based on soil conditions and expected market demand, information sharing through knowledge hubs, and providing training to farmers on agricultural practices and marketing strategies constitute innovative collaborative practices, which contribute to sustainability in FSCs and strengthen relationships both upstream and downstream. When applying innovative practices in a collaborative manner supported by technologies rather than individual efforts in silos, farmers can obtain greater market control, improve efficiency, reduce production costs, and contribute to sustainability. In addition, an electronic database that contains information such as land area, crop variety, expected yield, and harvest date helps to estimate the total available supply and align it with market demand. A direct communication channel with customers allows to inform customers about the availability of products and to manage orders, while reducing stock uncertainty and maximizing margins (Krishnan et al. 2021).

Technology is a vital enabler for sustainability in FSCs since it helps with prevention of waste and enables seamless information flows by means of data analyses, demand forecasting, or monitoring a myriad of data throughout the entire

FSC (Ciccullo et al. 2021). Already basic information and communication technologies improve decision-making of farmers (Ali and Kumar 2011). Also, advanced technologies like blockchain or artificial intelligence offer a great potential for more sustainable FSCs (Ashraf and Daus 2022). However, the comprehensive impact of such disruptive technologies is yet unexplored, with only a few applications in FSCs (Kumar et al. 2022).

One-third of global food product consumption is wasted along FSCs. Consequently, all the resources utilized in the processing and distribution of these food products are wasted as well, including the energy required for their production and transportation (Göbel et al. 2015). The negative effects impose a considerable burden on natural resources, ecosystems, or human health, especially in developing countries that often have poor production capabilities or ineffective processes along FSCs (Krishnan et al. 2020).

The food industry is seeking to adopt circular economy and sustainability aspects in their business to deal with the issues of food wastage and safety, higher demand and cost of food products, and higher energy and natural resources requirements. However, organizations lack capacities in utilizing the environmental and social sustainability considerations in value chain practices due to lack of awareness, resources, or infrastructure. Thus, it is important to understand and develop the circular economy aspects to improve the overall sustainability of food value chains in a developing economy context (Sharma et al. 2019).

A decreased food demand and hence lower prices can reduce food loss and waste and improve food security. Less food loss and waste can mitigate the need for more food production while decreasing material, water, land, and energy demand along supply chains. This would also lead to more household savings and efficiency in supply chains. However, a lower food demand is also associated with higher unemployment and lower GDP (PACE 2021).

2.4 Sustainability in Citrus Supply Chains

Citrus production in the Mediterranean region accounts for approximately 20% of the global output, primarily consisting of oranges and mandarin-like fruits used for both fresh consumption and agro-industrial purposes. This sector is a major source of income for many people in the Mediterranean. However, farmers face challenges related to environmental conditions such as drought, pests, and diseases. Despite its economic importance, traditional citrus cultivation practices in the region are labor-intensive (Duarte et al. 2016). The citrus sector displays diversity, with fruits cultivated across a range of farm sizes, including small, medium, large, and extremely large farms (FAO 2021).

Differences in agricultural conditions and farm characteristics between northern and southern Europe require tailored approaches for future scenarios. For example, southern countries have more numerous citrus tree crops, more small-scale farms, as well as different weather and soil conditions compared to northern countries. Since each farmer makes individual decisions on the regional level, different approaches are taken regarding pesticide use or application methods (Ramos et al. 2000).

Extreme temperature variations, drought, flooding, and the salinity of irrigation water and substrates, all of which are correlated with climate change, are some of the most detrimental elements affecting citrus productivity and the yield's commercial worth. High temperatures have an impact on citrus plants by increasing transpiration and photosynthesis, weakening their cell membranes, and increasing oxidative damage.

The Mediterranean region is characterized as a climate change hotspot, facing the threat of drought episodes leading to variable production and conflicts over irrigation water. These areas experience pronounced physical and ecological impacts of climate change, significantly affecting agriculture (Cos et al. 2022). Climatic conditions are highly relevant for crop production and crop water requirements, tightly linking the quantity and quality of citrus production to climatic aspects (Albrigo 2019). The geographical scope of citrus cultivation is constrained by low temperatures, primarily in coastal regions. Temperature variations impact various stages of plant development, including growth, blooming, fruit set, and overall quality. Consequently, water usage is critical for food security, employment, and economic development in the Mediterranean region, which faces dramatically increased irrigation requirements due to climate change or population growth. By implementing more efficient irrigation technologies, the savings potential could compensate to some degree for the increased water requirements, but water scarcity and droughts remain big threats in some scenarios for several Mediterranean countries (Daccache et al. 2014; Fader et al. 2016).

North African regions exhibit dispersed climate differences, resulting in varying crop water requirements and notable water deficits across different areas. In the upcoming decades, the Mediterranean is anticipated to experience increased aridity during the winter rainy season, potentially facing a 40% reduction in precipitation in the western region and 20–30% in the eastern region. Future projections suggest a possible rise in crop water requirements from 6% to 22%, with Egypt being particularly vulnerable. Additionally, the summer months are projected to become warmer by 3–4 °C (Tuel and Eltahir 2020). The occurrence of droughts in various regions is evident. This leads to agricultural stress, particularly impacting critical stages such as the flowering phase. Drought poses a growing challenge for agriculture, placing a substantial portion of the world's population at risk of food insecurity and hunger (Rojas 2021). Enhancing our knowledge of the ecology of citrus trees and fruit production may improve agricultural efficiency within the framework of sustainability.

To overcome such challenges, a transformation of the citrus sector accompanied by a redesign of supply chains is required. CSC design research has been active for some time, but only some models integrate social and environmental impacts to enhance decision-making in the design of networks, determining the physical configuration and infrastructure of the supply chain. Recently, a model was proposed to develop a sustainable citrus closed-loop supply chain CLSC network, considering the triple bottom line of sustainability and circularity strategy in manufacturing, distribution, inventory control, recycling, and locational decisions (Goodarzian et al. 2023). At the same time, Sicily's CSC is an excellent example of a circular economy approach, processing citrus pulp often considered waste. Similarly, the paper by (Chinnici et al. 2019) looked at the waste and by-products of the CSC and found technical and organizational ways to make the CSC more sustainable. Hence, putting circular economy ideas into the CSC can lead to less waste, better use of resources, more value, more environmentally friendly practices, and more market competition.

3 Research Method

Organizations must equip themselves for the future through development and implementation of resilient strategies. Scenario design proves to be a valuable method for decision-making in circumstances with uncertain outcomes. The scenario method is an appropriate approach for long-term planning, aiding decision-makers in navigating through uncertain situations (Gaspars-Wieloch 2019).

Scenarios generally constitute possible developments of a special area of observation. The scenario planning technique is based on network thinking (interrelationships between influencing factors) and multiple futures (possibility of several possible developments of each influencing factors). By thinking in complex networks in the light of a myriad of social and environmental problems, the development of a company as a system can no longer be considered separately from the development of cities, the environment, technology, or society, which also constitute interconnected subsystems within overall system. The idea of multiple futures is important to embrace because often it is the inconceivable rather than the supposedly probable that has become reality. This means that it is necessary to mentally prepare for possible changes in order to not be surprised by sudden developments (Gausemeier and Plass 2014).

In order to promote the adoption of innovation within CSCs and unleash the transformative potential for businesses, the scenario planning and management approach from Gausemeier and Plass (2014) was applied. It consists of five stages: (i) preparation, (ii) scenario field analysis, (iii) projections development, (iv) scenario development, and (v) scenario transfer.

The results of the first four phases are described in this chapter. The research design approach for these four phases is represented in Fig. 12.1.

Research of potential futures can be divided into explorative (What can happen?), predictive (What will happen?), and normative (how can a specific scenario be reached) studies (Lorenz and Veenhoff 2013). Exploratory scenarios were selected due to their capability to describe different developments of social, economic, and environmental factors in a participatory way. Key change factors (descriptors) were identified by means of expert workshops, and their potential impact was quantitatively assessed. A literature analysis on scenarios and sustainability in FSCs provided the groundwork for establishing projections, from which scenarios

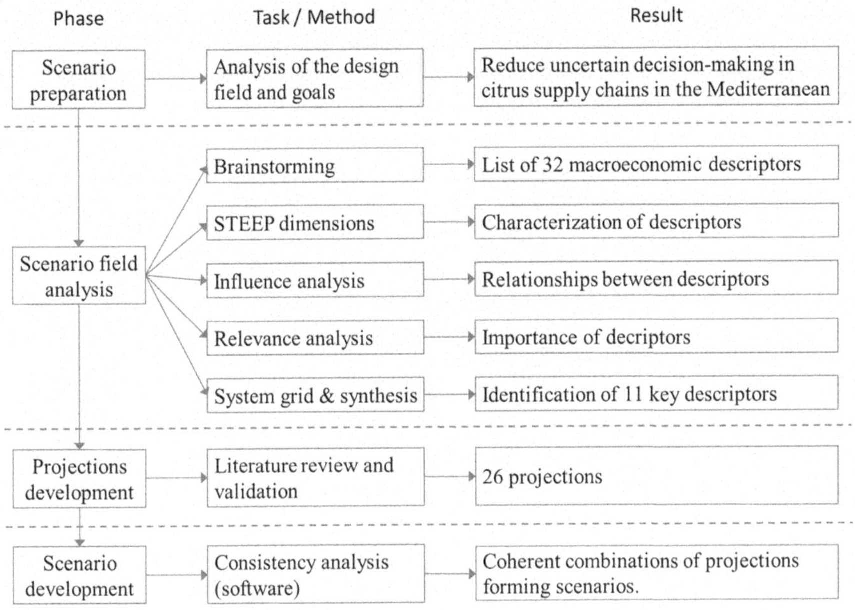

Fig. 12.1 Research design

were developed, reflecting different possible futures. The integration of various perspectives in a multidisciplinary team allows to generate profound scenarios that consist of a wide range of possible futures (Volkery et al. 2008). Thereby, the socioeconomic situation of the stakeholders, driving factors and challenges for citrus production, and digitization progress were taken into consideration, providing the necessary context for scenario design. By understanding the context of the situation, planners can make better decisions. The technique points out the impacts of potential future systems and projects consequences of different decisions (Amer et al. 2013). This approach was combined with STEEP (Social, Technological, Environmental, Economic, Political) and Cross-Impact Balances (Weimer-Jehle 2023) methods to develop scenarios for the citrus sector in the Mediterranean area.

3.1 *Scenario Preparation*

In the first phase, the system boundaries, a specific subject (a so-called design field), and goals of the scenario project relating to the design field were defined. The design field should be analyzed in terms of what should be achieved with the scenarios and what kind of strategies should be derived. Since companies in a specific sector are usually confronted with similar questions, the goal is to elaborate on challenges and develop improvement opportunities for the citrus sector as a whole, so that the stakeholders can prepare for the future and transform their businesses accordingly. For this study, the design field is defined as *citrus supply chains in the Mediterranean* with the goal to foster innovation and by-product valorization.

3.2 Scenario Field Analysis

The aim of this step was to identify characteristic key influencing factors (so called descriptors), which are relevant for the development of the scenario field. The scenario field describes what should be explained by the created scenarios. Besides aspects that directly affect the design field, these are also external citrus production and market conditions that describe the global environment. This is referred to as *system scenarios*. A system scenario consists of internal and external descriptors. The former can be steered towards own strategic goals, whereas the latter are not controllable and therefore represent general business limits.

To characterize the descriptors on the global scale, the STEEP dimensions were deployed. The STEEP method helps to identify external descriptors impacting organizations or systems. After listing macroeconomic descriptors that influence the scenario field by participatory brainstorming sessions, they were structured according to the STEEP dimensions. The brainstorming sessions took place in parallel at four different universities in Algeria, Egypt, Germany, and Tunisia. The results from each university were gathered and consolidated to avoid duplications or similar descriptors with different names. To ensure a comprehensive set of alternatives, these descriptors were consolidated without any judgement or evaluation. Due to the large number of potential descriptors, considering all of them would be too complex and bear the risk of getting lost in details. To filter out relevant key descriptors, subsequent influence analysis and relevance analysis were conducted in a workshop with 16 participants on February 8, 2023. Among the industrial participants, two representatives were from a citrus by-products processor in Egypt, one participant represented a lemon juice producer in Algeria, and one participant was an expert in technology adoption and management. The remaining participants were academic experts in the areas sustainable business models, strategy development, supply chain management, logistics, transportation modeling, operations management, mathematical programming and optimization, agricultural production and handling, circular economy, and environmental sciences.

The influence analysis evaluates the system behavior between descriptors on a scale between 0 (no influence) and 3 (strong influence). For each descriptor pair, direct relationships were evaluated in terms of how strongly (or quickly) one descriptor changes due to the direct influence of the other, and vice versa. The sum of the values of each row constitutes the active sum, which shows how strongly a descriptor influences the others. The sum of the values of each column constitutes the passive sum, which shows how strongly a descriptor is influenced by the others.

The relevance analysis evaluates on a binary scale for each descriptor pair which descriptor is more or less important for the design field. The sum of the values of each row constitutes the relevance sum, which shows how important a descriptor is for the design field. Hence, the result shows the strength with which the descriptors impact the investigated design field. Since these analyses are time consuming and couldn't be completed within one workshop day, participants finalized the influence matrix and the relevance matrix on an individual basis. The results were then consolidated, and deviations were discussed in a follow-up online meeting, so that a consensus was established.

The results of both the influence and relevance analysis are eventually combined in a system grid (Fig. 12.4). The location in the grid visualizes the ranking of active and passive sums, whereas the size of the bubbles represents the importance of each descriptor for the CSCs in the Mediterranean. Based on this, a synthesis of the descriptors regarding similar levels of detail and innovation potential resulted in a selection of descriptors that have a strong influence on other descriptors and are significant for CSCs in the Mediterranean.

3.3 Projections Development

For every descriptor, potential development paths (so called projections) were developed in phase three. The actual glimpse into the future happens in this phase, which is an essential step because the building blocks of the final scenarios are developed here. Two to three projections for each descriptor were obtained by means of literature analysis. Thereby, trends and challenges concerning the descriptors were derived and related to the citrus industry in the Mediterranean. Subsequently, the projections were evaluated and analyzed in a workshop with 11 academic participants on July 24, 2023. The participants covered the academic fields as stated for the first workshop. As a guideline to obtain a set of profound alternative projections, each projection had to fulfil the following quality criteria (Daus et al. 2018):

- Plausibility for the scenario team
- Dissimilarity: distinct from each other
- Completeness: comprehensive set of possible developments
- Relevance for the future
- Information content: add further value to the set of projections within a descriptor

As long as the projections meet these quality criteria, they can also contain extreme developments, which fosters creativity in the later stages of strategy development. Similar to the definition of descriptors, the development of projections requires analytical and creative thinking, capturing quantitatively measurable projections. To support creativity, the potential likelihood of occurrence or reasonable combinations of the projections into scenarios were neglected at this stage. The procedure to determine projections was supported by following several guidelines. (It is important to note that projections are of hypothetical nature. They intend to inspire forward-thinking and encourage proactive measures.)

- Extrapolate or simulate current developments
- Exaggerate projections (extreme developments)
- Accelerate current developments
- Incorporate surrounding developments
- Derive projections from current development processes

3.4 Scenario Development

The actual scenarios were developed in this phase. A scenario is a combination of future projections that fit together well. Consistency is the decisive factor for the credibility of projections into the future. To check the compatibility of the individual projections, a pairwise consistency assessment (Fig. 12.5.) was carried out by the scenario project team in the same workshop on July 24, 2023, by applying the Cross-Impact Balance analysis.

Hence, each projection was evaluated with every other projection in terms of their compatibility by assigning values between −2 (strongly restricting influence) and +2 (strongly promoting influence), whereas 0 means that the projections are independent of each other.

The interdependencies between the projections regulate which combinations of the various projections can form a projection bundle, which constitutes a selection of projections where exactly one projection per descriptor exists. Thus, a projection bundle can be regarded as a consistent scenario and has as many projections as descriptors. While the consistency check itself is straightforward and can be conducted manually, evaluating an entire cross-impact matrix demands frequent repetition of this check. To handle the complexity of 10,368 potential scenarios (see Sect. 4.3), the *ScenarioWizard* software was employed. The values from the consistency analysis matrix were transmitted into the software, resulting in the identification of five scenarios, which were eventually consolidated into three scenarios.

The final scenarios were then visualized (Fig. 12.7) through key distinguishing features represented by arrows. With this approach, interrelationships and fundamental differences between scenarios can be explained.

4 Results

This chapter outlines the results obtained from the first four stages of the scenario development process.

4.1 Descriptors

In the initial phase, descriptors (influencing factors) were gathered from each project partner to formulate a thorough list organized based on the STEEP dimensions. The initial consolidation of this compilation yielded 32 descriptors (Table 12.1). However, the objective is to narrow down to 10–20 key descriptors, ensuring the manageability of the scenario planning process. To achieve this, influence analysis

Table 12.1 Initial compilation of descriptors

STEEP	Descriptors
Social	Consumer preferences for products
	Labor availability
	Working conditions
	Sustainability awareness along the SC
	Action for sustainability qualification
	Consumer health and well-being
Technological	Transparency along the SC
	Technology adoption
	Forecasting capabilities
	Digitalization in the supply chain
	Technology Finance
Environmental	Packaging recovery
	Water salinity
	Fruit and by-product waste
	Production footprint
	Renewable source of energy
	Impact of climate change
	Natural disasters
	Water resources
	Pesticide use
	Soil quality
Economic	Yield development at farm and processor
	Market access (distribution channels)
	Efficiency of resources
	Profit margin
	Operational costs (raw materials, packaging, fertilizer, energy, labor etc.)
	Circular practices adoption
	Stakeholder integration (collaboration)
	Logistics processes
Political	Trade barriers (foreign policy)
	Regulations
	(Geo-)Political environment

(Sect. 4.1.1) and relevance analysis (Sect. 4.1.2) were carried out, and the outcomes were presented in a system grid (Sect. 4.1.3) to finalize the selection of key descriptors.

4.1.1 Influence Analysis

In an influence matrix (Fig. 12.2), direct relationships between descriptors are examined. The analysis for each descriptor pair reveals the extent to which one descriptor changes due to the direct influence of the other descriptor, and vice versa.

The active sum indicates the strength of a descriptor's influence on other descriptors, while the passive sum reflects how strongly a descriptor is influenced by others. The influence analysis offers insights into the systemic behavior of descriptors but lacks information on their impact strength on the investigated field. This gap is addressed through the relevance analysis.

Question: **How strong does descriptor i (row) influence descriptor j (column)** **0 = no influence** **1 = weak influence** **2 = moderate influence** **3 = strong influence**		**Social**						**Technological**					**...**	**Political**			
		Consumer preferences for products	Labor availability	Working conditions	Sustainability awareness along the SC	Action for sustainability qualification	Consumer health and well-being	Transparency along the SC	Technology adoption	Forecasting capabilities	Digitalisation in the supply chain	Technology Finance	...	Trade barriers (foreign policy)	Regulations	(Geo-)Political environment	Active sum
Social	Consumer preferences for products		3	2	1	2	3	2	1	3	1	3	...	0	1	0	41
	Labor availability	0		2	0	1	0	0	2	1	2	2	...	0	1	1	30
	Working conditions	1	3		2	3	1	0	2	0	2	1	...	2	3	1	38
	Sustainability awareness along the SC	2	1	3		3	3	2	3	1	2	2	...	2	3	1	62
	Action for sustainability qualification	1	2	3	3		3	2	2	1	2	3	...	1	2	0	59
	Consumer health and well-being	3	1	1	2	2		0	1	0	1	1	...	2	3	1	23
Technological	Transparency along the SC	2	1	1	3	1	0		2	2	1	1	...	1	3	1	43
	Technology adoption	1	1	2	2	1	2	3		3	3	1	...	1	2	0	62
	Forecasting capabilities	0	2	1	0	0	0	2	1		1	2	...	0	2	1	38
	Digitalisation in the supply chain	2	1	2	1	2	1	3	3	3		1	...	1	2	1	64
	Technology Finance	0	0	1	0	1	2	3	3	3	3		...	0	0	0	39
...	...	...	...	...	...	...	...	...	...	...	...	...	...	...	...	...	...
Political	Trade barriers (foreign policy)	0	0	0	0	2	0	0	0	0	0	0	...		2	3	18
	Regulations	0	1	3	3	3	3	1	0	0	0	2	...	3		2	38
	(Geo-)Political environment	0	0	0	0	0	1	0	2	0	2	1	...	3	2		22
	Passive sum	22	19	28	36	47	30	25	44	22	35	42	...	27	41	15	

Fig. 12.2 Influence matrix (abbreviated)

Question: **Is descriptor i (row) more important than descriptor j (column)** **0 = no (i is less important than j)** **1 = yes (i is more important than j)**		**Social**						**Technological**					**...**	**Political**			
		Consumer preferences for products	Labor availability	Working conditions	Sustainability awareness along the SC	Action for sustainability qualification	Consumer health and well-being	Transparency along the SC	Technology adoption	Forecasting capabilities	Digitalisation in the supply chain	Technology Finance	...	Trade barriers (foreign policy)	Regulations	(Geo-)Political environment	Relevance sum
Social	Consumer preferences for products		0	0	0	0	0	1	1	1	1	1	...	0	0	0	12
	Labor availability	1		0	0	0	0	1	1	1	1	1	...	0	0	0	15
	Working conditions	1	1		1	1	0	1	1	1	1	1	...	1	1	1	25
	Sustainability awareness along the SC	1	1	0		0	0	1	1	1	1	1	...	1	0	0	15
	Action for sustainability qualification	1	1	0	1		0	1	1	1	1	1	...	1	0	1	17
	Consumer health and well-being	1	1	1	1	1		1	1	1	1	1	...	1	1	1	30
Technological	Transparency along the SC	0	0	0	0	0	0		1	1	1	1	...	0	0	1	8
	Technology adoption	0	0	0	0	0	0	0		1	1	0	...	0	0	1	5
	Forecasting capabilities	0	0	0	0	0	0	0	0		0	0	...	0	0	1	1
	Digitalisation in the supply chain	0	0	0	0	0	0	0	0	1		1	...	0	0	1	5
	Technology Finance	0	0	0	0	0	0	0	1	1	0		...	0	0	1	8
...	...	...	...	...	...	...	...	...	...	...	...	...	...	...	...	...	...
Political	Trade barriers (foreign policy)	1	1	0	0	0	0	1	1	1	1	1	...		0	1	12
	Regulations	1	1	0	1	1	0	1	1	1	1	1	...	1		1	17
	(Geo-)Political environment	1	1	0	1	0	0	0	0	0	0	0	...	0	0		5

Fig. 12.3 Relevance matrix (abbreviated)

4.1.2 Relevance Analysis

Unlike the influence matrix, the relevance matrix (Fig. 12.3) conducts pairwise comparisons of descriptors in a single direction. For the second direction (below the black diagonal separation line), inverted values are utilized. Moreover, the evaluation of descriptors involves providing only yes (1) and no (0) responses.

The relevance sum indicates the importance (or significance) of a descriptor for CSCs in the Mediterranean and ranks the descriptors accordingly.

4.1.3 System Grid

To determine key descriptors with the most significant impact on the future of CSCs in the Mediterranean, the active sum and passive sum from the influence analysis, along with the relevance sum from the relevance analysis, can be illustrated in a system grid (Fig. 12.4). Both axes are scaled based on ranks for clarity in presentation. The bubble sizes reflect the outcomes of the relevance analysis, where a larger bubble signifies a greater influence on CSCs.

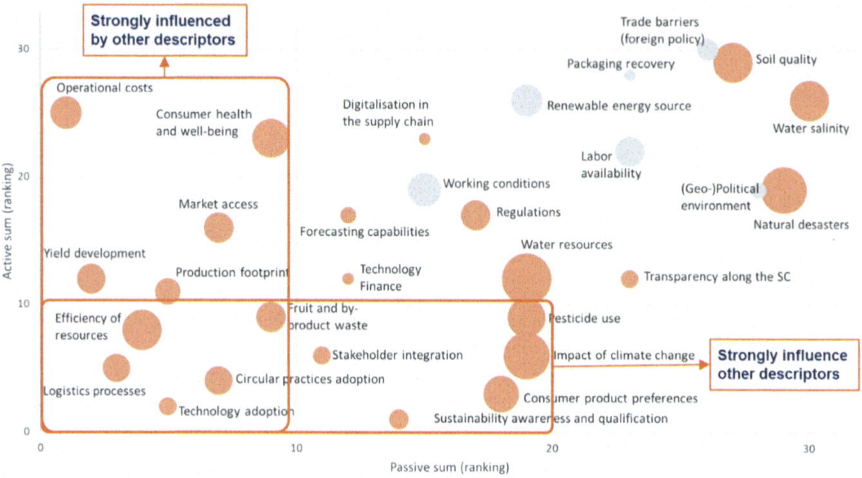

Fig. 12.4 System grid (selection of key descriptors)

The key descriptors are identified by a large bubble size (indicating high relevance for the CSC) and a high-ranking position (reflecting strong activity in the system network of descriptors). The chosen ranking range of 1–10 fulfills these criteria, encompassing a total of 15 descriptors. However, this is a subjective range definition. Some descriptors, while not within this range, are close or also large, suggesting their relevance. Following a workshop decision within the project team, nine additional descriptors were included in the scenario development, resulting in a total of 24 key descriptors. To streamline and attain the desired 10–20 descriptors, the collection was reduced through the consolidation of similar key descriptors, along with the provision of reasoning (Table 12.2).

The consolidation process yields ten key descriptors. Ultimately, recognizing *geopolitical environment* as a crucial factor, given its impact on trade policies, transportation routes, and overall stability in the Mediterranean, 11 key descriptors (Table 12.3) are considered for further analysis. Political decisions can significantly influence cross-border movements, tariffs, and market access, thereby affecting the reliability and cost-effectiveness of the CSC. It is imperative to monitor geopolitical dynamics to anticipate potential disruptions and make informed decisions.

4.2 *Projections*

Projections outline potential future development paths and were established for each descriptor. The time horizon for which these projections are applicable is approximately 10 years. These projections were formulated through a literature review and insights gathered from experts, resulting in a comprehensive catalog with

Table 12.2 Consolidation of key descriptors

STEEP	Key descriptor	Consolidation	Reasoning
Social	Consumer product preferences	Sustainability awareness along the CSC	Preference for sustainable products and willingness to pay for it generally comes from sustainability awareness
	Sustainability awareness		
	Consumer health and well-being	Social health and well-being	Includes not only consumers, but also labor along the CSC
Technological	Transparency in the SC	Digitalization in the CSC	Digitalization drives responsiveness (real-time data), collaboration, traceability (logistics, inventory, materials), predictive analytics of internal and external data
	Forecasting capabilities		
	Digitalization in the SC		
	Technology adoption	Technology adoption capabilities	Capabilities to adopt technologies include financing possibilities, capacities, competences
	Technology Finance		
Environmental	Fruit and by-product waste	Valorization of secondary resources	Includes fruit and by-product waste, wastewater, energy
	Production footprint	Environmental performance	Efficient use of resources and capacities along the supply chain
	Pesticide use		
	Impact of climate change	Natural citrus growing conditions	Given surrounding conditions and impacts of it High relevance (bubble size) based on relevance matrix
	Natural disasters		
	Water resources		
	Water salinity		
	Soil quality		
Economic	Fruit and by-product waste	Efficiency of production means	Resource-related: All describe the use of land, labor, capital etc. that can be used to produce products, which also effects the profit margin
	Efficiency of resources		
	Circular practices adoption		

(continued)

Table 12.2 (continued)

STEEP	Key descriptor	Consolidation	Reasoning
	Operational costs	Logistical order fulfilment	Process-related: Describes the 6Rs—Product, time, place, quantity, quality, cost. Also integrating stakeholders in the processes fosters collaboration and industrial symbiosis
	Stakeholder integration		
	Logistics processes		
Political	Fruit and by-product waste	Regulations	Regulations are the state's attempt to intervene in private activities and protect domestic industries (policy rules, enforcement of compliance, economic controls, taxation
	Regulations		

Table 12.3 Final key descriptors

STEEP	Final key descriptors
Social	Sustainability awareness along the CSC
	Social health and well-being
Technological	Digitalization in the supply chain
	Technology adoption capabilities
Environmental	Valorization of secondary resources
	Environmental performance
	Natural citrus growing conditions
Economic	Efficiency of production means
	Logistical order fulfilment
Political	Regulations
	Geopolitical environment

elaborated 26 projections summarized in Table 12.4. Subsequently, all projections were validated in a workshop, making sure that the projections meet the required quality criteria described in Sect. 3.3. Each key descriptor was introduced briefly, followed by a description of the corresponding projections. Due to the large volume of projections developed, the full descriptions only for the projections for the descriptor *technology adoption capabilities* are presented here as an example (Sect. 4.2.1). The full catalogue can be found in the *report on proposed citrus supply chain scenarios* (Daus 2023).

Social: Social Health and Well-Being

The Harmony Initiative projection emphasizes the interconnectedness between agricultural practices and of social cohesion within the Mediterranean CSCs. Envisaging collaboration approaches, fosters solidarity among stakeholders and promotes shared responsibility for sustainable development. A Resilience Alliance revolves around overcoming the lack of resilience in CSCs by focusing on building adaptive

Table 12.4 Summary of projection

STEEP	Key descriptor	Projection	Reference
Social	Social health and well-being	Harmony initiative	Consolidation of linkages with social agriculture and critical consumption movements is necessary (Fonte and Cucco 2017)
		Resilience Alliance	Lack of consensus on defining and implementing resilience in food supply chains (Stone et al. 2015), focusing on overall food security, instead of individual company performance (Stone and Rahimifard 2018)
	Sustainability awareness in the citrus supply chain	Eco-Educators	Improvement of education is required to adapt to challenges, contributing to closing the skills gap in the food sector (Jonas Lazaro-Mojica and Rebeca Fernandez 2021)
		Sustainable Empowerment Network	Support of sustainable FSC practices through the provision of critical resources and collaboration for partners along the supply chain (Adams et al. 2021)
Technological	Technology adoption capabilities	Tech Integration Hub	Importance of incorporating the needs and perceptions of farmers during technology design and development to enhance adoption as well as provide political incentives (Yokamo 2020)
		Innovation Sandbox	Many countries require technological innovation in the food industry and policy support, fostering the development of novel food products and processes (Castillo-Vergara et al. 2021)
		Slow Adopters	Despite the presence of technological advancements, the adoption of technology in agricultural practices remains unaccomplished, particularly in less-developed countries (Ugochukwu and Phillips 2018)
	Digitalization in the supply chain	Digital Citrus Network	Integrating technological advancements for a digital transformation at all levels of FSCs can enhance optimization and promote increased sustainability throughout the entire supply chain, meeting the increasing demand for safe and nutritious food (Galih Nugroho et al. 2023)
		Citrus E-Commerce Platform	E-commerce logistics can have a strong positive impact on customer satisfaction and last-mile food logistics (Yi Jiang et al. 2021)

(continued)

Table 12.4 (continued)

STEEP	Key descriptor	Projection	Reference
Environmental	Valorization of secondary resources	Circular Citrus Bioeconomy	Moving towards a regenerative bioeconomy in the context of circular economy requires adoption of sustainability and efficiency principles as well as innovative technologies (Hamam et al. 2021)
		Waste-to-Energy Initiatives	Conversion technologies for food waste provide a potential solution for effectively managing food waste and transforming it into valuable by-products (Jeevahan et al. 2018)
	Environmental performance	Sustainable Farming Certification	Proof of quality assurance becomes increasingly important and often correspond to high export volumes of agricultural and food products (Gawron and Theuvsen 2009)
		Production Footprint Reduction	Significant reductions of carbon emissions in food systems can be achieved by effective process control of carbon emissions, package design, or waste treatment (Xu et al. 2015)
	Natural citrus growing conditions	Climate-Resilient Citrus Cultivation	Adapting and mitigating climate change in agriculture must align with global food production demands, lacking a standard framework for implementing both practices (Reddy 2014)
		Organic Citrus Farming	Organic farming, while yielding less than conventional farming, is more profitable and environmentally friendly, with equally or more nutritious foods and fewer pesticide residues. Still, it requires policy instruments to overcome barriers for development and implementation (Reganold and Wachter 2016)
		Unfavorable Growing Conditions	Climate change might make foodborne illnesses and chemicals contaminating our food more common and severe, while extreme weather events threaten food safety and require adaptation strategies (Duchenne-Moutien and Neetoo 2021)
Economic	Efficiency of production means	Technology-Driven Efficiency Gains	Technological innovation is the fundamental approach for enhancing industrial efficiency and fostering a low-carbon economy in the food processing industry, addressing environmental and resource use challenges (Xuxian Yan and Xianjun Li 2015)

(continued)

Table 12.4 (continued)

STEEP	Key descriptor	Projection	Reference
		Circular Gains and Collaboration	The agro-food sector will face business scenarios, where resource efficiency and innovative technologies offer opportunities for sustainability and economic viability by reducing food loss and waste, which positively affects profits and well-being (Esposito et al. 2020)
	Logistical order fulfilment	Smart Logistics and Real-time Tracking	Smart solutions in agribusiness can enhance logistics by preventing waste, cold chain disruptions, and food contamination, ensuring comprehensive control, efficiency and transparency along the FSC (Capello et al. 2016)
		Collaborative Supply Chain Networks	An effective cooperative approach in food supply networks and inclusion of vendors in delivery planning benefits delivery transport costs and service levels, while meeting uncertain retailer demands (Accorsi et al. 2018)
		Last-Mile Delivery Optimization	Advanced last mile solutions for urban FSCs, help minimize costs, avoid traffic, and decrease emissions, thereby providing guidance to city authorities for the adoption of more sustainable concepts for urban last mile distribution (Leyerer et al. 2020)
Political	Regulations	Sustainable Agricultural Policies	Enhancing policy network effectiveness involves clearly defining policy objectives and increased participation of environmental and developmental organizations (Lovec 2015), supporting policy changes in Southern and Eastern Mediterranean countries to improve water and food security (Ferragina and Canitano 2015)
		Food Safety and Quality Standards	Regulations serve as a crucial tool to boost organic fruit and vegetable production and enhance exports to the growing EU market, contributing to food security and the protection of consumers and farmers (Corona 2015)
		Trade Agreements and Market Access	Incentives for organization and certification of farmers can mitigate difficulties in accessing markets, while ensuring socially and environmentally sustainable agricultural practices (King-Okumu and Aboukheira 2015)

(continued)

Table 12.4 (continued)

STEEP	Key descriptor	Projection	Reference
	Geopolitical environment	Favorable Political Conditions	The trend in global politics fosters multipolarity in the Mediterranean and causes concerns of structural divisions. However, the reduced U.S. interest and the EU's inability to engage have created a gap in the Mediterranean region, filled by other powers trying to strengthen relationships and engage in the region (Cristiani 2020)
		Unfavorable Political Conditions	Rising tension between countries in the Mediterranean nourish geopolitical considerations regarding balance of power, sovereignty, and energy security, challenging the status quo and leading to the creation of strategic military-energy alliances and bilateral deals with other regional players, while also affecting energy security of the EU (Proedrou 2023)

capacity, fostering community engagement, and enhancing working conditions, creating a resilient network that not only withstands uncertainties but actively contributes to food security.

Social: Sustainability Awareness in the Citrus Supply Chain
Eco-Educators outlines a pathway dedicated to the improvement of continuous learning and informed choices by prioritizing sustainability education and training, knowledge dissemination, stakeholder engagement, and the adoption of sustainable practices. A Sustainable Empowerment Network focuses on uplifting marginalized stakeholders, emphasizing sustainable livelihoods, social equity, and inclusivity. This projection envisions a future where the empowerment of all participants contributes to the overall sustainability of the citrus production and distribution system.

Technological: Technology Adoption Capabilities
The Tech Integration Hub involves the establishment of digital training networks, focused on capacity building and tailored to stakeholders' needs. It aims to provide stakeholders with access to essential digital resources and technologies in a tech-savvy environment. An Innovation Sandbox foresees digital advancements, encourages experimentation, and embraces an agile mindset within Mediterranean CSCs. It envisions a dynamic digital innovation ecosystem through technology scouting and entrepreneurship, driving sustained efficiency and innovation. The Slow Adopters path addresses the challenges faced by certain stakeholders in embracing digital-driven change. It underlines resistance and adoption barriers, including the fear of change and the scarcity of skilled labor.

Technological: Digitalization in the Supply Chain
A Digital Citrus Network represents a future characterized by enhanced transparency and traceability across Mediterranean CSCs. Through robust data integration and predictive analytics, this trajectory envisions a seamlessly interconnected and optimized supply chain that offers stakeholders real-time insights into production and distribution operations. In the Citrus E-Commerce Platform projection, online marketplaces play a pivotal role in enhancing customer engagement and order fulfillment. This trajectory anticipates a user-friendly platform that not only enhances accessibility but also maximizes customer satisfaction.

Environmental: Valorization of Secondary Resources
A Circular Citrus Bioeconomy charts a course towards a regenerative bioeconomy in which citrus waste is transformed into valuable products, while maximizing efficiency and minimizing environmental impact. By emphasizing waste and byproducts valorization, resource recovery, and the development of bioproducts, this path envisions the establishment of closed-loop systems and industrial symbiosis concepts. Waste-to-Energy Initiatives focus on leveraging waste for renewable energy projects. Emphasizing biomass conversion plants, this path envisions a future where citrus waste becomes a valuable resource for generating energy. The initiatives aim to contribute to both sustainability and energy efficiency within CSCs.

Environmental: Environmental Performance
The Sustainable Farming Certification projection focuses on the promotion of environmentally conscious farming practices and conservation agriculture principles by adhering to sustainability standards, implementing best practices, and fostering a culture of continuous improvement. A Production Footprint Reduction strategy aims to minimize environmental impact by focusing on identifying emission hotspots, implementing logistics optimization strategies, and incorporating carbon offset initiatives. In this development, citrus production significantly reduces its ecological footprint.

Environmental: Natural Citrus Growing Conditions
The Climate-Resilient Citrus Cultivation path guides towards sustainable citrus production in the face of climate challenges within the Mediterranean CSCs. It emphasizes the implementation of risk management strategies, the cultivation of resilient citrus varieties, and the adoption of precision agriculture, which helps adapt to changing climatic conditions. Organic Citrus Farming centers on the promotion of organic practices, addressing key areas such as pest control, soil, and nutrition management. By providing incentive programs and policy support, this path envisions a future where stakeholders are encouraged to adopt organic principles, leveraging the growing organic market.

Unfavorable Growing Conditions address the challenges posed by severe climate change impacts. With present issues such as less agricultural land, decreased citrus yield, and concerns about pests and diseases, this development poses a threat to Mediterranean CSCs and calls for mitigation strategies.

Economic: Efficiency of Production Means
Technology-Driven Efficiency Gains describe income sources generated by the deployment of advanced technologies. With a focus on optimized citrus yield, resource allocation, and productivity gains through supply chain integration, automation, and robotics, this trajectory leverages innovative technologies to enhance overall productivity and sustainability in citrus production and distribution. In contrast, the Circular Gains and Collaboration projection emphasizes income sources generated by efficient resource management practices, by-products valorization, and collaborative resource pooling, highlighting a future where circularity and collaboration are central.

Economic: Logistical Order Fulfillment
The Smart Logistics and Real-time Tracking projection charts a course where the collection of real-time data is instrumental for supply chain and product visibility, ensuring better decision-making and overall improvement in the management of logistics processes. Collaborative Supply Chain Networks focus on fostering cooperation to achieve economic gains. Through sharing supply chain information, collaborative process improvement can take place, while cross-docking solutions enhance efficiency in logistics operations. Last-Mile Delivery Optimization projection centers on refining the final stage of Mediterranean CSCs by focusing on route optimization and alternative delivery methods to serve better the increasing customer requirements.

Political: Regulations
The Sustainable Agricultural Policies projection outlines a path for shaping sustainable farming policies, environmental regulations, and water management regulations that address environmental concerns and ensure overall sustainability of citrus cultivation. Food Safety and Quality Standards emphasize high standards within CSCs, ensure the safety and quality of citrus products by means of quality control measures, certification and labeling, and requirements for product traceability. It also involves adherence to consumer protection laws, creating a trustworthy framework that safeguards the integrity and quality of citrus products. With Trade Agreements and Market Access, smooth interactions between stakeholders can be facilitated by reducing trade barriers, harmonizing product compliance schemes, and streamlining customs procedures. The establishment of agreements can simplify cross-border transactions, ensuring compliance and accessibility for citrus products in diverse markets.

Political: Geopolitical Environment
Favorable Political Conditions highlight the importance of political stability and collaborative environments that support growth and the sustainable development of citrus production. In contrast, Unfavorable Political Conditions address challenges posed by uncertainties and conflicts, causing disruptions and threatening societies.

4.2.1 Descriptor Example: Technology Adoption Capabilities

The adoption of new technologies is often a complex process that is influenced by a variety of factors, such as the cost of new technologies, the lack of infrastructure, the availability of skilled labor, the fear of change, or the potential benefits. However, the adoption of new technologies can be accelerated, e.g., by governmental support, the availability of financing, and the development of new business models, but it will require a concerted effort by all stakeholders (Gausemeier and Plass 2014). Adoption of technologies has the potential to revolutionize the citrus industry. These technologies could make citrus production more efficient, sustainable, and profitable. As a result, CSCs can become more resilient and competitive, and could better meet the challenges of the future.

Projection 1: Tech Integration Hub
In this projection, a Tech Integration Hub is established to enhance technology adoption capabilities and digital training networks throughout the CSC. The hub serves as a central platform for knowledge sharing to enhance the technology literacy and skills of citrus growers, workers, and other stakeholders. Collaboration and capacity-building initiatives are implemented to foster a culture of continuous learning, especially in rural areas, ensuring equitable access to digital resources and connectivity (Ghisi and da Silva 2001). The projection includes a comprehensive set of selective technologies, ranging from the integration of IoT sensors, to the adoption of precision agriculture techniques for optimized resource management (Ruzzante et al. 2021). For example, precision agriculture can be used to target fertilizer and pesticide applications more precisely, which can help to reduce water pollution (Yokamo 2020). Stakeholders choose technologies that are most suited to their specific needs and local agricultural and cultural contexts (Ruzzante et al. 2021). This leads to a more gradual and sustainable adoption of new technologies, which helps to minimize the risks and maximize the benefits. These developments overcome common technology adoption barriers like scalability, interoperability, lack of expertise, and regulations (Mohammed et al. 2023; Salemink et al. 2017).

Projection 2: Innovation Sandbox
This projection fosters technology adoption capabilities by creating an Innovation Sandbox as a digital innovation ecosystem. The sandbox provides a multi-actor supportive environment for experimentation, collaboration, and the development of a multidisciplinary innovation culture within the CSC. Stakeholders are encouraged to adopt an agile mindset, explore new technologies through scouting initiatives, and take calculated risks to drive technological advancements (Wolfert et al. 2023). This development accelerates the adoption of new technologies, ranging from the implementation of pilot projects for emerging technologies like artificial intelligence and robotics, to the promotion of entrepreneurship and start-ups within the citrus industry. This leads to increased efficiency and productivity, as well as improved quality and safety of citrus products. As a result, CSCs become more resilient to shocks and disruptions. This development path emphasizes the importance of innovation and risk-taking in technology adoption processes, fostering technology-

driven capabilities, and the creation of an innovation ecosystem within the CSC. (Masi et al. 2022; Nature Food 2022; Pierpaoli et al. 2013) Within this development path, the citrus industry is on the brink of a gradual technological revolution and innovation. New technologies have the potential to improve the sustainability and profitability of citrus production. As these technologies become more widely adopted, the citrus industry will be transformed.

Projection 3: Slow Adopters
In this projection, citrus growers and processors are slow to adopt technologies. This is due to several barriers in the agricultural sector, such as cultural beliefs, ethical concerns, regulatory issues, and lack of knowledge (Ugochukwu and Phillips 2018) The perception of farmers toward new technologies can be a major barrier, as some farmers may be reluctant to adopt new technologies that they are unfamiliar with (Mwangi and Kariuki 2015). High cost of new technologies can be a major obstacle, especially for small farmers. A lack of required skills can also be a problem, as new technologies often require specialized training and technical skills. Government policies or the complexity of agricultural technologies might furthermore hinder technology adoption (Kumari et al. 2018).

As a result, CSCs become less efficient and productive, and are more vulnerable to shocks and disruptions. This leads to increased prices for citrus products, and a decline in the competitiveness of the Mediterranean citrus industry. The developments in this projection assume that new technologies will continue to play an increasingly important role in the agricultural sector. However, their adoption includes several barriers and requires farmers to have a good understanding of their own operations and the potential benefits of new technologies. This would lead to a slower adoption rate for some farmers, as it requires them to take the time to learn about new technologies and to assess their potential benefits. This more pessimistic view states that, if citrus growers and processors are unable to adopt new technologies rapidly, they will be at a competitive disadvantage and may struggle to remain profitable.

4.3 Scenario Development

A scenario is composed of selecting one projection for each descriptor. Theoretically, the 26 projections of the 11 descriptors can be combined into a total of 10,368 potential scenarios ($2 \times 2 \times 3 \times 2 \times 2 \times 2 \times 3 \times 2 \times 3 \times 3 \times 2 = 10{,}368$). However, not all these scenarios are coherent due to the interdependencies between the descriptors. The credibility of these future scenarios is depending on their consistency, meaning the absence of contradictions among individual projections. Hence, the primary challenges in scenario development involve assessing the credibility of various combinations of projections and aggregating coherent combinations into a scenario. This methodology is termed *consistency analysis*.

Question:
How compatible (consistent) is projection i (row) with projection j (column)?

-2: strongly restricting influence (inconsistency)
-1: restricting influence
0: no influence (independent of each other)
+1: promoting influence
+2: strongly promoting influence (consistency)

Dimension	Key descriptors	Projections	Nr.	A1	A2	B1	B2	C1	C2	C3	D1	D2	...	J1	J2	J3	K1	K2
			Dimension	Social				Technological					...	Political				
			Key descriptors	Social health and well-being		Sustainability awareness along the citrus supply chain		Technology adoption capabilities			Digitalisation in the supply chain		...	Regulations			Geopolitical environment	
			Projections	Harmony Initiative	Resilience Alliance	Eco-Educators	Sustainable Empowerment Network	Tech Integration Hub	Innovation Sandbox	Slow adopters	Digital Citrus Network	Citrus E-Commerce Platform	...	Sustainable Agricultural Policies	Food Safety and Quality Standards	Trade Agreements and Market Access	Favourable	Unfavourable
Social	Social health and well-being	Harmony Initiative	A1			2	1	2	0	0	1	1	...	1	1	0	0	0
		Resilience Alliance	A2			2	2	1	1	-2	2	2	...	1	1	0	0	0
	Sustainability awareness along citrus supply chains	Eco-Educators	B1	2	2			2	1	-1	0	0	...	1	1	0	1	-1
		Sustainable Empowerment Network	B2	2	2			1	1	-1	0	0	...	1	1	1	0	0
Technological	Technology adoption capabilities	Tech Integration Hub	C1	2	2	1	1				2	1	...	0	1	0	0	0
		Innovation Sandbox	C2	0	1	1	1				2	1	...	1	0	0	0	0
		Slow adopters	C3	-1	-1	0	0				-2	-1	...	0	0	0	0	0
	Digitalisation in the supply chain	Digital Citrus Network	D1	1	2	0	1	1	1	-2			...	0	1	2	1	-1
		Citrus E-Commerce Platform	D2	0	0	1	1	2	0	-1			...	0	1	2	0	0
...	...	...	...	...	...	...	...	...	...	...	...	...	...	...	...	...	...	...
Political	Regulations	Sustainable Agricultural Policies	J1	1	2	2	2	2	2	1	2	2	...				1	1
		Food Safety and Quality Standards	J2	2	2	2	2	1	2	-1	2	0	...				1	1
		Trade Agreements and Market Access	J3	2	2	2	2	1	1	-2	2	2	...				2	2
	Geopolitical environment	Favourable	K1	2	2	1	1	2	2	-2	2	2	...	2	2	2		
		Unfavourable	K2	-2	-2	-2	-2	-2	-2	1	-2	-2	...	-1	-1	-2		

Fig. 12.5 Consistency analysis matrix (abbreviated)

4.3.1 Pairwise Consistency Analysis

The assessment of consistency for each pair of projections was conducted in a workshop by members of the scenario project team. The results are shown in Fig. 12.5. Although this is a subjective approach, the discussions surrounding the synchronization of alternative projections contribute additional value to the scenario project. Each projection was evaluated against every other projection in terms of their compatibility by means of a cross-impact matrix. Their interdependencies regulate which combinations can be regarded as consistent scenarios.

The identification of inconsistencies involves examining the influence relationships in the cross-impact matrix. For example, the cross-impact value of the matrix cell at the intersection B1 → C1 is +2, meaning that the projection *Eco-Educators* has a strongly promoting influence on the projection *Tech Integration Hub*. This comparison is performed for every projection pair.

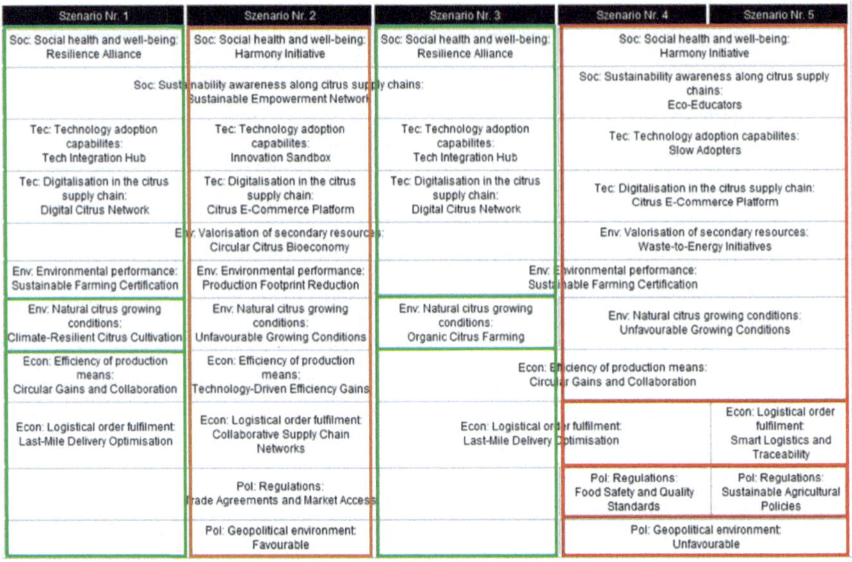

Fig. 12.6 Scenarios (obtained from ScenarioWizard software)

4.3.2 Scenarios

After transmitting the values from the consistency analysis into the ScenarioWizard software, five scenarios were obtained (Fig. 12.6).

When comparing the five scenarios, it becomes evident that certain scenarios closely resemble each other, differing only in one or two projections. Consequently, scenarios 1 and 3, as well as scenarios 4 and 5, were merged. This consolidation yields three distinct final scenarios—an optimistic (green), a pessimistic (red), and a neutral (orange) scenario.

To visualize the differences between these three scenarios, the key distinguishing features (arrows) were used in Fig. 12.7. The arrows represent selected key descriptors whose projections are at the arrowheads. The x- and y-axes represent selected STEEP dimensions that make the differences between the scenarios particularly clear.

Based on the projections assigned to scenarios by the software, the main prevailing differentiators were determined as environmental and technological factors, and to a lesser extent also social considerations.

The current CSCs can be associated with the negative (red) scenario, labeled as "Citrus Evolution Call: business as usual in food supply chains." In this scenario, unfavorable natural citrus growing conditions and customer engagement through simple e-commerce solutions are embedded in a social environment based on collaboration, cohesion, and well-being. To transform the current state towards more

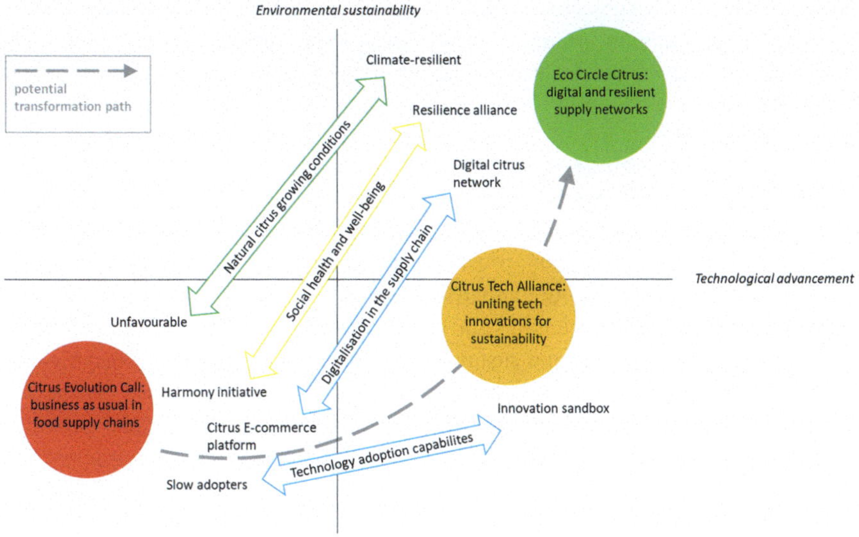

Fig. 12.7 Differences between scenarios and proposed transformation path

sustainability and technology adoption, a potential transformation path serves as a guideline.

To achieve the positive (green) scenario termed "Eco Circle Citrus: digital and resilient supply networks," climate-resilient growing conditions and a transparent digital citrus network should be promoted. These projections would also foster resilient CSCs, addressing social challenges such as poor working conditions or social inequalities, which are still present in the negative scenario. Especially technological advancement seems to play a crucial role in the transformation path. Therefore, a transformation towards the positive scenario should take an intermediate step from the technological point of view. This includes enhancing technology adoption capabilities by creating an *Innovation Sandbox* as a digital innovation ecosystem, which provides a multi-actor supportive environment for experimentation, collaboration, and the development of a multidisciplinary innovation culture within the CSCs. In this pathway, the existing slow adoption of technologies is enhanced, fostering an environment where stakeholders embrace an agile mindset. This involves actively exploring new technologies through scouting initiatives and being willing to take calculated risks to drive technological advancements.

5 Discussion

Although attempts are made to put the development of scenarios on a sound basis with the help of literature research, expert surveys, workshops, and other methods, scenarios are always subjective in nature. Looking into the future is also always associated with many uncertainties and complex interdependencies between many variables. Scenarios depend, for example, on national and local conditions. While several studies covered scenarios on the European level (Huan-Niemi et al. 2010; Meyer 2007), other studies took a global or mixed perspective (Ambler-Edwards et al. 2008; Benton 2019; Hamilton et al. 2020; Huan-Niemi et al. 2010; Hubert et al. 2010; Sundbo 2016). A FSC scenario deemed realistic or probable for a specific region does not have to be necessarily true for another region.

Nevertheless, gathering insights from other regions or similar conditions can help to frame the overall picture and assess how own strategic decision fit into the global context. Exploring scenarios for CSCs contributes to the broader discourse on sustainable food systems and sustainability literature by providing insights into potential development paths for this specific sector. This can guide businesses in making strategic decisions with regard to an uncertain future, and provides policymakers with a robust foundation to develop regulations that support the regional citrus sector.

This study extends previous research, making novel contributions to the literature in several ways. Firstly, it responds to the request for increased awareness for environmental and social responsibility in FSC transformation (Melkonyan et al. 2019) by incorporating these aspects as crucial descriptors into the scenario generation process. The evaluation of the scenario team, which resulted in the selection of these descriptors as key descriptors, underlines their importance for Mediterranean CSCs. Secondly, the findings enhance the connection between CSCs, FSCs, and sustainable supply chain management, addressing the importance of the supply chain management function to help businesses create more sustainable products and processes (UNGC 2015). Thirdly, the created scenarios facilitate the handling of both existing and emerging risks in CSCs, contributing to the enhancement of global food security (Hamilton et al. 2020) and the reduction of famine, especially in developing countries. Fourthly, this study adds to the available literature on scenarios in FSCs, particularly closing the gap of absent scenarios for CSCs in the Mediterranean.

Innovative Business Strategies

Based on expert assessments, the study conducted by Joshi et al. (2023) examined the use of innovation to enhance the ability of food supply networks to withstand and recover from disruptions, which may strengthen the resilience of supply chains, including sophisticated logistics, traceability, green technology, and sustainable practices. The authors call for more innovations in FSCs to succeed in a post-pandemic scenario. In particular, business strategy innovations and technological innovations stand out as the preferred options to enhance resilience in FSCs. Positive effects on the end-of-life of products can be credited to advancements of digital technologies in data management, real-time decision-making, and innovative practices. Additionally, implementing e-commerce operations can support agri-food companies to widen their scope. Our study supports these insights in two ways. On the one hand, CSC scenarios provide a sound basis for the development of innovative business strategies. On the other hand, the scenarios identify that technological innovations (technology adoption) are a significant driver for achieving the positive scenario *Eco Circle Citrus*. Nevertheless, our results also implicate that e-commerce is positioned in a *business-as-usual* environment. According to this, advanced digitalization in the FSC requires a more sophisticated approach, meaning the establishment of a comprehensive digital network of stakeholders, which facilitates information sharing, collaboration, and circular operations.

Waste Management

The notion of food waste plays an important role in view of sustainable resource use in FSCs. Melkonyan et al. (Melkonyan et al. 2019) tackle this problem from the sharing society angle, highlighting business models that focus on saving food at the consumption stage of the FSC. Adding to that view, the results of this study advocate for avoiding waste in the first place by aligning demand and supply, e.g., through deployment of advanced data analytics. In a similar vein, Hubert et al. (2010) propose deploying technologies for growing food that can mitigate poverty by connecting the urban demand side with local production.

In addition, waste not only occurs at the consumption side but throughout the FSC. Therefore, it is not enough to only share food and align supply and demand. Instead, waste occurring at earlier stages in the supply chain should be transformed into further valuable products demanded by consumers, emphasizing the urgent need for circular economy concepts across FSCs.

Technology Adoption and Collaboration

The role of technology is underlined in several FSC scenario studies (Hamilton et al. 2020; Hubert et al. 2010; Meyer 2007; Sundbo 2016). Technologies play an ambiguous role in FSCs. Whilst technologies support the transformation towards sustainable FSCs and have enhanced productivity, increased automation and technologies also create an uncertain environment (Meyer 2007) and pose new challenges due to their high degree of connectivity, dependency on certain technologies, or potential cyber-attacks (Hamilton et al. 2020). Addressing this concern, our results show that technology adoption requires proper education tailored to individual needs, highlighted in the *Tech Integration Hub* projection. Through training and capacity building, the benefits of technologies outweigh the challenges associated with it. The important role of technologies is represented in Fig. 12.7, where technology adoption capabilities play a central role in the proposed transformation path from the current business as usual, over the neutral *Citrus Tech Alliance* scenario, reaching the desired positive scenario *Eco Circle Citrus*. Hence, technologies should play an important role in the transformation of food systems.

Similarly, Sundbo (2016) develops a scenario at global, national, and regional levels focused on high-tech food production for the year 2025. In contrast, our results show that citrus production (at least in the Mediterranean CSCs) is still a long way off that path. Although the large enterprises in the Mediterranean citrus sector deploy semi-automated production lines, it is far away from high-tech, meaning that the focus is often on the own premises, while upstream and downstream processes are barely considered. Therefore, we argue that a successful digitalization and automation of food production needs to consider the entire supply chain, including harvesting, procurement, production, and distribution, while paying attention to social and environmental sustainability. This systemic view on the CSCs facilitates necessary information flows between stakeholders and requires close collaboration on eye-level. FSCs should be created through collaborative efforts that involve coordination with food producers, trading firms, corporations, and external parties, enhancing the agility and resilience of the value chains (Joshi et al. 2023). This view is also in line with Benton (2019), who suggests to move from agricultural productivity to the total productivity of the food system. The supply chain is only as resilient as the weakest link. Hence, knowledge sharing and technology adoption throughout the CSCs is an important factor, elevating less advanced players. Consequently, the entire supply chain will benefit and increase competitiveness, resulting in a win-win situation while improving food security.

The importance of collaboration was also emphasized by Meyer (2007), who suggested a close collaboration between governments and supply chain associates, and by Hamilton (2020), who called for the development of projections in cooperation between research, industry, and policy.

Policies

As underlined by Hubert et al. (2010), besides obvious technological challenges, also institutional barriers constitute major obstacles to entering local, regional, and global markets. Technology has the potential to overcome institutional barriers by enabling the production of high-quality, safe, and nutritious food.

Not surprisingly, also other studies highlighted the role of policies on the food sector (Huan-Niemi et al. 2010; Meyer 2007). Huan-Niemi et al. (2010) compared a business-as-usual scenario with, among others, the removal of subsidies and tariffs in the agricultural sector in the EU and globally, with the outlook to liberalize global trade. Yet, more than a decade later agricultural tariffs still affect imports, decreasing international trade (Cipollina and Salvatici 2020). The recent demonstrations by farmers in several European countries against subsidy cuts also illustrate the continuing importance of these issues.

Likewise, global agricultural trade liberalization was also seen as a crucial factor in both our neutral and positive scenarios, which involves reducing trade barriers, streamlining customs procedures, and establishing agreements to simplify cross-border transactions. Even though reduction of subsidies and trade liberalization could lead to less domestic production and stronger competition from abroad (Huan-Niemi et al. 2010), we argue that a free market is efficient in itself, revealing inefficient players while transferring resources from high-cost to low-cost producers, who serve the needs of the consumers in sustainable and innovative ways for the price they are willing to pay. Supporting this claim, policies developed to promote affordable and sustainable food often lead to the opposite effect, having a negative impact on human health and the environment (Benton 2019).

6 Conclusion and Outlook

This chapter presents scenarios for the Mediterranean citrus industry, incorporating considerations of socioeconomic factors, environmental challenges, geopolitical reflections, and digitalization possibilities across CSCs. The identification of influencing factors and potential development paths was conducted through literature analysis and insights from various workshops with a multidisciplinary team of experts. The scenarios aim to align business priorities with the regional and global citrus industry environment, fostering innovation adoption, and unlocking transformation potential in the citrus sector. Three alternative scenarios were generated—positive, negative, and neutral. They provide a comprehensive exploration of challenges and opportunities in CSCs, emphasizing the necessity for innovation, collaboration, technology adoption, and strategic planning within the principles of circular economy and sustainability. Adopting novel strategies and practices is pivotal for addressing global challenges such as food security and environmental threats, paving the way for a resilient and efficient citrus production and distribution system. This study advocates for a collaborative innovation adoption approach within the citrus sector by focusing on technological advancement as well as biological and technological circularity within the frames of a sustainable business strategy. The identification of key differentiators between the three scenarios on social, technological, environmental, economic, and political dimensions offers added value for business and policy. Proposed transformational pathway offers a guideline for developing business strategies and industrial policies.

To enhance the robustness of the scenarios, critical challenges can be introduced, and further aspects may be considered. This could involve conducting sensitivity analyses and incorporating a wider range of perspectives to ensure a more comprehensive and objective understanding of the potential future developments in the citrus industry.

The scenarios created elaborate on possible future paths and therefore form a sound basis for the development of strategies for an uncertain future. Hence, the next step (scenario transfer) will involve analyzing the scenarios to evaluate their potential for success, discovering possible threats to today's businesses in the citrus sector, and assessing their impacts and probability of occurrence. Subsequent analysis will involve the development of sustainable scenario-based business model concepts and strategies for the Mediterranean citrus sector, considering the adoption of circular economy principles and technological advancement. Also, defining indicators to monitor the transformation progress towards scenario achievement should be targeted.

References

Abu Seman NA, Govindan K, Mardani A, Zakuan N, Mat Saman MZ, Hooker RE, Ozkul S (2019) The mediating effect of green innovation on the relationship between green supply chain management and environmental performance. J Clean Prod 229:115–127. https://doi.org/10.1016/j.jclepro.2019.03.211

Accorsi R, Baruffaldi G, Manzini R, Tufano A (2018) On the design of cooperative vendors' networks in retail food supply chains: a logistics-driven approach. Int J Logist Res Appl 21(1):35–52

Adams D, Donovan J, Topple C (2021) Achieving sustainability in food manufacturing operations and their supply chains: key insights from a systematic literature review. Sustain Prod Consumption 28:1491–1499. https://doi.org/10.1016/j.spc.2021.08.019

Albrigo LG (2019) Citrus, Crop production science in horticulture, No. 25, 2nd edn. CAB International, Oxford

Ali J, Kumar S (2011) Information and communication technologies (ICTs) and farmers' decision-making across the agricultural supply chain. Int J Inf Manag 31:149–159. https://doi.org/10.1016/j.ijinfomgt.2010.07.008

Ambler-Edwards S, Bailey KS, Kiff A, Lang T, Lee R, Marsden TK, Simons DW, Tibbs H (2008) Thinking about the future of food: the Chatham House food supply scenarios. In: Chatham House Food Supply Project.

Amer M, Daim TU, Jetter A (2013) A review of scenario planning. Futures 46:23–40. https://doi.org/10.1016/j.futures.2012.10.003

Amorim P, Costa AM, Almada-Lobo B (2014) Influence of consumer purchasing behaviour on the production planning of perishable food. OR Spectr 36:669–692. https://doi.org/10.1007/s00291-013-0324-9

Ashraf SRB, Daus D (2022) Integration of Blockchain technology in the citrus food supply chain: a systematic literature review. Accessed 9 June 2023

Banaitė D, Tamošiūnienė R (2016) Sustainable development: the circular economy indicators' selection model. JSSI 6:315–323. https://doi.org/10.9770/jssi.2016.6.2(10)

Banaitė D (2016) Towards circular economy: analysis of indicators in the context of sustainable development. Soc Transform Contemporary Soc: 142–150

Beh L-S, Ghobadian A, He Q, Gallear D, O'Regan N (2016) Second-life retailing: a reverse supply chain perspective. SCM 21:259–272. https://doi.org/10.1108/SCM-07-2015-0296

Benton TG (2019) Using scenario analyses to address the future of food. EFSA J 17. https://doi.org/10.2903/j.efsa.2019.e170703

Bouchery Y, Corbett CJ, Fransoo JC, Tan T (eds) (2017) Sustainable supply chains: a research-based textbook on operations and strategy. Springer Series in Supply Chain Management, vol 4. Springer International Publishing, Cham. https://doi.org/10.1007/978-3-319-29791-0

Bragdon SH, Smith C (2015) Small-scale farmer innovation. Geneva. http://www.quno.org/sites/default/files/resources/SSF%20Innovation%20WEB.pdf

Capello F, Toja M, Trapani N (2016) A real-time monitoring service based on industrial internet of things to manage agrifood logistics. Conference: ILS2016, Information Systems, Logistics and Supply Chain

Caputo P, Ducoli C, Clementi M (2014) Strategies and tools for eco-efficient local food supply scenarios. Sustain For 6:631–651

Carter CR, Liane Easton P (2011) Sustainable supply chain management: evolution and future directions. Int J Phys Dist Log Manag 41:46–62. https://doi.org/10.1108/09600031111101420

Castillo-Vergara M, Quispe-Fuentes I, Poblete J (2021) Technological innovation in the food industry: a bibliometric analysis. Eng Econ:197–209. https://doi.org/10.5755/j01.ee.32.3.26000

Chinnici G, Carla Z, Hamam M, Pecorino B, Mario D (2019) A model of circular economy of citrus industry. 19th International Multidisciplinary Scientific GeoConference SGEM 2019 in Vienna. https://doi.org/10.5593/sgem2019V/4.2

Cicculló F, Cagliano R, Bartezzaghi G, Perego A (2021) Implementing the circular economy paradigm in the agri-food supply chain: the role of food waste prevention technologies. Resour Conserv Recycl 164:105114. https://doi.org/10.1016/j.resconrec.2020.105114
Cipollina M, Salvatici L (2020) On the effects of EU trade policy: agricultural tariffs still matter. Eur Rev Agric Econ 47:1367
Corona D (2015) Allowing or banning genetically modified food and feed? Morocco in search of a synthesis between conflicting regulatory models. In: Paciello M (ed) Building sustainable agriculture for food security in the Euro-Mediterranean: challenges and policy options
Cos J, Doblas-Reyes F, Jury M, Marcos R, Bretonnière P-A, Samsó M (2022) The Mediterranean climate change hotspot in the CMIP5 and CMIP6 projections. Earth Syst Dyn 13:321–340. https://doi.org/10.5194/esd-13-321-2022
Cristiani D (2020) Making Sense of the Geopolitical Pluralisation of the Mediterranean. https://revistaidees.cat/en/making-sense-of-the-geopolitical-pluralisation-of-the-mediterranean/
Daccache A, Ciurana JS, Rodriguez Diaz JA, Knox JW (2014) Water and energy footprint of irrigated agriculture in the Mediterranean region. Environ Res Lett 9:124014. https://doi.org/10.1088/1748-9326/9/12/124014
Daus D (2023) Report on proposed citrus supply chain scenarios, online. https://impulse-sustainable-supply-chains.net/wp-content/uploads/2023/12/ImPUlSe_D3.2_Report_on_proposed_Citrus_Supply_Chain_Scenarios_final.pdf
Daus D, Barros AC, Kalaitzi D, Muerza V, Marchiori I (2018) Alternative development paths for supply chains in 2030. In: Proceedings of the Hamburg International Conference of Logistics (HICL) – 25. The Road to a Digitalized Supply Chain Management., pp 121–141
Duarte A, Fernandes J, Bernardes J, Miguel G (2016) Citrus as a component of the Mediterranean diet. J Spatial Org Dyn IV:289–303
Duchenne-Moutien RA, Neetoo H (2021) Climate change and emerging food safety issues: a review. J Food Prot 84:1884–1897. https://doi.org/10.4315/JFP-21-141
Eskandarpour M, Dejax P, Miemczyk J, Péton O (2015) Sustainable supply chain network design: an optimization-oriented review. Omega 54:11–32. https://doi.org/10.1016/j.omega.2015.01.006
Esposito B, Sessa MR, Sica D, Malandrino O (2020) Towards circular economy in the agri-food sector. A systematic literature review. Sustainability 12:7401. https://doi.org/10.3390/su12187401
Fader M, Shi S, von Bloh W, Bondeau A, Cramer W (2016) Mediterranean irrigation under climate change: more efficient irrigation needed to compensate for increases in irrigation water requirements. Hydrol Earth Syst Sci 20:953–973. https://doi.org/10.5194/hess-20-953-2016
FAO (2021) Making agrifood systems more resilient to shocks and stresses. The state of food and agriculture, vol 2021. Food and Agriculture Organization of the United Nations, Rome
FAO, IFAD, UNICEF, WFP, WHO (2020) The state of food security and nutrition in the world 2020: transforming food systems for affordable healthy diets. FAO, Rome
Fernandes N (2020) Economic effects of coronavirus outbreak (COVID-19) on the world economy. SSRN J. https://doi.org/10.2139/ssrn.3557504
Ferragina E, Canitano G (2015) Geopolitical implications of water and food security in southern and eastern Mediterranean countries. In: Paciello M (ed) Building sustainable agriculture for food security in the Euro-Mediterranean: challenges and policy options
Fonte M, Cucco I (2017) Cooperatives and alternative food networks in Italy. The long road towards a social economy in agriculture. J Rural Stud 53:291–302. https://doi.org/10.1016/j.jrurstud.2017.01.019
Galih Nugroho F, Tedjakusuma D, Lo AR, Pamungkas DH, Kinardi SA (2023) Review of the application of digital transformation in food industry, vol 13. J Curr Sci Technol, p 774
Gaspars-Wieloch H (2019) Role of scenario planning and probabilities in economic decision problems – literature review and new conclusions. In: Proceedings of 6th international scientific conference contemporary issues in business, management and economics engineering '2019
Gausemeier J, Plass C (2014) Zukunftsorientierte Unternehmensgestaltung: strategien, Geschäftsprozesse und IT-Systeme für die Produktion von morgen, 2nd edn. Hanser, München

Gawron J-C, Theuvsen L (2009) Certification schemes in the European agri-food sector. Outlook Agric 38:9–14. https://doi.org/10.5367/000000009787762716

Ghisi F, da Silva AL (2001) The information technology on food supply chain management. In: PICMET '01. Portland international conference on management of engineering and technology. Proceedings Vol. 1: book of summaries, IEEE Cat. No.01CH37199

Goodarzian F, Ghasemi P, Santibanez Gonzalez E, Tirkolaee EB (2023) A sustainable-circular citrus closed-loop supply chain configuration: Pareto-based algorithms. J Environ Manag 328:116892. https://doi.org/10.1016/j.jenvman.2022.116892

Göbel C, Langen N, Blumenthal A, Teitscheid P, Ritter G (2015) Cutting Food Waste through Cooperation along the Food Supply Chain. Sustainability 7:1429–1445. https://doi.org/10.3390/su7021429

Guerrieri V, Lorenzoni G, Straub L, Werning I (2022) Macroeconomic implications of COVID-19: Can negative supply shocks cause demand shortages? Am Econ Rev 112(5):1437–1474

Hamam M, Chinnici G, Di Vita G, Pappalardo G, Pecorino B, Maesano G, D'Amico M (2021) Circular economy models in agro-food systems: a review. Sustain For 13:3453. https://doi.org/10.3390/su13063453

Hamilton H, Henry R, Rounsevell M, Moran D, Cossar F, Allen K, Boden L, Alexander P (2020) Exploring global food system shocks, scenarios and outcomes. Futures 123:102601. https://doi.org/10.1016/j.futures.2020.102601

Huan-Niemi E, Niemi J, Niemi J (2010) Global food production under alternative scenarios. Int Food Agribu Manag Rev 13:117–135

Hubert B, Rosegrant MW, van Boekel MA, Ortiz R (2010) The future of food: scenarios for 2050. Crop Sci 50. https://doi.org/10.2135/cropsci2009.09.0530

Ibn-Mohammed T, Mustapha KB, Godsell J, Adamu Z, Babatunde KA, Akintade DD, Acquaye A, Fujii H, Ndiaye MM, Yamoah FA, Koh SCL (2021) A critical analysis of the impacts of COVID-19 on the global economy and ecosystems and opportunities for circular economy strategies. Resour Conserv Recycl 164:105169. https://doi.org/10.1016/j.resconrec.2020.105169

Jeevahan J, Anderson A, Sriram V, Durairaj R, Britto Joseph G, Mageshwaran G (2018) Waste into energy conversion technologies and conversion of food wastes into the potential products: a review. Int J Ambient Energy. https://doi.org/10.1080/01430750.2018.1537939

Jonas Lazaro-Mojica, Rebeca Fernandez (2021) Review paper on the future of the food sector through education, capacity building, knowledge translation and open innovation. Curr Opinion Food Sci 38:162

Joshi S, Sharma M, Ekren BY, Kazancoglu Y, Luthra S, Prasad M (2023) Assessing supply chain innovations for building resilient food supply chains: an emerging economy perspective. Sustain For 15:4924. https://doi.org/10.3390/su15064924

King-Okumu C, Aboukheira AA (2015) The challenges to achieve sustainable exports of fruit from small-Holdings in the Southern Mediterranean. In: Paciello M (ed) Building sustainable agriculture for food security in the Euro-Mediterranean: challenges and policy options

Krishnan R, Agarwal R, Bajada C, Arshinder K (2020) Redesigning a food supply chain for environmental sustainability – an analysis of resource use and recovery. J Clean Prod 242:118374. https://doi.org/10.1016/j.jclepro.2019.118374

Krishnan R, Yen P, Agarwal R, Arshinder K, Bajada C (2021) Collaborative innovation and sustainability in the food supply chain- evidence from farmer producer organisations. Resour Conserv Recycl 168:105253. https://doi.org/10.1016/j.resconrec.2020.105253

Kumar A, Mangla SK, Kumar P (2022) An integrated literature review on sustainable food supply chains: exploring research themes and future directions. Sci Total Environ 821:153411. https://doi.org/10.1016/j.scitotenv.2022.153411

Kumari S, Jeble S, Patil Y (2018) Barriers to technology adoption in agriculture-based industry and its integration into technology acceptance model. Int J Agric Resour Gov Ecol. https://doi.org/10.1504/IJARGE.2018.10019353

Leyerer M, Sonneberg M-O, Heumann M, Breitner MH (2020) Shortening the last mile in urban areas: optimizing a smart logistics concept for e-grocery operations. Smart Cities 3:585–603. https://doi.org/10.3390/smartcities3030031

Lorenz U, Veenhoff S (2013) Integrated scenarios of sustainable food production and consumption in Germany. Sustain Sci Pract Policy 9:92–104. https://doi.org/10.1080/15487733.2013.11908118

Lovec M (2015) The European Union's common agricultural policy reforms and the sustainability of agro-food Systems in the Euro-Mediterranean Region: how to get trade and development Back on the agenda? In: Paciello M (ed) Building sustainable agriculture for food security in the Euro-Mediterranean: challenges and policy options

Masi M, de Rosa M, Vecchio Y, Bartoli L, Adinolfi F (2022) The long way to innovation adoption: insights from precision agriculture. Agric Econ 10:1–17. https://doi.org/10.1186/s40100-022-00236-5

Melkonyan A, Gruchmann T, Huerta A, Krumme K (2019) Scenario planning for sustainable food supply chains. In: Saini M (ed) Innovative logistics services and sustainable lifestyles. Springer International Publishing, Cham, pp 211–233

Meyer R (2007) Comparison of scenarios on futures of European food chains. Trends Food Sci Technol 18:540–545. https://doi.org/10.1016/j.tifs.2007.02.008

Mohammed A, Potdar V, Quaddus M, Hui W (2023) Blockchain adoption in food supply chains: a systematic literature review on enablers, benefits, and barriers. IEEE Access 11:14236–14255. https://doi.org/10.1109/ACCESS.2023.3236666

Mwangi M, Kariuki (2015) Factors determining adoption of new agricultural technology by smallholder farmers in developing countries. J Econ Sustain Dev 6:208–216

Nature Food (2022) Innovation in fruit and vegetable supply chains. Nat Food 3:387–388. https://doi.org/10.1038/s43016-022-00548-1

PACE (2021) Circular Economy Action Agenda: Food. https://pacecircular.org/sites/default/files/2021-02/circular-economy-action-agenda-food_0.pdf

Pierpaoli E, Carli G, Pignatti E, Canavari M (2013) Drivers of precision agriculture technologies adoption: a literature review. Procedia Technol 8:61–69. https://doi.org/10.1016/j.protcy.2013.11.010

Proedrou F (2023) A geopolitical account of the Eastern Mediterranean conundrum: sovereignty, balance of power and energy security considerations. Camb Rev Int Aff 36:679–696. https://doi.org/10.1080/09557571.2021.1897088

Ramos C, Carbonell G, Garcia BJM, Tarazona JV (2000) Ecological risk assessment of pesticides in the Mediterranean region. The need for crop-specific scenarios. Sci Total Environ 247:269–278. https://doi.org/10.1016/s0048-9697(99)00496-9

Reddy P (2014) Climate resilient agriculture for ensuring food security. Springer India

Reganold J, Wachter JM (2016) Organic agriculture in the twenty-first century, vol 2. Nat Plants, p 15221

Rojas O (2021) Next generation agricultural stress index system (ASIS) for agricultural drought monitoring. Remote Sens 13:959. https://doi.org/10.3390/rs13050959

Ruzzante S, Labarta R, Bilton A (2021) Adoption of agricultural technology in the developing world: a meta-analysis of the empirical literature. World Dev 146:105599. https://doi.org/10.1016/j.worlddev.2021.105599

Salemink K, Strijker D, Bosworth G (2017) Rural development in the digital age: a systematic literature review on unequal ICT availability, adoption, and use in rural areas. J Rural Stud 54:360–371. https://doi.org/10.1016/j.jrurstud.2015.09.001

Sarkis J, Cohen MJ, Dewick P, Schröder P (2020) A brave new world: lessons from the COVID-19 pandemic for transitioning to sustainable supply and production. Resour Conserv Recycl 159:104894. https://doi.org/10.1016/j.resconrec.2020.104894

Schimmenti E, Borsellino V, Galati A (2013) Growth of citrus production among the Euro-Mediterranean countries: political implications and empirical findings. Span J Agric Res 11:561. https://doi.org/10.5424/sjar/2013113-3422

Sharma K, Mahato N, Cho MH, Lee YR (2017) Converting citrus wastes into value-added products: economic and environmently friendly approaches. Nutrition 34:29–46. https://doi.org/10.1016/j.nut.2016.09.006

Sharma YK, Mangla SK, Patil PP, Liu S (2019) When challenges impede the process: for circular economy driven sustainability practices in food supply chain. Manag Decis 57:995–1017. https://doi.org/10.1108/MD-09-2018-1056

Skawińska E, Zalewski RI (2018) Circular economy as a management model in the paradigm of sustainable development. Management 22:217–233. https://doi.org/10.2478/manment-2018-0034

Stone J, Rahimifard S (2018) Resilience in agri-food supply chains: a critical analysis of the literature and synthesis of a novel framework. SCM 23:207–238. https://doi.org/10.1108/SCM-06-2017-0201

Stone J, Rahimifard S, Woolley E (2015) An overview of resilience factors in food supply chains. Loughborough University. Conference contribution. https://hdl.handle.net/2134/19826

Sundbo J (2016) Food scenarios 2025: drivers of change between global and regional. Futures 83:75–87. https://doi.org/10.1016/j.futures.2016.03.003

Tuel A, Eltahir EAB (2020) Why is the Mediterranean a climate change hot spot? J Clim 33:5829–5843. https://doi.org/10.1175/JCLI-D-19-0910.1

Ugochukwu AI, Phillips PWB (2018) Technology adoption by agricultural producers: a review of the literature. In: From agriscience to agribusiness, pp 361–377. https://doi.org/10.1007/978-3-319-67958-7_17

UNEP (2017) Food and food waste. https://www.unep.org/explore-topics/resource-efficiency/what-we-do/sustainable-lifestyles/food-and-food-waste. Accessed 2 Jan 2023

UNGC (2015) Supply chain sustainability: a practical guide for continuous improvement, 2nd edn

United Nations (2022) World population prospects 2022: summary of results. United Nations, New York

Vlajic JV, Mijailovic R, Bogdanova M (2018) Creating loops with value recovery: empirical study of fresh food supply chains. Prod Plan Control 29:522–538. https://doi.org/10.1080/09537287.2018.1449264

Volkery A, Ribeiro T, Henrichs T, Hoogeveen Y (2008) Your vision or my model? Lessons from participatory land use scenario development on a European scale. Syst Pract Action Res 21:459–477. https://doi.org/10.1007/s11213-008-9104-x

Wan Ahmad WNK, Rezaei J, Tavasszy LA, de Brito MP (2016) Commitment to and preparedness for sustainable supply chain management in the oil and gas industry. J Environ Manag 180:202–213. https://doi.org/10.1016/j.jenvman.2016.04.056

wbcsd (2021) Circular Transition Indicators V2.0: Metrics for business, by business, Geneva, Beijing, Delhi, London, New York, Singapore. https://www.wbcsd.org/contentwbc/download/11256/166026/1

Weimer-Jehle W (2023) Cross-impact balances (CIB) for scenario analysis: fundamentals and implementation, Contributions to management science, 1st edn. Springer Nature Switzerland, Cham

Wolfert S, Verdouw C, van Wassenaer L, Dolfsma W, Klerkx L (2023) Digital innovation ecosystems in agri-food: design principles and organizational framework. Agric Syst 204:103558. https://doi.org/10.1016/j.agsy.2022.103558

World Manufacturing Foundation (2018) The 2018 World Manufacturing Forum Report: Recommendations for the Future of Manufacturing. www.worldmanufacturingforum.org

Xu Z, Sun D-W, Zeng X-A, Liu D, Pu H (2015) Research developments in methods to reduce the carbon footprint of the food system: a review. Crit Rev Food Sci Nutr 55:1270–1286. https://doi.org/10.1080/10408398.2013.821593

Xuxian Yan, Xianjun Li (2015) The impact of technological innovation on industrial efficiency and food manufacturing industry. Adv J Food Sci Technol. https://doi.org/10.19026/ajfst.7.1326

Yi Jiang, Polin Lai, Chia-Hsun Chang, Kum Fai Yuen, Sihan Li, Xinchen Wang (2021) Sustainable management for fresh food e-Commerce logistics services. Sustainability https://doi.org/10.3390/su13063456

Yokamo S (2020) Adoption of improved agricultural technologies in developing countries. Lit Rev 4:183

Zailani S, Govindan K, Iranmanesh M, Shaharudin MR, Sia Chong Y (2015) Green innovation adoption in automotive supply chain: the Malaysian case. J Clean Prod 108:1115–1122. https://doi.org/10.1016/j.jclepro.2015.06.039

Zhu Z, Chu F, Dolgui A, Chu C, Zhou W, Piramuthu S (2018) Recent advances and opportunities in sustainable food supply chain: a model-oriented review. Int J Prod Res 56:5700–5722. https://doi.org/10.1080/00207543.2018.1425014

Chapter 13
Choice-Based Conjoint Analysis for Orange Juice Types: A Study in the South Mediterranean Region

Heba Yusuf, Hadi Fors, and Nermine Harraz

1 Introduction

Orange juice consumption is a key factor in the growth of the citrus-byproduct industry in the global economy. The orange juice trade is leading globally as opposed to the sale of fresh oranges (Allegra et al. 2019). However, there have been shifts in the global consumption of orange juice; it is losing its market share over the other kinds of juices and drinks with functional ingredients such as, e.g., probiotic drinks, energy drinks, sports drinks, and meal replacers. Moreover, the presence of concerns on production sustainability and organic farming contributes to this shift (Fava Neves et al. 2020). Myriad solutions, such as the use of citrus processing waste (CPW), can contribute to solving the decreased consumption of traditional orange juice, aiming at ensuring food security, enhancing economic conditions of the citrus supply chain, and securing its circularity. CPW generated during juice extraction can be used to produce probiotics, which are high in nutritional value (Cristóbal et al. 2018). The probiotics prepared from the CPW have potential health benefits for the immune system and the gut, as they reduce side effects after using antibiotics, reduce symptoms associated with irritable bowel syndrome, help in reducing lactose intolerance, and have antimicrobial and anticancer characteristics (Zaki and Naeem 2021). The probiotics are nowadays added to food and to orange juice, among other drinks, to produce "functional probiotic drinks." Using CPW to produce probiotics achieves three goals simultaneously: first, coping with fluctuation of orange juice demand due to consumers' need to buy differentiated functional drinks with health

H. Yusuf (✉) · H. Fors · N. Harraz
Production Engineering Department, Industrial Engineering Division, Faculty of Engineering, Alexandria University, Alexandria, Egypt
e-mail: hebatullah.yusuf@alexu.edu.eg; hadifors@alexu.edu.eg; nharraz@alexu.edu.eg

A. Melkonyan-Gottschalk, D. Daus (eds.), *Transformation Towards Circular Food Systems*, Greening of Industry Networks Studies 13,
https://doi.org/10.1007/978-3-031-63793-3_13

benefits; second, promoting circular economy; and third, sustainability of the citrus supply chain.

Yet, there are market challenges due to lack of information on consumer acceptance and willingness to pay for products developed from CPW (Teigiserova et al. 2019; Eliopoulos et al. 2022; Zuin and Ramin 2018). Sustainability is crucial for food and beverage supply chains (Adams et al. 2023), as consumers are becoming more conscious of the consequences of their purchasing decisions in terms of environmental and social impact (Toussaint et al. 2021). Pesticide residues are another challenge facing the orange juice industry, which urges the need for organic farming (Drábová et al. 2022). However, organic products are generally more expensive than conventionally grown crops, due to the higher production costs associated with organic farming. Accordingly, examining consumer preferences and willingness to pay for orange juice is needed to assess the economic viability of producing new orange drinks utilizing probiotics produced from CPW.

Given the various ways to collect data, there are two main methods for measuring consumer preferences: stated preference and revealed preference methods.

- *Stated preference data* is a survey-based technique for collecting data. The data collected are on the key purchase drivers, mainly the price, to estimate price-response functions. The main approaches used to collect and analyze data are conjoint analysis and discrete choice analysis (Paczkowski 2018). The discrete choice analysis is also known as "choice-based conjoint analysis" (Raghavarao et al. 2010).
- *Revealed preference data* is a consumer transactions-based technique for collecting data stored in large databases. This method has some drawbacks, such as lack of variability, which is very important for regression models; obsoleteness of data, as collected data may be old; and the data don't show how consumers make purchase decisions while comparing a product with another substitute (Paczkowski 2018); therefore, the stated preference data is used in this study to avoid the abovementioned drawbacks.

Conjoint analysis has been used to understand how consumers develop preferences for products or services (Sekaran and Bougie 2016). It is a statistical technique used in many fields, including marketing, product development, and operations research. Many studies have been conducted to measure consumer preferences for food and drinks. Gül and Akpinar (2012) have conducted market research using choice-based conjoint analysis, collecting 1152 responses, targeting fruit juice drinkers in Turkey. The aim of the study was to figure out the consumers' preferences for fruit juice to help firms in juice product development in Turkey. The investigated attributes were product types (orange, sour cherry, etc.), price level, international quality and food safety certificates, being organic, production methods, and containing sugar. They found that the most important factor in consumers' purchasing preferences was the product type, followed by the price, and having international quality and food safety certificates. They concluded that the fruit juice that maximizes consumer satisfaction is "the orange juice which is 100% fruit juice, without sugar additive, organic, has international quality and food safety certificates, and with affordable price." Kumar and Babu (2021) have conducted market research by collecting 500 responses, targeting orange juice drinkers in Andhra Pradesh, India. They used a conjoint analysis using a 20-point scale to rate each product profile. The product profile is a combination of attributes to be measured with certain levels. The aim of the study was to examine consumers' preferences for a combination of orange juice attributes and provide a market simulation analysis for a new product with desired attribute levels based on the results of conjoint analysis. The investigated attributes were flavor, preservatives, sweetener, brand, taste, pulp concentration, packaging, production method of orange, and price. They found that the "production method of orange" was the most important attribute, followed by the "brand" being available in the market and the pulp concentration. The product's "price" was the least important attribute by the consumers, showing that orange juice demand is inelastic in the investigated market. Annunziata and Vecchio (2013) developed a conjoint analysis to understand consumer preferences regarding functional foods' attributes, namely base product, health claim, price, and brand. From the reviewed studies, it can be concluded that consumers are concerned with the healthier options.

There is scarcity in investigations measuring consumer preference and willingness to pay for orange juice using conjoint study, especially in the Mediterranean basin. There are three drivers for conducting this research: (1) No similar studies conducted in this targeted region; (2) Investigating consumers' inclination toward functional drinks containing probiotics, as these functional drinks are less familiar to consumers in the region; and (3) Exploring consumers' preference for sustainability measures in supply chain indicated by the sustainability labels.

The aim of this study is to conduct market research in three South Mediterranean countries: Algeria, Egypt, and Tunisia to explore the consumer preferences and willingness to pay considering the type of orange juice, price, packaging, and sustainability labels. For this, a choice-based conjoint study is conducted. Three types of orange juice are included: orange probiotic drinks, nectar orange drinks, and pasteurized not from the concentrate (P-NFC) orange juice. The price levels are set based on the market prices for orange juice in the three countries. Two packaging alternatives are considered: plastic bottles and Tetra Pak carton bottles. The sustainability labels investigated are related to organic, environmental, and fairtrade certifications.

The results of this research help in analyzing consumer preferences and willingness to pay regarding orange juice attributes in the three countries. Accordingly, juice producers' marketing decisions could be guided by the outputs of this study, and the viability of improving the citrus and its byproducts supply chain in this region could be assessed.

The steps to conduct the choice-based conjoint study are illustrated in the following section. Section 3 presents the results of the three surveys and the insights on attributes' importance, utility values, and marginal willingness to pay. Finally, the study is concluded in Sect. 4.

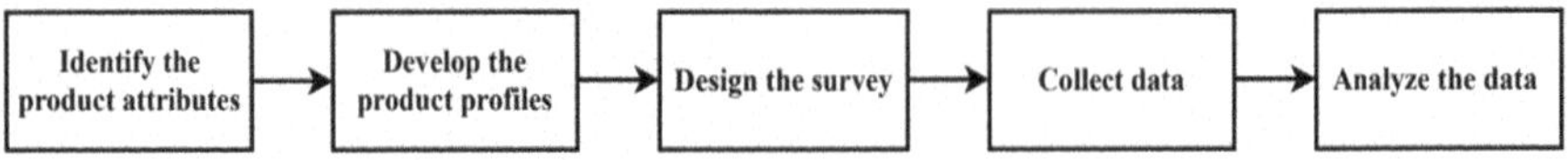

Fig. 13.1 Followed steps to conduct choice-based conjoint study

2 Methodology

Figure 13.1 illustrates the followed steps to conduct the choice-based conjoint study to explore the consumer preferences for a combination of orange juice attributes in three South Mediterranean countries (Algeria, Egypt, and Tunisia). The details of each step are illustrated in the following subsections.

2.1 Identification of the Product Attributes

The choice of the orange juice attributes included in the market research is based on the aim of the study. These attributes are the type of orange juice, price, packaging, and sustainability labels. Three sustainability labels are indicated, corresponding to organic, fairtrade, environmental certifications (Li and Kallas 2021). The health claim related to each type of juice is presented to the consumer (Mohebalian et al. 2012). The levels of each attribute are set up based on observations from the target markets. Orange juice attributes and their corresponding levels are illustrated in Table 13.1.

2.2 Development of the Product Profiles (Products with Specific Features)

Once the attributes and levels are determined, a set of product profiles is generated to be used in the discrete-choice conjoint study. A group of features presents the profile, which is a combination of attributes with various levels. It should be noted that some attributes' levels are prohibited from appearing in the product profile. For example, the lowest price level is not allowed to appear with the organic certified products, the highest price level is intended not to appear with non-organic certified products, and the organic certification is prohibited with nectar orange drink. In this study, each respondent is asked to give her/his preference (choice) for a couple of choice sets. Each choice set consists of two product profiles and an option of "no choice." An example of a choice set presented to a respondent is illustrated in Fig. 13.2. Due to the difficulty of introducing considerable number of product profiles to a respondent, incomplete block designs are applied (Rao 2013). There have been 160 blocks generated; each block contains 10 choice sets. The number of choice sets

Table 13.1 Orange juice attributes and corresponding levels

Attribute	Level (Egypt)	Level (Tunisia)	Level (Algeria)	Description
Type of juice	P-NFC			Healthy, good taste, higher price, 100% juice
	Orange nectar			Not healthy, good taste, lower price, not 100% juice
	Probiotic drink			Healthier, good taste, higher price, probiotics prepared from citrus processing waste
Prices per liter	34 EGP per liter	5.7 TND per liter	250 DZD per liter	Market price
	39 EGP per liter	6.7 TND per liter	290 DZD per liter	Market price
	44 EGP per liter	7.7 TND per liter	330 DZD per liter	Market price
	49 EGP per liter	8.7 TND per liter	370 DZD per liter	Market price
Organic certification	Organic certified			No pesticides, no fertilizers, and no synthetic additives
	Not organic certified			The product may hold pesticide residues
Fairtrade certified	Fair trade certified			Ensure worker's safety, fair payment, and product data is recorded from farm to store
	Not fair trade certified			No guarantee that the product is ethically sourced

(continued)

Table 13.1 (continued)

Attribute	Level (Egypt)	Level (Tunisia)	Level (Algeria)	Description
Environmental certification	Environmentally certified			Product does not harm environment
	Not environmentally certified			No guarantee that the product does not harm environment
Packaging	Plastic bottle			1 liter plastic bottle (transparent with black cap) 
	Tetra Pak			1 liter white carton bottle 

Fig. 13.2 Example of choice set

in a block that is presented to a respondent is calculated using the following formula (Kumar and Babu 2021):

$$\text{Number of choice sets} = [(1 + \text{Total number of attribute levels}) \\ - \text{Number of attributes}]$$

2.3 *Designing the Survey*

To conduct the choice-based conjoint study, a survey is designed with the aid of the advanced survey platform "Conjoint.ly." The survey includes demographic questions to ensure that the sample is representative of the target population and the choice sets, which are presented to the respondent to decide the preferences.

2.4 Collecting Data

The study data is collected online or through face-to-face interviews with the respondents in the three countries. The interviews are moderated in the three most spoken languages in these countries, Arabic, English, and French. The sample size used in this study is more than 200 respondents from each country, as recommended by Conjoint.ly. The sample size in this type of study must be large enough to provide reliable estimates for part-worths, which are the level utilities for conjoint attributes, and to detect significant differences between attribute levels. It is recommended to use specialized software to figure out the proper sample size for a conjoint analysis study (Rao 2013).

2.5 Analyzing Data

In conjoint study, utility is a key concept for measuring value, which assesses unique subjective preferences for each respondent (Kumar and Babu 2021). Attribute importance is a measure of the relative importance of each attribute in the study. The higher the value of attribute importance, the higher the impact of the attribute on consumer decision. The level part-worth or utility value is a measure of the attribute level's influence on the consumer's decision to select an alternative. Marginal Willingness to Pay (MWTP) is the amount of money that customers are willing to pay for a particular feature of a product over the price they are already paying for the baseline product (Conjointly 2022). The term "marginal" refers to the fact that MWTP is always compared to a baseline product. After collecting the data using Conjoint.ly, the results of the three surveys are analyzed to obtain: (1) attribute importance, (2) utility value for each level, and (3) marginal willingness for each feature.

3 Results and Discussions

Table 13.2 summarizes the socioeconomic profile of respondents from Algeria, Egypt, and Tunisia. The gender distribution for the three samples nearly reflects the actual gender distribution for each of the three countries. Most of the participants are aged between 20 and 35 years in Algerian and Tunisian cases, while most of the participants are aged between 30 and 50 years in Egypt. For the three countries, the largest percent of participants have "University degree." For Egypt and Tunisia, the average annual household income ranges for most of the participants between EGP 60,000 and 500,000 for Egypt and TND 2400 and 30,000 for Tunisia.

To have a comprehensive insight on the consumer preference, three indicators are assessed: attributes' importance, utility values, and marginal willingness to

Table 13.2 Socioeconomic profile of respondents from the three countries

	Algeria	Egypt	Tunisia
Number of responses	209	201	200
Age distribution			
15–20	9.1%	6.5%	11.5%
20–35	73.7%	22.3%	46%
35–50	9.6%	42.3%	17%
Over 50	7.6%	28.9%	25.5%
Gender distribution			
Male	31.1%	50.2%	50.5%
Female	68.9%	49.8%	49.5%
Educational level			
Preparatory degree	4.8%	0.5%	5%
Secondary degree	8.1%	9.5%	15.5%
Technical secondary degree	1.5%	1%	6%
Post secondary degree	3.8%	2%	8.5%
University degree	49.3%	50.7%	29%
High diploma degree	8.6%	10.9%	2.5%
Master's degree	17.7%	15.9%	15%
Doctor of philosophy PhD	3.3%	8.5%	12%
X None of the above	2.9%	1%	6.5%
Average annual household income			
Less than 240,000 DZD Less than 60,000 EGP Less than 2400TND	25.4%	20%	1%
240,000–480,000 DZD 60,000–500,000 EGP 2400–30,000 TND	22.5%	77.5%	63%
More than 480,000 DZD More than 500,000 EGP More than 30,000 TND	23.4%	2.5%	17%
Refuse to answer	28.7%	0%	19%

pay. Figures 13.3, 13.4, and 13.5 show attributes' importance for orange juice in Algeria, Egypt, and Tunisia, respectively. It is clear from the figures that the most important orange juice attribute for Algerian and Tunisian consumers is price per liter, followed by type of juice, then organic certification. While Egyptian consumers give higher importance to the organic certification over the type of juice. This shows that consumers are conscious about the advantages of organic products. It is also noted that the environmental certification comes in the fourth place for the Algerian consumers, while the Egyptian and Tunisian consumers give higher importance to the fair-trade certificate over the environmental certification. All three countries respondents are less concerned with the type of packaging.

Figures 13.6, 13.7, and 13.8 illustrate the utility values of attributes' levels in Algeria, Egypt, and Tunisia, respectively. Goodness of fit is indicated by

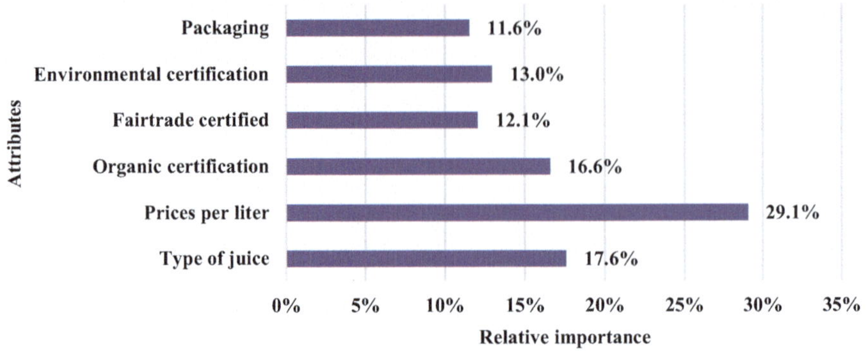

Fig. 13.3 Attribute importance for Algerian orange juice drinkers

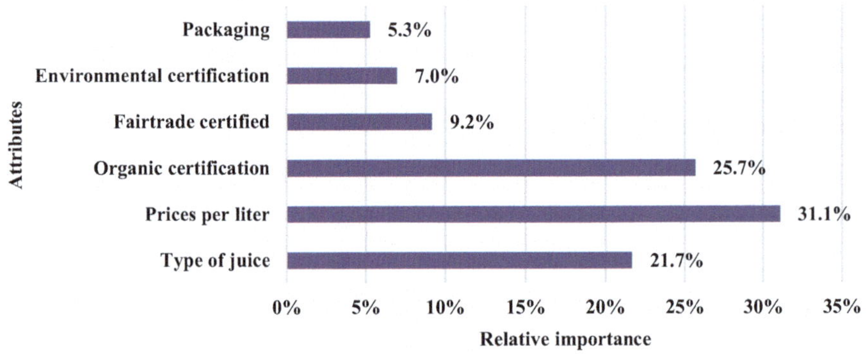

Fig. 13.4 Attribute importance for Egyptian orange juice drinkers

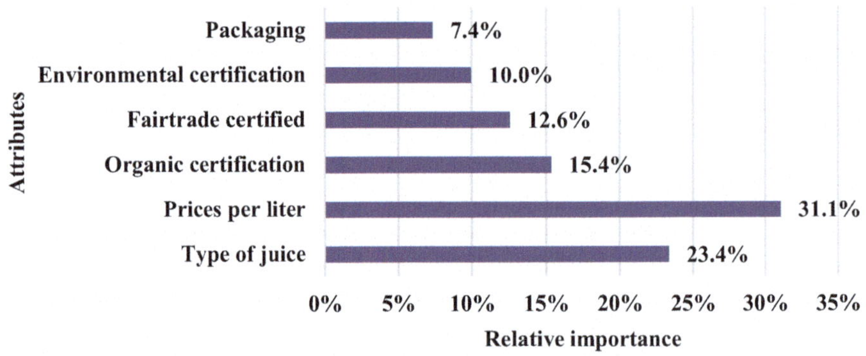

Fig. 13.5 Attribute importance for Tunisian orange juice drinkers

McFadden's pseudo-R square. The most favorable attribute level is indicated by the greatest positive utility value, while the smallest negative utility value indicates the least favorable attribute level. For Algerian consumers, the preferred attribute level

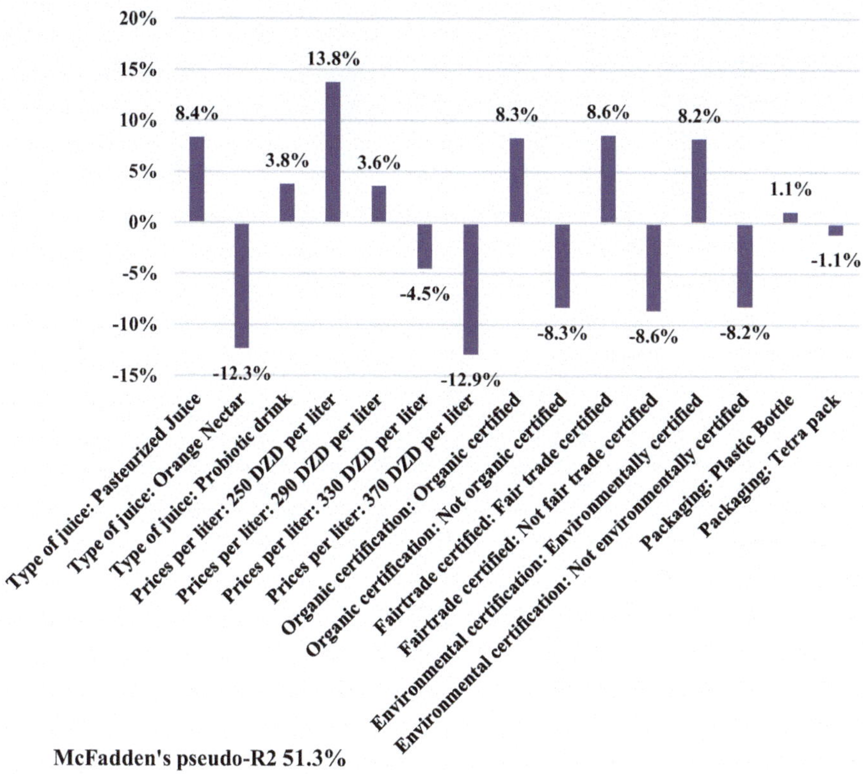

Fig. 13.6 Utility values for Algerian orange juice drinkers

is “price level DZD 250 per liter,” and the least preferred one is “price level DZD 370 per liter.” While for Egyptian consumers, the most preferred attribute level is the “organic certification: organically certified,” and the least preferred one is “type of juice: orange nectar.” For Tunisian consumers, the most preferred attribute level is “price level 5.7 TND per liter,” and the least preferred one is “price level 8.7 TND per liter.” Regarding the type of juice, the Algerian consumers preferred the pasteurized juice, and the Egyptian consumers almost have the same preference for probiotic drinks and pasteurized juice (P-NFC). The Tunisian consumers gave priority to the probiotic drink, knowing that the orange juice with added probiotic is a hypothetically suggested functional drink for the aim of the study and is not yet on the market. This gives an interesting insight that the consumers are health and innovation conscious. Orange nectar was the least preferred type in all three countries.

Figures 13.9, 13.10, and 13.11 present MWTP for each feature of orange juice in Algeria, Egypt, and Tunisia, respectively. As illustrated in Fig. 13.9, the Algerian consumers need the price to be reduced by DZD 27.01 for orange nectar drink and by DZD 2.66 for probiotic orange drink, in case of not considering the pasteurized

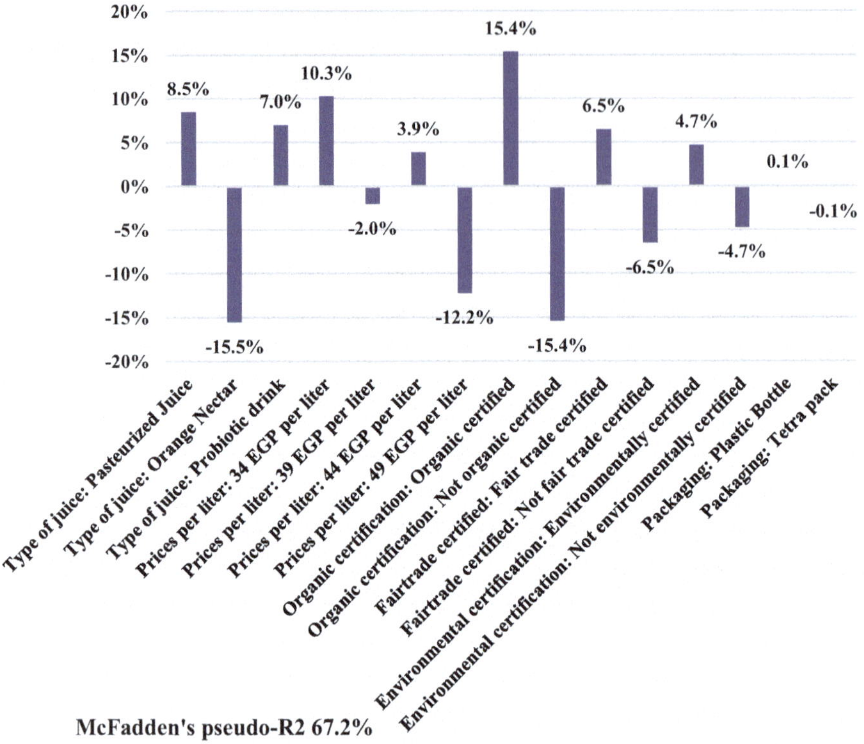

Fig. 13.7 Utility values for Egyptian orange juice drinkers

juice. Also, it is less preferred to have the orange juice not organic certified, not fairtrade certified, not environmental certified, and with Tetra Pak packaging.

For the Egyptian consumers, as demonstrated in Fig. 13.10, the price should be reduced by EGP 0.54 for the orange nectar drink in case of not choosing the pasteurized juice, and they are willing to pay EGP 0.44 more for the probiotic orange drink than the pasteurized juice. Furthermore, it is less favored to have the orange juice not organic certified, not fairtrade certified, not environmental certified, and with Tetra Pak packaging.

Figure 13.11 shows the MTWP for Tunisian consumers. The price should be reduced by TND 0.53 for orange nectar drink in case of not deciding the pasteurized juice, and they are willing to pay TND 0.82 more for probiotic orange drink than the pasteurized juice. Moreover, the orange juice with the following attributes is less preferred: not organic certified, not fairtrade certified, not environmental certified, and with Tetra Pak packaging.

The above findings show that the price of orange juice is the most important attribute for orange juice consumers in Algeria, Egypt, and Tunisia, revealing that the demand for orange juice is elastic in these markets. These results are different than the studies of (Gül and Akpinar 2012) and (Kumar and Babu 2021). In Gül and

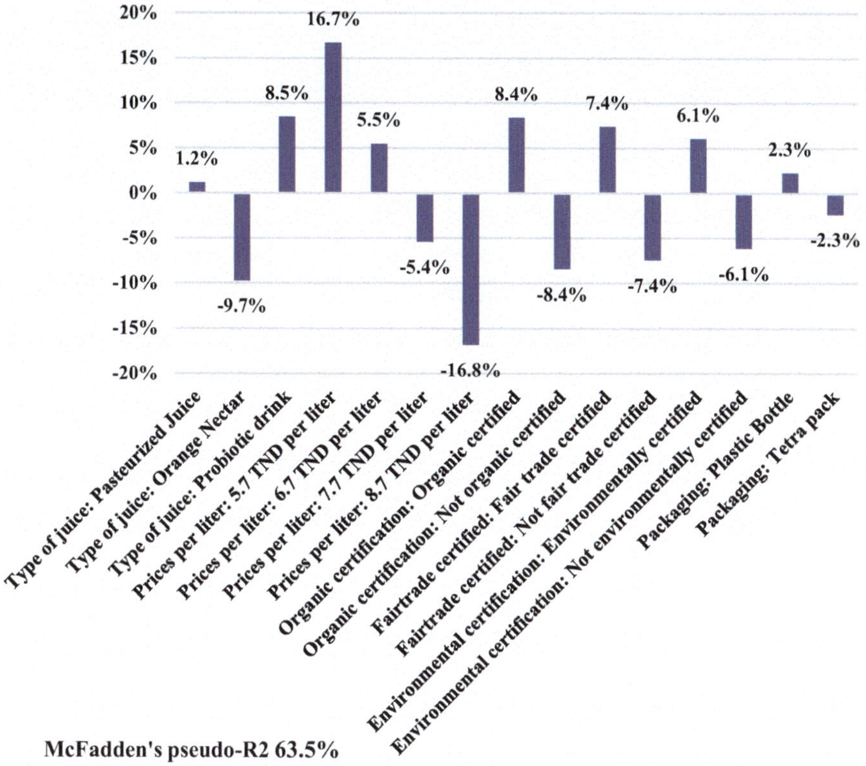

Fig. 13.8 Utility values for Tunisian orange juice drinkers

Akpinar (2012), revealing that even though the price was an important attribute, it was not the most important one. On the contrary, Kumar and Babu (2021) showed that the price had lower priority over other attributes due to COVID pandemic. The consumers in Algeria, Egypt, and Tunisia showed interest for organic certified orange juice, which is somewhat like the results provided by (Gül and Akpinar 2012) where consumers showed preference for organic orange juice.

4 Conclusion

A choice-based conjoint study was conducted in Algeria, Egypt, and Tunisia to explore the consumer preferences and willingness to pay for orange juice with different features. Price, packaging, and sustainability labels with different levels and designs were examined for three orange juice types, namely, P-NFC, nectar, and orange juice with added probiotics. The results of the study show that: (1) price comes first as the most important attribute for consumers in the three countries; (2)

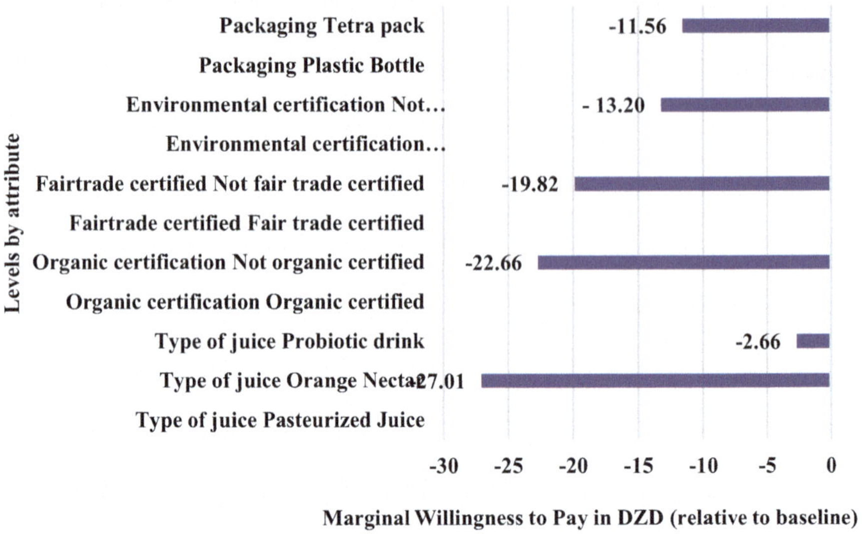

Fig. 13.9 MWTP in DZD for each feature of orange juice in Algeria

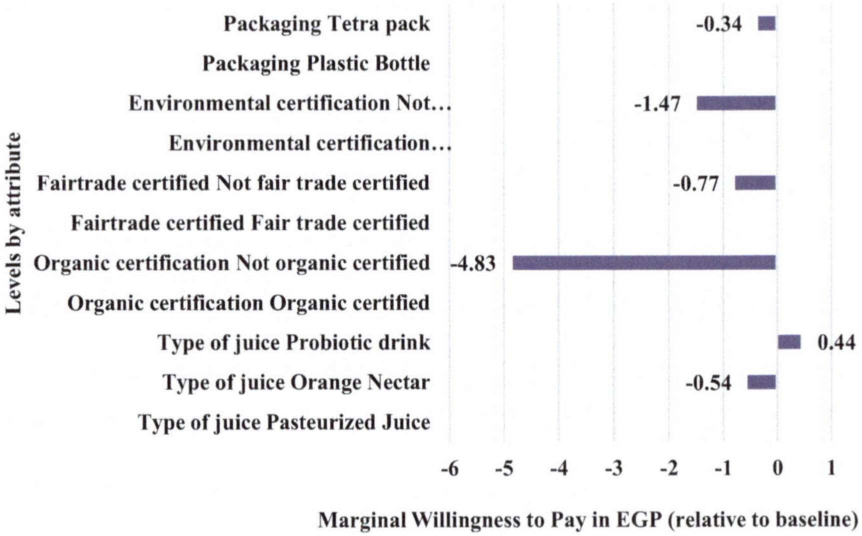

Fig. 13.10 MWTP in EGP for each feature of orange juice in Egypt

Algerian consumers favored the pasteurized juice; (3) consumers in Egypt almost have the same preference for probiotic added drink and P-NFC juice. The probiotic-added juice had the highest preference over the other juice types for the Tunisian consumers; (4) Egyptian and Tunisian consumers are willing to pay more for the probiotic drink; (5) the consumers in Algeria, Egypt, and Tunisia's interest for

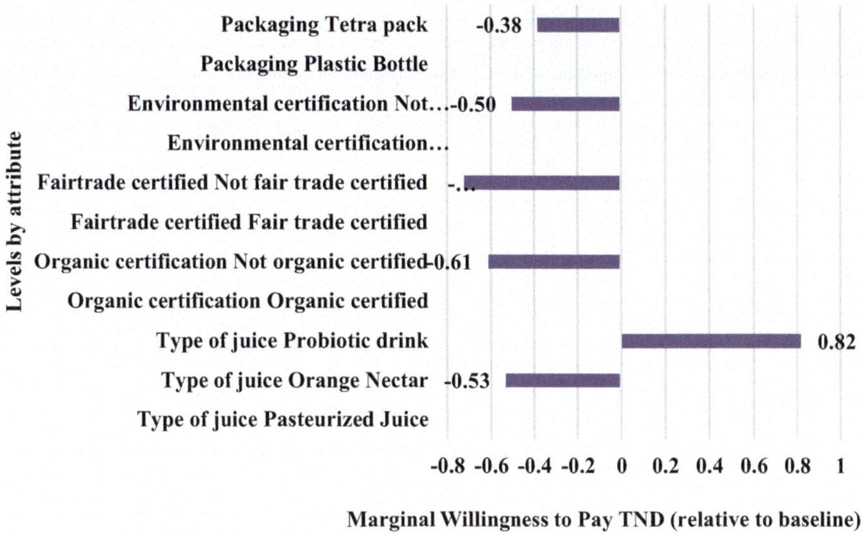

Fig. 13.11 MWTP in TND for each feature of orange juice in Tunisia

sustainability labels comes after the price and type of orange juice. The insights from this study can guide citrus supply chain stakeholders in their marketing decisions and assess the viability of improving the citrus as well as the citrus byproduct supply chains.

The authors of this work perceive extending this work to exploring the consumers' preferences for probiotic-added orange drink sensory attributes such as, e.g., taste, acid, astringent, carbonation, strength of flavor, sweetness, and thickness viscosity to have more accurate insights on price sensitivity and consumer willingness to pay for innovative functional products.

Acknowledgments This work was supported by the Academy of Scientific Research and Technology (ASRT)—Egypt, PRIMA call 2020 section 2 project no.26 entitled "Innovation in the By-Product Supply Chain of Citrus in Mediterranean area" (ImPUISe). The authors would like to thank the colleagues and ImPUlse partners from University of Sfax—Tunisia and University of Abou Bekr Belkaid—Algeria for their support in collecting the market research data.

References

Adams D, Donovan J, Topple C (2023) Sustainability in large food and beverage companies and their supply chains: an investigation into key drivers and barriers affecting sustainability strategies. Bus Strateg Environ 32(4):1451–1463. https://doi.org/10.1002/bse.3198

Allegra V, Zarbà C, La Via G, Zarbà AS (2019) Why the new orange juice consumption model favors global trade and growth in orange production. Br Food J 121(9):1954–1968. https://doi.org/10.1108/BFJ-05-2019-0316

Annunziata A, Vecchio R (2013) Consumer perception of functional foods: a conjoint analysis with probiotics. Food Qual Prefer 28(1):348–355. https://doi.org/10.1016/j.foodqual.2012.10.009

Conjointly (2022) Interpreting marginal willingness to pay. Conjointly

Cristóbal J, Caldeira C, Corrado S, Sala S (2018) Techno-economic and profitability analysis of food waste biorefineries at European level. Bioresour Technol 259:244–252. https://doi.org/10.1016/j.biortech.2018.03.016

Drábová L, Mráz P, Krátký F, Uttl L, Vacková P, Schusterova D, Zadražilová B, Kadlec V, Kocourek V, Hajšlová J (2022) Assessment of pesticide residues in citrus fruit on the Czech market. Food Addit Contam Part A Chem Anal Control Expo Risk Assess 39(2):311–319. https://doi.org/10.1080/19440049.2021.2001579

Eliopoulos C, Markou G, Langousi I, Arapoglou D (2022) Reintegration of food industry by-products: potential applications. In: Foods, vol 11, Issue 22. MDPI. https://doi.org/10.3390/foods11223743

Fava Neves M, Vinícius, Trombin G, Marques VN, Franco Martinez L (2020) Global orange juice market: a 16-year summary and opportunities for creating value. Trop Plant Pathol 45:166–174. https://doi.org/10.1007/s40858-020-00378-1

Gül M, Akpinar MG (2012) The analysis of households' purchasing preferences for fruit juice in Turkey. J Food Agric Environ 10(3–4):119–123

Kumar KNR, Babu SC (2021) An analysis of consumers' preferences for orange juice in India during covid-19. Stud Agric Econ 123(3):131–140. https://doi.org/10.7896/j.2151

Li S, Kallas Z (2021) Meta-analysis of consumers' willingness to pay for sustainable food products. In: Appetite, vol 163. Academic. https://doi.org/10.1016/j.appet.2021.105239

Mohebalian PM, Cernusca MM, Aguilar FX (2012) Discovering Niche Markets for Elderberry Juice in the United States. HortTechnology 22(4):556–566. https://doi.org/10.21273/HORTTECH.22.4.556

Paczkowski WR (2018) Pricing analytics: models and advanced quantitative techniques for product pricing, 1st edn. Routledge. https://doi.org/10.4324/9781315178349

Raghavarao D, Wiley JB, Chitturi P (2010) Choice-based conjoint analysis: models and designs, 1st edn. Chapman and Hall/CRC. https://doi.org/10.1201/9781420099973

Rao VR (2013) Applied conjoint analysis. https://doi.org/10.1007/978-3-540-87753-0

Sekaran U, Bougie R (2016) Research methods for business: a skill building approach. John Wiley & Sons

Teigiserova DA, Hamelin L, Thomsen M (2019) Review of high-value food waste and food residues biorefineries with focus on unavoidable wastes from processing. Resour Conserv Recycl 149:413–426. https://doi.org/10.1016/j.resconrec.2019.05.003

Toussaint M, Cabanelas P, González-Alvarado TE (2021) What about the consumer choice? The influence of social sustainability on consumer's purchasing behavior in the Food Value Chain. Eur Res Manag Bus Econ 27(1). https://doi.org/10.1016/j.iedeen.2020.100134

Zaki N, Naeem M (2021) Antioxidant, antimicrobial and anticancer activities of citrus peels to improve the shelf life of yoghurt drink. Egypt J Food Sci. https://doi.org/10.21608/ejfs.2021.58310.1092

Zuin VG, Ramin LZ (2018) Green and sustainable separation of natural products from agro-industrial waste: challenges, potentialities, and perspectives on emerging approaches. In: Topics in current chemistry, Issue 1, vol 376. Springer International Publishing. https://doi.org/10.1007/s41061-017-0182-z

Chapter 14
The Way Forward: Scientific, Practical, Political, and Managerial Impact of the Book

Ani Melkonyan-Gottschalk, Denis Daus, Emad Alzubi, and Ahmed Kassem

This book explores the application of integrated sustainability assessment methods in both operational and strategic decision-making processes within food supply chains (FSCs), with a specific focus on citrus supply chains (CSCs). It emphasizes evidence-based practices and encourages decision-makers to implement *data-driven approaches to enhance the sustainability, resilience, and circularity of FSCs.* To facilitate the implementation of data-driven solutions, the *advanced data analytics* methodologies have been envisaged to forecast crop yields accurately, *optimize supply chain processes*, and minimize *crop wastage, while opening new markets for citrus by-products.* These data-driven solutions empower stakeholders with valuable insights to make informed decisions and achieve operational excellence, meanwhile offering healthy and sustainable nutrition to the consumers. The development of *consumer-driven, efficient, and innovative supply chains* is crucial for the growth of citrus and *fruit consumption in Europe.* Furthermore, the information on the different *roles along the supply chain* can help in understanding the dynamics of the industry, inform decision-makers about leveraging benefits, and identify areas for collaborative innovation mechanisms to enable circular transformation.

A. Melkonyan-Gottschalk (✉)
Clausthal University of Technology, Clausthal-Zellerfeld, Germany
e-mail: ani.melkonyan-gottschalk@uni-due.de

D. Daus
Centre for Logistics and Traffic (ZLV), University of Duisburg-Essen (UDE), Duisburg, Germany
e-mail: denis.daus@uni-due.de

E. Alzubi · A. Kassem
Transportssysteme und -logistik, Fakultät für Ingenieurwissenschaften, Universität Duisburg–Essen, Duisburg, Germany
e-mail: emad.alzubi@stud.uni-due.de; ahmed.kassem@uni-due.de

A. Melkonyan-Gottschalk, D. Daus (eds.), *Transformation Towards Circular Food Systems*, Greening of Industry Networks Studies 13,
https://doi.org/10.1007/978-3-031-63793-3_14

Thus, the book delves into the implementation of circular economy strategies, illustrating how these approaches can significantly enhance sustainability benefits within FSCs, such as redirecting food loss and waste (FLW) and by-products generated through the different stages in FSCs to create value out of it.

FLW is an important key performance indicator in any FSC. Therefore, stakeholders should take action to reduce the discarded amount of FLW to enhance the economic performance (Alzubi and Akkerman 2022), since the competitiveness within FSC will be improved (Eriksson et al. 2014). FLW quantification is important, when thinking toward adopting circular economy options. Attempts toward quantifying FLW in several FSCs exist in the literature in particular stages. For example, Dusch Silva (2018) analyzed the orange supply chain and found around 20% of oranges produced in Brazil are discarded at the farm stage for several causes, such as workers availability, and citrus canker. Similarly, Alhiary (2012) claimed that around 23% of fruits and vegetables is categorized as loss and waste on the farm in Jordan, while around 18% lattice is discarded annually in this stage in Libya.

In the processing stage, the percentages of FLW vary between 5% and 82.5%. For instance, Fornereto Soldan et al. (2021) studied the sweet pepper processing stage and found the percentage of FLW can vary from 5% to 30%, where the highest amount of FLW in the processing stage was found to be 82.5% for olive in the 2-phase extraction process (Nasir and Markham 2008). Other works quantified the FLW in the processing stage such as in date processing with 10–20% (Attia et al. 2021), 70% for broccoli (Gudiño et al. 2022), 30–50% in carrot processing (Badjona et al. 2019), 12% in cucumber (Sai Prasanna and Mitra 2020), 25% for garlic (dos Santos et al. 2022), onion with 37% (Sharma et al. 2016), watermelon with 60% (Zia et al. 2021), tomato with 20–30% (Andreou et al. 2020), wheat with 25–40% (Kanoija et al. 2018), orange 50% (Siddiqui et al. 2022), and mandarin 50–60% (Yun and Liu 2022). Other research quantified FLW in transportation, such as Alzubi et al. (2022) estimated the FLW for citrus in transportation at 16%, peach at 25% (Sasaki et al. 2021), and the same ratio for strawberry (Sasaki et al. 2022), while Alzubi et al. (2024) estimated the FLW in the whole CSC as fresh fruit from 46% to 56%.

Considering these amounts of FLW generated within the different FSC stages, it is essential to promote FSCs and integrate the circular economy. FLW valorization options are also discussed for the different FSCs. For instance, citrus loss and waste can be recycled to produce fertilizer (Alzubi et al. 2024), essential oil (Shaw et al. 2023), or biodiesel (Rizzioli et al. 2023). Furthermore, wheat straw can also be used in paper production, as discovered by Agrawal et al. 2023, while wheat bran can be directed to produce sugar (Glaser et al. 2023) or lignin (Yafetto et al. 2023). Tomato residual pulp can be used in milk production process to enhance its fatty acid profile (Correddu et al. 2023). However, circularity options are different and need further investigation to explore their applicability according to the available resources, technologies, and technical parameters (Alzubi and Noche 2022).

Beyond theoretical discussions, the book provides practical insights into the characteristics and functioning of CSCs and real-world case studies that highlight successful sustainable practices implemented by various stakeholders. These examples serve as inspirations for readers to initiate positive changes within their own organizations, countries, and cities through implementation of interdisciplinary and systemic transformation pathways within food systems.

Considering the farming stage of the supply chain (Part II), ongoing drought poses a significant threat to citrus production, emphasizing the importance of efficient irrigation for both quantitative and qualitative crop outcomes. Hence, to decrease dependency on irrigation and to enhance the resilience and productivity of citrus orchards amidst climate change challenges, adopting sustainable practices is required, including improved water management, soil conditions, and irrigation systems. To combat crop-threatening insects and increase crop yield, farmers use pesticides while often neglecting the risks associated with them. Improper application, including excessive use and inadequate protective measures, poses risks to human health and the environment. Low education levels among citrus growers contribute to misuse, and the absence of legislation further increases the issue.

The book also shows the high practical potential of technology adoption and advanced data analytics (Part III), leading to innovative solutions that improve the sustainability of processes, for example by eliminating the need for heating, centrifugation, or ultrafiltration cleaning solutions. Sustainability in agricultural practices is further fostered by shifting operations to a circular model. For example, integrating a composting unit in an orange juice supply chain can reduce fertilizer usage and mitigate environmental impacts (global warming potential, ozone depletion, eutrophication of freshwater, eutrophication of marine) throughout the supply chain while improving farmers' profitability. Also, yield forecasting techniques (or their absence) have a high impact on the sustainability of agricultural practices, especially in sectors with perishable goods. In combination with artificial intelligence and precise crop monitoring, yield forecasting can be an effective tool for enhanced productivity, guiding decisions to balance food quality, environmental integrity, and cost-effectiveness. This emphasizes the importance of integrating traditional knowledge with innovative approaches. In supply chain planning, accurate yield forecasts have an influence on product prices, pesticide/fertilizer usage, workforce management, product allocations, logistical planning, and capacity planning. However, to excel machine learning algorithms, extensive data and vast information on external factors affecting crop yield are required.

Beyond technological drivers, there are social, environmental, economic, and political parameters supporting a circular transformation. These parameters underline the development of comprehensive strategies for transformational pathways, and close collaborations. The formulation of such strategies requires quality assurance measures, greener technologies, exploitation of the market potential, strategic alliances, and additional valorization paths for competitive supply chains that serve customers' needs. Thus, taking the consumer perspective in Part V, price, packaging, and sustainability labels play an important role in the choice of products. Although price seems to be the most important attribute, consumers are also willing to pay higher prices for juice with certain characteristics (e.g., probiotic), underlining a region-specific marketing approach.

Based on these insights, the impacts of the book are as following:

1. *Foster better access for the farmers, food processors, and distributors to the market and to the developed technological and sustainable business innovations*:

   The authors aim to increase awareness amongst key stakeholders about socioeconomic impacts of innovation adoption along citrus (by-product) supply chain through this book. Firstly, to gain insights into sustainability within CSCs, it's essential to engage with multi-stakeholders. This involves sharing knowledge and experiences among various participants in the supply chain, including producers, distributors, and retailers. Secondly, the book provided state-of-the-art insights to inform and reform the entire supply chain, promoting product and process innovations, expansion to new markets, and agri-food product diversification in respect to economic, environmental, and social sustainability. The book supports the actors engaged in citrus (by-products) supply chains in identifying joint cooperation opportunities on overcoming barriers of commercialization and industrial use of wasted products by utilizing innovative digital technologies and new business models.
2. *Enhance sustainability of the small-scale farming systems and preservation of natural resources*:

   Implementation of smart solutions, in addition to numerous business and organizational improvements, will dramatically increase efficiency in water and energy consumption. Effective use of data and decision support systems is the key to efficiency in FSCs. Automation of field data capture and analysis provides a reliable way to implement sustainability assessment. One of the major challenges to enhance sustainability is the fact that assessing the sustainability of current practices is overly difficult, costly, and time consuming. These assessment methods, even when implemented, are not useful for operational decision making. The book included newly developed integrated sustainability assessment methods that can be tightly coupled with operational and strategic decision making along the FSCs.
3. *Enhance innovation capacities and appropriate technology transfer of existing or new infrastructures*:

   For enhancement of innovations, information sharing by participatory strategies is the key to involve stakeholders in the co-creation of solutions in a gender inclusive way, supported by the latest digital transformation technologies. It should address stakeholder needs and limitations considering investment scarcities, gender, age, and inclusiveness constraints, facilitate knowledge, awareness of transparency, massive changes in culture and behavior, new services and jobs, particularly for smallholders. Specific research on the collection, sorting, and pre-treatment processes for ensuring a high-quality secondary raw material for energy generation or new by-products was carried out. Completely integrated processes for each type of the by-products, as well as the improvements achieved in food processing, logistics, and recovery stages have been demonstrated with operational trials.

In essence, the book serves as a valuable resource for stakeholders along the value chain, from raw material suppliers to end consumers, providing them with insights and strategies to navigate the complexities of sustainable and circular supply chains in the citrus industry. It underscores the importance of transformational change, not just adaptation, in addressing environmental challenges and building a more resilient and holistic agricultural system. The chapters provide a deeper understanding of how these concepts can be practically applied in the real world, making it a valuable for the stakeholders involved in the citrus supply chain and those they could be interested in.

The current book also emphasizes the enhancement of sustainability in small-scale farming systems and the preservation of natural resources. It advocates for the implementation of smart solutions, which, along with numerous business and organizational improvements, will dramatically increase efficiency in water and energy consumption. The effective use of data and decision support systems is identified as the key to efficiency in food supply chains. Automation of field data capture and analysis provides a reliable way to implement sustainability assessment.

Major challenges that individuals may encounter when enhancing sustainability include data availability, costs, the availability of experts, and the time-consuming nature of current practices. These challenges often hinder the operational decision-making process, even when applying existing methods. Therefore, the book advocates for the integration of sustainability assessment methods that can be closely aligned with both operational and strategic decision-making processes throughout FSCs. It also focuses on enhancing innovation capacities and the appropriate technology transfer of existing or new infrastructures. Additionally, it highlights the importance of information sharing through participatory strategies to involve stakeholders in the co-creation of solutions, highlighting a clear managerial implication at each stage of the business decision-making process.

Various case studies carried out with different focus areas of each participating country exhibit distinct approaches to sustainability across CSCs. Egypt and the United States prioritize agricultural practices, Italy emphasizes life cycle assessment in the farming stage, Brazil and China balance efforts across farming and processing stages, while India and Mexico stand out for successfully transforming waste into valuable products using innovative technologies. No matter which approach in which region, achieving increased sustainability requires ongoing collaboration, socio-technical innovation, and knowledge-sharing of best practices. This is required to align business priorities with the regional and global environment, fostering innovation adoption, and unlocking transformation potential in the food sector. Adopting novel strategies and practices is crucial for addressing global challenges like food security and environmental threats. Finally, by emphasizing pathways for technological advancement and biological and technological circularity, this book should serve as a practical guide for developing business strategies and industrial policies in FSCs.

References

Agrawal S, Sharma D, Nagpal R, Kaur A, Bhardwaj N, Mahajan R (2023) Valorisation of wheat straw into paper with improved quality characteristics using ultrafiltered xylano-pectinolytic pulping approach. 3 Biotech 13(3):106. https://doi.org/10.1007/s13205-023-03526-x

Alhiary M (2012) Estimation of post harvest losses of some fruit crops in Jordan. Available online at https://www.researchgate.net/publication/261440921_Estimation_of_post_harvest_losses_of_some_fruit_crops_in_Jordan#fullTextFileContent. Updated on 2012, checked on 1/29/2024

Alzubi E, Akkerman R (2022) Sustainable supply chain management practices in developing countries: an empirical study of Jordanian manufacturing companies. Cleaner Prod Lett 2:100005. https://doi.org/10.1016/j.clpl.2022.100005

Alzubi E, Noche B (2022) A multi-objective model to find the sustainable location for citrus hub. Sustainability 14(21):14463. https://doi.org/10.3390/su142114463

Alzubi E, Kassem A, Noche B (2022) A comparative life cycle assessment: polystyrene or polypropylene packaging crates to reduce citrus loss and waste in transportation? Sustainability 14(19):12644. https://doi.org/10.3390/su141912644

Alzubi E, Kassem A, Melkonyan A, Noche B (2024) Enhancing economic-social sustainability through a closed-loop citrus supply chain: a life cycle cost analysis. Resour Conserv Recyl Adv 21:200199. https://doi.org/10.1016/j.rcradv.2023.200199

Andreou V, Dimopoulos G, Dermesonlouoglou E, Taoukis P (2020) Application of pulsed electric fields to improve product yield and waste valorization in industrial tomato processing. J Food Eng 270:109778. https://doi.org/10.1016/j.jfoodeng.2019.109778

Attia AI, Reda FM, Patra AK, Elnesr SS, Attia YA, Alagawany M (2021) Date (Phoenix dactylifera L.) by-products: chemical composition, nutritive value and applications in poultry nutrition, an updating review. Animals 11(4). https://doi.org/10.3390/ani11041133

Badjona A, Adubofuor J, Amoah I, Diako C (2019) Valorisation of carrot and pineapple pomaces for rock buns development. Sci Afr 6:e00160. https://doi.org/10.1016/j.sciaf.2019.e00160

Correddu F, Caratzu MF, Lunesu MF, Carta S, Pulina G, Nudda A (2023) Grape, pomegranate, olive, and tomato by-products fed to dairy ruminants improve Milk fatty acid profile without depressing Milk production. Foods (Basel, Switzerland) 12(4). https://doi.org/10.3390/foods12040865

dos Santos PCM, Da Silva LMR, Magalhaes FEA, Cunha FET, Ferreira MJG, de Figueiredo EAT (2022) Garlic (Allium sativum L.) peel extracts: from industrial by-product to food additive. Appl Food Res 2(2):100186. https://doi.org/10.1016/j.afres.2022.100186

Dusch Silva S (2018) SQUEEZED Behind the scenes of the juice industry. CIR

Eriksson M, Strid I, Hansson P-A (2014) Waste of organic and conventional meat and dairy products—a case study from Swedish retail. Resour Conserv Recycl 83:44–52. https://doi.org/10.1016/j.resconrec.2013.11.011

Fornereto Soldan AC, Arvelos S, Watanabe ÉO, Hori CE (2021) Supercritical fluid extraction of oleoresin from Capsicum annuum industrial waste. J Clean Prod 297:126593. https://doi.org/10.1016/j.jclepro.2021.126593

Glaser SJ, Al-Rudainy B, Hatti-Kaul R, Galbe M (2023) Wheat bran fractionation: effect of steam explosion and hydrotropic extraction conditions on the recovery of sugars and lignin. Ind Crops Prod 195:116405. https://doi.org/10.1016/j.indcrop.2023.116405

Gudiño I, Martín A, Casquete R, Prieto MH, Ayuso MC, Córdoba MG (2022) Evaluation of broccoli (Brassica oleracea var. italica) crop by-products as sources of bioactive compounds. Sci Hortic 304:111284. https://doi.org/10.1016/j.scienta.2022.111284

Kanoija V, Kushwaha NL, Reshi M, Rouf MA, Muzaffar H (2018) Products and byproducts of wheat milling process. Int J Chem Stud 6(4):990–993

Nasir NG, Markham J (2008) Recycling solid waste from the olive oil extraction process. Available online at https://agrifutures.com.au/wp-content/uploads/publications/08-165.pdf. Checked on 1/29/2024

Rizzioli F, Benedetti V, Patuzzi F, Baratieri M, Bolzonella D, Battista F (2023) Valorization of orange peels in a biorefinery loop: recovery of limonene and production of volatile fatty acids and activated carbon. Biomass Conv Bioref. https://doi.org/10.1007/s13399-023-03738-4

Sai Prasanna N, Mitra J (2020) Isolation and characterization of cellulose nanocrystals from Cucumis sativus peels. Carbohydr Polym 247:116706. https://doi.org/10.1016/j.carbpol.2020.116706

Sasaki Y, Orikasa T, Nakamura N, Hayashi K, Yasaka Y, Makino N et al (2021) Life cycle assessment of peach transportation considering trade-off between food loss and environmental impact. Int J Life Cycle Assess 26(4):822–837. https://doi.org/10.1007/s11367-020-01832-7

Sasaki Y, Orikasa T, Nakamura N, Hayashi K, Yasaka Y, Makino N et al (2022) Optimal packaging for strawberry transportation: evaluation and modeling of the relationship between food loss reduction and environmental impact. J Food Eng 314:110767. https://doi.org/10.1016/j.jfoodeng.2021.110767

Sharma K, Mahato N, Nile SH, Lee ET, Lee YR (2016) Economical and environmentally-friendly approaches for usage of onion (Allium cepa L.) waste. Food Funct 7(8):3354–3369. https://doi.org/10.1039/C6FO00251J

Shaw D, Tripathi AD, Paul V, Agarwal A, Mishra PK, Kumar M (2023) Valorization of essential oils from citrus peel powder using hydro-distillation. Sustain Chem Pharm 32:101036. https://doi.org/10.1016/j.scp.2023.101036

Siddiqui SA, Pahmeyer MJ, Assadpour E, Jafari SM (2022) Extraction and purification of d-limonene from orange peel wastes: recent advances. Ind Crops Prod 177:114484. https://doi.org/10.1016/j.indcrop.2021.114484

Yafetto L, Odamtten GT, Wiafe-Kwagyan M (2023) Valorization of agro-industrial wastes into animal feed through microbial fermentation: a review of the global and Ghanaian case. Heliyon 9(4):e14814. https://doi.org/10.1016/j.heliyon.2023.e14814

Yun D, Liu J (2022) Recent advances on the development of food packaging films based on citrus processing wastes: a review. J Agric Food Res 9:100316. https://doi.org/10.1016/j.jafr.2022.100316

Zia S, Khan MR, Shabbir MA, Aadil RM (2021) An update on functional, nutraceutical and industrial applications of watermelon by-products: a comprehensive review. Trends Food Sci Technol 114:275–291. https://doi.org/10.1016/j.tifs.2021.05.039

Index

A. Melkonyan-Gottschalk, D. Daus (eds.), *Transformation Towards Circular Food Systems*, Greening of Industry Networks Studies 13,
https://doi.org/10.1007/978-3-031-63793-3

The manufacturer's authorised representative in the EU is Springer Nature Customer Service Centre GmbH, Europaplatz 3, 69115 Heidelberg, Germany. If you have any concerns regarding our products, please contact ProductSafety@springernature.com

Printed and bound by CPI Group (UK) Ltd, Croydon, CR0 4YY
01/10/2026
02225969-0013